Mathematics Education in Brazil

Alessandro Jacques Ribeiro
Lulu Healy
Rute Elizabete de Souza Rosa Borba
Solange Hassan Ahmad Ali Fernandes
Editors

Mathematics Education in Brazil

Panorama of Current Research

Springer

Editors
Alessandro Jacques Ribeiro
Center for Mathematics, Computing and Cognition
Federal University of ABC
Santo André, SP, Brazil

Lulu Healy
School of Education, Communication and Society
King's College
London, UK

Rute Elizabete de Souza Rosa Borba
Graduate Program in Mathematics & Technology Education
Federal University of Pernambuco
Recife, PE, Brazil

Solange Hassan Ahmad Ali Fernandes
Graduate Program in Mathematics Education
Anhanguera University of São Paulo
São Paulo, SP, Brazil

ISBN 978-3-030-06663-5 ISBN 978-3-319-93455-6 (eBook)
https://doi.org/10.1007/978-3-319-93455-6

Softcover re-print of the Hardcover 1st edition 2018

This Springer imprint is published by the registered company Springer Nature Switzerland AG
The registered company address is: Gewerbestrasse 11, 6330 Cham, Switzerland

Preface

This book aims to present a panorama of research in mathematics education developed in Brazil to the international academic community. It is the result of a collaboration of members of the Brazilian Society of Mathematics Education (SBEM—Sociedade Brasileira de Educação Matemática).

Since its founding in January 1988, the Brazilian Society of Mathematics Education (SBEM) has been organised to discuss issues related to teaching and learning in mathematics, with a view to contributing to the consolidation of mathematics education as an area of knowledge. Mathematics education in Brazil is still relatively a new field of research, which arose initially from concerns that emerged from the programme of expansion of mathematics teaching in the beginning of the 1950s. These concerns led to the first National Congress of Mathematics Teaching, followed by the creation of what became known as the Circles of Mathematics Teachers and the Brazilian Association of Mathematics Teachers and Researchers.

The creation of SBEM began to be structured in 1985, during the 6th Inter-American Conference on Mathematics Education (VI CIAEM), held in Guadalajara, Mexico, and the organisation came into being in its own right following the realisation of the first National Meeting of Mathematics Education (I ENEM) in the city of São Paulo in 1987. From its conception, SBEM has played an important role in the development of Brazilian mathematics education. As a scientific association, it expanded its area of activity, with the creation of regional directories in almost all states of the federation, and has been instrumental in the organisation of 12 National Meetings, six International Seminars on Research in Mathematics Education (SIPEM) and dozens of regional meetings. SBEM maintains two journals, *Mathematics Education in Journal (EMR)* and the *International Journal for Research in Mathematics Education (RIPEM).*

At the beginning of the twenty-first century, conscious of the growing number of researchers in mathematics education across Brazil, together with the size and diversity of the country, SBEM recognised a need to create mechanisms that would both enable collaborations between researchers throughout Brazil to be built and sustained and open channels through which to communicate the country's ongoing research concerns beyond its borders. This motivated the conception of what has become a triannual event, the International Seminar of Research in Mathematics (Seminário Internacional de Pesquisa em Educação Matemática—SIPEM), an event which serves to congregate the Brazilian mathematics education research community together with international collaborators and to simultaneously showcase, review and plan national research efforts.

The first SIPEM took place in 2000. The aims of the seminar were (and continue to be) fivefold: (1) to encourage exchange between local and regional groups, based in different states of Brazil and in different countries and dedicated to researching the area of mathematics education; (2) to disseminate Brazilian research to the community of mathematics educators within and beyond Brazil; (3) to provide a forum for meetings between researchers in mathematics education, opening possibilities to share, collaborate and plan research; (4) to promote the formation of integrated Brazilian-based research groups that work together on problems of national interest, with collaboration from foreign researchers (5) to support advances in research in mathematics education.

Six editions of SIPEM have now taken place. A total of 319 registered researchers participated in SIPEM IV, where 169 scientific papers were presented and discussed. During the event, plenaries, lectures and panels were also conducted, a number of new books related to the area of mathematics education were launched and meetings for discussions of topics involving education were organised.

The organisation of the conference around the national working groups has proved successful in forging research collaborations around particular themes. Members of the groups have worked both at and between the events on a variety of joint projects, including the organisation of books and special editions of research journals. Not surprisingly, in great part, these projects and the outcomes associated with them have been disseminated more widely within Brazil than beyond, not least because they have tended to be published in Portuguese. The idea of publishing a book aimed more specifically at the international community emerged during SIPEM VI. This book is a result of the discussions that began there, and its main focus is on the research of the 13 national working groups of the SBEM that participated in SIPEM VI.

These working groups are organised around the following themes: Mathematics Education in the Early Years and Primary Education (Y1–Y5); Mathematics Education in the Middle School (Y6–Y9); Mathematics Education in the High School (Y10–Y12); Mathematics Education at the University Level; History of Mathematics, Culture and Mathematics Education; Digital Technologies and Distance Education; Teacher Education; Assessment and Mathematics Education; Cognitive and Linguistic Processes in Mathematics Education; Mathematical Modelling; Philosophy of Mathematics Education; Teaching Probability and Statistics; and Difference, Inclusion and Mathematics Education. Twelve of these thirteen groups have participated in all the editions of SIPEM, while for the thirteenth group SIPEM VI occasioned its first participation. The seminars are organised around the working groups, with refereed papers forming the basis of the working group discussions during the event and the planning of the joint research activities that occur in the interim years between seminars. In this way, the papers presented at the triennial research seminars can be considered as a representative sample of the concerns of the Brazilian mathematics education community at a particular moment in time.

The 6th International Seminar of Research in Mathematics Education (SIPEM VI) took place from the 15th to the 19th of November 2015, and the contributions presented at this event provide the means we have chosen in this book to reflect upon the current concerns of the Brazilian mathematics education community and how the themes addressed within the national working groups have developed over time. The book contains a total of 14 chapters. Following an invited chapter from Professor Ubiratan D'Ambrósio, an internationally renowned researcher in the field and a founding member of the Brazilian Mathematics Education community, are chapters corresponding to each of the 13 working groups who met during SIPEM VI.

In his chapter, D'Ambrosio discusses how mathematics education—in Brazil and other parts of the world—should be thought in a new way, i.e. centred on humanity. He defends the idea that mathematics, which he sees as the most universal mode of thought, is intrinsically related to the future of humankind, and thus, mathematicians and mathematics educators are responsible for developing concepts and techniques that contribute to the survival, rather than destruction, of civilisation. Mathematics, he says, should enable, not endanger, the pursuit of human well-being and full citizenship, as well as offering means for understanding, explaining and transforming reality. It is fitting that most of the challenges posed in D'Ambrosio's chapter resurface in the remaining 13 chapters of the book in forms that are specific to the different areas of interest that currently characterise the field of mathematics education in Brazil. Indeed, tracing the ideas he raises through the rest of the book is one of the ways that readers might gain a sense of some of the characteristic flavours of Brazilian research, as well as the ways in which these flavourings vary in their intensity, impact and amplitude.

The chapters prepared by members of the 13 SBEM working group have a similar structure, with each chapter emphasising the major themes and research questions, theoretical and methodological approaches and emerging results, along with considerations of the tensions and issues that characterise research in the particular area of interest. In this way, the aim is that the book presents a panorama of research in the area of mathematics education in Brazil at a particular moment in time. Its chapters reflect the diversity of research themes of the national working groups, contemplating questions related to the teaching and learning of mathematics at all education levels—from early years to university mathematics—and encompassing issues of history, philosophy, language and cognition, digital technology, inclusion, assessment, teacher education, mathematical modelling and statistics education. We hope that the book will convey not only a sense of some of the special flavours of Brazilian research in mathematics education, but also the vibrancy and dynamism of our community, as a well as offering an overview of recent research tendencies and results from research programmes, much of which was previously available only in Portuguese.

Santo André, SP, Brazil — Alessandro Jacques Ribeiro
London, UK — Lulu Healy
Recife, PE, Brazil — Rute Elizabete de Souza Rosa Borba
São Paulo, SP, Brazil — Solange Hassan Ahmad Ali Fernandes

Acknowledgements

We would like to acknowledge and thank all those who have contributed to make this book possible.

First of all, we are indebted to the Brazilian Society of Mathematics Education (SBEM—Sociedade Brasileira de Educação Matemática) for its support at each stage of the book's development, from conception to completion. We reserve special mention for the current and previous members and Executive Board of SBEM (Diretoria Nacional Executiva—DNE) who have contributed to the consolidation of the International Seminar for Research in Mathematics Education (Seminário Internacional de Pesquisa em Educação Matemática—SIPEM) and to the strengthening of the working groups that compose the triannual events and work in the years between the seminars.

We would also like to thank the members of the working groups who accepted the challenge of writing the chapters that appear in the book, as well as the critical readers who kindly read different versions of the chapters and suggested essential changes.

Finally, we are grateful to all the associates of SBEM for extending to us their trust and confidence in the realisation of this project.

Contents

Contributors

Lourdes Maria Werle de Almeida Universidade Estadual de Londrina, Londrina, Brazil

Cristiane de Arimatéa Rocha Universidade Federal de Pernambuco, Recife, Brazil

Tânia Baier Universidade de Blumenau—FURB, Blumenau, Brazil

Marcelo Bairral Universidade Federal Rural do Rio de Janeiro, Seropédica, Brazil

Barbara L. Bianchini Pontifícia Universidade Católica de São Paulo, São Paulo, SP, Brazil

Maria Apaecida Viggiani Bicudo Universidade Estadual Paulista, Campus de Rio Claro, São Paulo, Brazil

Marcelo Borba Universidade Estadual Paulista, Rio Claro, Brazil

Rute Borba Universidade Federal de Pernambuco, Recife, Brazil

Airton Carrião Universidade Federal de Minas Gerais, Belo Horizonte, Brazil

José Ivanildo Felisberto de Carvalho Universidade Federal de Pernambuco, Recife, Brazil

Sônia Maria Clareto Universidade Federal de Juiz de Fora, Juiz de Fora, Brazil

Márcia Cristina de Costa Trindade Cyrino Universidade Estadual de Londrina, Londrina, Brazil

Edda Curi Universidade Cruzeiro do Sul, São Paulo, Brazil

Ubiratan D'Ambrosio Universidade Estadual de Campinas, Campinas, SP, Brazil

Universidade Anhanguera de São Paulo, São Paulo, SP, Brazil

Jader Otávio Dalto Universidade Tecnológica Federal do Paraná, Cornélio Procópio, Brazil

Solange Hassan Ahmad Ali Fernandes Anhanguera University of São Paulo, São Paulo, SP, Brazil

Ana Cristina Ferreira Universidade Federal de Ouro Preto, Ouro Preto, Brazil

Cláudia Regina Flores Universidade Federal de Santa Catarina, Florianópolis, SC, Brazil

Verônica Gitirana Universidade Federal de Pernambuco, Recife, Brazil

Elenilton Vieira Godoy Federal University of Paraná, Curitiba, PR, Brazil

Claudia Lisete Oliveira Groenwald Universidade Luterana do Brasil, Canoas, Brazil

Gilda Guimarães Universidade Federal de Pernambuco, Recife, Brazil

Lulu Healy School of Education, Communication and Society, King's College, London, UK

Sonia B. C. Igliori Pontifícia Universidade Católica de São Paulo, São Paulo, SP, Brazil

Carmen Teresa Kaiber Universidade Luterana do Brasil, Canoas, Brazil

Tiago Emanuel Klüber Universidade Estadual do Oeste do Paraná, Cascavel, Brazil

Sintria Labres Lautert Universidade Federal de Pernambuco, Recife, Brazil

Vinício de Macedo Santos University of São Paulo, São Paulo, SP, Brazil

Silvia Dias Alcântara Machado Pontifícia Universidade Católica de São Paulo, São Paulo, Brazil

Fabiane Guimarães Vieira Marcondes Instituto Federal de São Paulo, São José dos Campos, SP, Brazil

Renata Cristina Geromel Meneghetti Universidade de São Paulo, São Paulo, Brazil

Cristiano Muniz Universidade de Brasília, Brasília, Brazil

Adair Mendes Nacarato Universidade São Francisco, Itatiba, Brazil

Lilian Nasser Universidade Federal do Rio de Janeiro, Rio de Janeiro, RJ, Brazil

Clélia Maria Ignatius Nogueira Universidade Estadual do Oeste do Paraná e Universidade Estadual de Maringá, Cascavel, Maringá, PR, Brazil

Cristiane Coppe de Oliveira Universidade Federal de Uberlândia, Uberlândia, MG, Brazil

Maria Cristina Araújo de Oliveira Universidade Federal de Juiz de Fora, Juiz de Fora, MG, Brazil

Lourdes Onuchic Universidade Estadual Paulista Júlio de Mesquita Filho, São Paulo, SP, Brazil

Daniel Clark Orey Universidade Federal de Ouro Preto, Ouro Preto, MG, Brazil

Maria Isabel Ramalho Ortigão Universidade do Estado do Rio de Janeiro, Rio de Janeiro, Brazil

Cármen Lúcia Brancaglion Passos Universidade Federal de São Carlos, São Carlos, Brazil

Miriam Godoy Penteado Universidade Estadual Paulista, Rio Claro, SP, Brazil

Célia Maria Carolino Pires Universidade Cruzeiro do Sul, São Paulo, SP, Brazil

Mauren Porciúncula Universidade Federal do Rio Grande, Rio Grande, Brazil

Alessandro Jacques Ribeiro Center of Mathematics, Computing and Cognition, Federal University of ABC, Santo André, SP, Brazil

Maurício Rosa Universidade Federal do Rio Grande do Sul, Porto Alegre, Brazil

Suzi Samá Universidade Federal do Rio Grande, Rio Grande, Brazil

João Ricardo Viola dos Santos Universidade Federal de Mato Grosso do Sul, Campo Grande, Brazil

Marcio Antonio da Silva Federal University of Mato Grosso do Sul, Campo Grande, MS, Brazil

Elizabeth Gomes Souza Universidade Federal do Pará, Belém, Brazil

Alina Galvão Spinillo Universidade Federal de Pernambuco, Recife, Brazil

Leo Akio Yokoyama Colégio de Aplicação da Universidade Federal do Rio de Janeiro, Rio de Janeiro, Brazil

Chapter 1
To Think in a New Way in Mathematics Education

Ubiratan D'Ambrosio

We have to learn to think in a new way. ... There lies before us, if we choose, continual progress in happiness, knowledge, and wisdom. Shall we, instead, choose death, because we cannot forget our quarrels? We appeal as human beings to human beings: Remember your humanity, and forget the rest. If you can do so, the way lies open to a new Paradise; if you cannot, there lies before you the risk of universal death.

The Russell-Einstein Manifesto, 1955

Abstract In this chapter, mathematics educators are encouraged to think about a broader conception of mathematics, one that aims towards human well-being, freedom and choice, good social relations, security, and peace of mind. It is argued that, since as mathematics educators we influence the generations that will oversee future world affairs, it is our responsibility to prepare them to shape a new civilization, in which social justice and peace with dignity are privileged. This involves a rethinking of the ways in which mathematics is taught and practiced: rather than stressing only principles and techniques, it is also important to reflect on the ethics of mathematics, its uses for good or for evil. Part of this rethinking involves reflecting on why mathematics is taught and while the declared intention is often that it will be useful for everyday life, it is also important to recognize that our most successful students may design lethal weapons or reinforce the practices of brutal capitalism. History tells us that mathematics is the dorsal spine of modern civilization, hence mathematics and mathematics education have everything to do with the state of the world. In short, this chapter considers how, in an era of increasing globalization in all sectors of society, the ethics of respect, solidarity, and cooperation might permeate all aspects of mathematics education.

U. D'Ambrosio (✉)
Universidade Estadual de Campinas, Campinas, SP, Brazil

Universidade Anhanguera de São Paulo, São Paulo, SP, Brazil
e-mail: ubi@usp.br

A. J. Ribeiro et al. (eds.), *Mathematics Education in Brazil*,
https://doi.org/10.1007/978-3-319-93455-6_1

1.1 Introduction

Right at the beginning of my career as a mathematician, I was impressed by the appeal of the two eminent mathematicians of the twentieth century. They were concerned with the world rapidly approaching doomsday in view of the insane nuclear race. Their appeal was partially successful, in the sense of avoiding a final nuclear confrontation by the two then superpowers, the USA and the USSR. But it was and continues to be ignored as an appeal to human beings to remember their humanity.

Half a century later, the survival of the species is threatened by a societal breakdown, as is strongly expressed by distinguished astrophysicist Lord Martin Rees, FRS:

> The main threats to sustained human existence now come from people, not from nature. Ecological shocks that irreversibly degrade the biosphere could be triggered by the unsustainable demands of a growing world population. Fast-spreading pandemics would cause havoc in the megacities of the developing world. And political tensions will probably stem from scarcity of resources, aggravated by climate change. Equally worrying are the imponderable downsides of powerful new cyber-, bio-, and nanotechnologies. Indeed, we're entering an era when a few individuals could, via error or terror, trigger societal breakdown (Rees, 2013).

Mathematics is present in all the stages of the evolution of the human species, since pre-historic times (D'Ambrosio & Almeida, 2017) advances in mathematics are associated with progress. As historian Mary Lefkowitz says: "The evolution of general mathematical theories from those basics [the mathematics of Egyptians, Sumerians, and others] *is the real basis of Western thought*" (emphasis is mine).[1]

Mathematics educators will miss their most important ethical responsibility without a clear understanding of how mathematics and mathematics education are related to this tragic perspective for the future of humanity. It is a gross mistake to ignore that the transmission of mathematics consists only of principles and techniques. The same is true for all fields of knowledge. Indeed, it is more important to reflect on the ethics of mathematics, its uses for good or for evil, than in merely transmitting contents. Although mathematics is taught with the declared intention that it will be useful for everyday life, mathematics educators cannot ignore the fact that their most successful students may be engineers who design lethal weapons or economists that reinforce the practices of brutal capitalism.

The main issues affecting society nowadays can be synthesized in:

- National security, personal security
- Government/politics
- Economics: social and environmental impact
- Relations among nations
- Relations among social classes

---

[1] This is quoted in Ken Ringle: A Professor's Collision Course: Classics Scholar Mary Lefkowitz Shakes Up Academia with Her Critique of Afrocentrism. *The Washington Post*, June 11, 1996, pp. A1, A3.

- People's welfare
- The preservation of natural and cultural resources

Mathematics, mathematicians, and mathematics educators are deeply involved with all these issues. History tells us that the technological, industrial, military, economic, and political complexes have developed thanks to mathematical material and intellectual instruments. It is also clear that mathematics has been relying on these complexes for the material bases of its continuing progress.

It is widely recognized that mathematics is the most universal mode of thought and that survival with dignity is the most universal problem facing mankind. It is expected that scientists, mathematicians, and mathematics educators be concerned with the most universal problem, that is, survival with dignity, and also have much familiarity with the most universal mode of thought, that is, mathematics. It is absolutely natural to expect that they, mathematicians and mathematics educators, look into the relations between these two universals, that is, into the role of mathematicians and math educators in the pursuit of a civilization with dignity for all, in which inequity, arrogance, and bigotry have no place. This means, to achieve a world in peace (Pugwash Manifesto, 1955).

My current concerns about research and practice in mathematics education fit into my broad interest in the human condition as related to the history of natural evolution (from the Cosmos to the future of the human species) and to the history of ideas, particularly the history of the explanations of creation and of evolution. Reflections on how mathematics contributes for human well-being must permeate the lectures of mathematics teachers and also of teacher educators in both pre-service and in-service courses and in professional development programs.

Human well-being in the broad sense is the main focus of both our pedagogical role and the reason for scientific advances. In this chapter, I discuss my views on the role of mathematics and mathematics education moving into a future focused on the broad concept of human well-being as conceived by the ICSU:

> Human well-being: A context- and situation-dependent state, comprising basic material for a good life, freedom and choice, health and bodily well -being, good social relations, security, peace of mind, and spiritual experience (ICSU, 2010).

For nearly 50 years now, I have been discussing proposals of mathematics for social justice, with different formulations, as the main focus of mathematics education. I believe that mathematics educators should be educators who regard mathematics as an important instrument to prepare future generations to live in a world with peace and human dignity for all.

In 1976, in my controversial discussion paper on why do we teach mathematics, presented at the Third International Congress of Mathematical Education (ICME-3), in Karlsruhe, Germany, I stated:

> We see the educational process as the conjugation of global socio-economic aspects aiming at the betterment of the quality of life. In this conjugation intervene, the same as in the technological process, the philosophy to which society subscribes, as well as considerations about manpower and available material resources (D'Ambrosio, 1976; p. 224).

Almost 20 years later, in 1993, at the 15th annual conference of the North American Chapter of the Psychology of Mathematics Education (PME-NA), in Pacific Grove, California, I went further in these ideas:

> Although the main concern of this meeting is Mathematics Education, I believe I will be allowed to subordinate my comments to a higher objective: the survival of civilization on Earth with dignity for all. This is not merely jargonizing. The world is threatened, not only by aggressions against nature and the environment. We are equally concerned with increasing violations of human dignity. We face more and more cases of life under fear, hatred and violation of the basic principles upon which civilization rests (D'Ambrosio, 1993; p. 31).

In other writings, I have asked for a new thinking in mathematics education. My objective, in this chapter, is to stress the fact that our most urgent concern is to teach mathematics for social justice as it is understood in the broad sense of attaining the kind of human well-being that the ICSU wrote about in 2010.

No one can deny that mathematics is an essential intellectual and material instrument for full citizenship. Higher education and research is responsible for preparing teachers and for proposing innovations in the curriculum and in methodological appropriation in education. The rapid development of digital technologies in information and communication calls for profound innovation in methodology.

I have often referred to mathematics as the imprint of modern society, for good and evil. But paradoxically mathematics has been the main instrument in weaponry and in economics. And the consequence has been wars, greedy capitalism, uncontrollable consumerism, and aggression against people, killing and suppressing dignity. I understand the violation of social justice in this broad conception. As political scientist Glenn D. Paige argues:

> The structure of society does not depend upon lethality. There are no social relationships that require actual or threatened killing to sustain or change them. No relationships of dominance or exclusion—boundaries, forms of government, property, gender, race, ethnicity, class, or systems of spiritual or secular belief—require killing to support or challenge them. This does not assume that such a society is unbounded, undifferentiated, or conflict-free, but only that its structure and processes do not derive from or depend upon killing. There are no vocations, legitimate or illegitimate, whose purpose is to kill (Paige, 2002; p. 30).

Similarly, the Charter For a World Without Violence, endorsed by Nobel laureates, ends with the appeal:

> To address all forms of violence we encourage scientific research in the fields of human interaction and dialogue, and we invite participation from the academic, scientific and religious communities to aid us in the transition to non-violent, and non-killing societies.[2]

Mathematicians and mathematics educators are among those invited to participate in creating the transition to nonviolent and non-killing societies. How do we respond to this appeal?

[2] *http://www.nobelforpeace-summits.org/charter-for-a-world-without-violence/*

1.2 The State of the World and Mathematics and Mathematics Education

Mathematician Mikhail Gromov, the 2009 Abel Prize laureate, says:

> Earth will run out of the basic resources, and we cannot predict what will happen after that. We will run out of water, air, soil, rare metals, not to mention oil. Everything will essentially come to an end within fifty years. What will happen after that? I am scared. It may be okay if we find solutions, but if we don't then everything may come to an end very quickly. Mathematics may help to solve the problem, but if we are not successful, there will not be any mathematics left, I am afraid! (Gromov, 2010; p. 401)

I am also afraid. What kind of world are we leaving to the future generations? The future may not even be. All our proposals for better educating the future generations may be voided. The tensions within our contemporary societies, both intranational and international, add to the feeling of fright and fear. As mathematicians and mathematics educators we have a responsibility for the future. We have to find ways to both recognize and respond to this responsibility.

The *Conference on Sustainable Development (Rio+20)*, organized by the United Nations in June 4–6, 2012, in Rio de Janeiro, had the participation of all the disciplines. Mathematics played a transdisciplinarian role in the discussions. Indeed, mathematics is deeply involved in the interdisciplinary research that took place in preparation for the conference.

Christiane Rousseau, then Vice-President of the Executive Committee of International Mathematics Union (IMU), herself a pure mathematician, announced the endorsement given by IMU to the Conference Rio+20 and sponsored the broad project Mathematics of Planet Earth (MPE) 2013. In her endorsement she stated:

> **A complex planet**. Earth is a planet with dynamic processes in the mantle, oceans and atmosphere creating climate. Mathematical sciences provide tools to understand these dynamic processes and measure the variation of its dynamics due to natural interaction or anthropogenic forcing.
>
> **A biologically diverse planet**. Earth is the home of life supporting systems which, through evolution, have generated biodiversity. Living species interact with ecosystems; new species appear or disappear; and species migrate to spread spatially. Mathematical modeling and analysis help us measure and understand what has happened in the past, predict what may happen in the future, and manage the present to promote more desirable future outcomes.
>
> **A planet structured by civilization**. Humans have developed systems of great complexity, including those for communication, transportation, energy, water, and health care. These man-made systems are the ones over which we exert the most direct control; however, we must often analyze massive amounts of data from monitoring systems, deal with multi-scale and/or distributed systems, and identify and quantify sources of uncertainty.
>
> **A planet at risk**. Human activity has increased to the point where it influences the global climate, impacts planet ability to feed its inhabitants, and threatens stability of these systems. We need to understand how the earth's systems interact. How does human activity affect climate, climate affect agriculture, agriculture affect water quality? How do the availability of food and water affect human health and migration and human migration and development affect animal migration? These are highly interconnected, complex adaptive

> systems. The mathematics of such systems is still poorly understood, while modeling them must draw expertise from many disciplines.
>
> MPE is being organized by a variety of committees with world-wide membership, including an "Umbrella" Leadership Committee and committees focusing on workshops, public awareness, education, etc. These committees coordinate activities among the many participating institutes and societies.[3]

As mathematics educators, we must be aware of this endorsement and all the issues raised therein. A number of mathematics educators have developed projects to involve schools with the major objective of showing the students local specificities in the Planet Earth. As an example, we have The Project Parallel Globe. Although conceived independently of the project MPE, the objective of this project is to recreate, for students, important stages in the development of mathematics, which are related to the observation of the skies.[4] The project intends to assist students in visualizing their position on the Earth surface in relation to the position occupied by other countries, including observing the way the Sun illuminates different regions of the Earth in real time. This visualization helps students to understand time zones and the alternation of the seasons on the planet. This tool of didactic practice is rich in mathematics content, as it allies observation, concrete experimentation with reflection, and data collection. In the development of the project, it became fundamental to share the results from countries of different longitude and latitude, thus helping to clarify the semantic and symbolic differences of the distinctions North-South, top-bottom, over-under, up-down in different languages and cultures.

For his dissertation, in 2013, Marcelo Ferreira Paiva conducted research with students of the 8th and 9th grades of elementary school aiming at involving them in developing an understanding of their position on the Earth in relation to other locations on the planet (Paiva, 2013). He joined the Globo Local Project in which students observe their position at different latitudes and longitudes during the Equinoxes and Solstices 2011. Paiva showed that the integrated study of mathematics and astronomy can help students to realize the importance of the production of knowledge for understanding the world in which they are inserted. To reach that goal, activities were developed using the software Stellarium, a gnomon, and the Globo Local Paralelo—GLP, among others (see Footnote 1). The intrinsic objective is that students realize that by expanding their knowledge they become critical and reflective citizens, aware of the need to care for the planet Earth.

Intercultural awareness has obvious political implications for a democratic perception of globalization, and it supports respect for difference and the recognition that all nations are part of the same global system, one that is threatened. The main objective of the project is to convey to students the message that civilization is

---

[3] http://mpe.dimacs.rutgers.edu/mpe2013-overview/ See also http://www.mpe2013.org, with very interesting links.

[4] A detailed explanation of the project is at the site https://www.academia.edu/8500143/The_Parallel_Globe_and_the_Globo_Local_Project

threatened, that all nations share different but common environmental conditions, and that mathematics is an important instrument for monitoring these conditions.

A remarkable conference on Visions in Mathematics—Towards 2000 was held in Tel Aviv, Israel, from August 25 to September 3, 1999. It was one of the most remarkable mathematical meetings in recent years and united some of the leading mathematicians worldwide. The goals of the conference were to discuss the importance, the methods, the past and the future of mathematics as we enter the twenty-first century and to consider the connection between mathematics and related areas. In this conference, Gromov delivered an address where he pointed to new directions for the development of mathematics, ones resulting from the sociocultural context rather than the conceptual necessities and details intrinsic to established mathematics theories. We need another mathematics.

In his remarkable chapter on Spaces and questions, M. Gromov makes very deep considerations about the nature of mathematics, inspired by historical reflections. He says.

> *Nature and naturality of questions.* Here are (brief, incomplete, personal and ambiguous) remarks intended to make clearer, at least terminologically, the issues raised during discussions we had at the meeting. "Natural" may refer to the structure or nature of mathematics (granted this exists for the sake of argument), or to "natural" for human nature. We divide the former into (pure) **mathematics, logic and philosophy**, and the latter, according to (internal or external) reward stimuli, into **intellectual, emotional, and social**. How can we trust our mind overwhelmed by intellectual, emotional, and social ideas to come up with true mathematics, logic and philosophy questions? (Gromov, 2010; p. 122)

and concludes the paragraph saying, "As for myself, I love unnatural, crazily unnatural problems but you stumble upon them so rarely!" (Gromov, 2010; p. 122).

The ideas of Gromov clearly indicate that the new generation of scientists, engineers, and, obviously, mathematicians will need broader attitudes towards mathematics.

Although these considerations are primarily addressed to research mathematicians, it is undeniable that they pose an even great challenge to mathematics educators. It is questionable if we should insist in consuming school time and energy teaching obsolete contents instead of moving more rapidly into the new concepts of mathematics, as suggested by Gromov and others (2010). Indeed, we may be doing a disservice to education and to future generations. Mathematics educators should invite their students to explore and to divagate into new ideas. These ideas may even be incomplete, only hints. Teachers should give students a possibility of exercising their creativity, their fantasy. I cite the eminent Norwegian mathematician Sophus Lie (1842–1899), when he said to a friend:

> Without Fantasy one would never become a mathematician, and what gave me a place among the mathematicians of our day, despite my lack of knowledge and form, was the audacity of my thinking (Stubhaug, 2000; p. 143).

It is undeniable that a new face of mathematics, designed to deal with the mounting menaces to civilization, is a strong motivation to study mathematics. The digital natives feel that the traditional mathematics that still dominates the curricula is obsolete, boring, and useless. I am convinced that this is the main cause of the bad results in tests.

The same question is applicable to the new physics, the new biology, and other scientific fields. The eminent physicist Richard Feynman denounces, in the preface of his well-known lectures, the damages caused to students when teachers insist in presenting obsolete theories:

> They [the students] have heard a lot about how interesting and exciting physics is—the theory of relativity, quantum mechanics, and other modern ideas. By the end of two years of our previous [i.e., traditional] course, many would be very discouraged because there were really very few grand, new, modern ideas presented to them. They were made to study inclined planes, electrostatics, and so forth, and after two years it was quite stultifying (Feynman, Leighton, & Sands, 1963; p. 3).

The challenging problems require, besides new mathematical techniques, the training of a new generation of researchers in the mathematical sciences. Again, citing Gromov (1998):

> We shall need for this the creation of a new breed of mathematical professionals able to mediate between pure mathematics and applied science. The cross-fertilization of ideas is crucial for the health of the science and mathematics (Gromov, 1998; pp. 846–847)

The mediation of mathematics and the other sciences is a fundamental theme. The concept of integrating mathematics with sciences and applications is the motivation of the concept of STEM in education. This concept was proposed in a pioneering symposium organized by the National Science Foundation of the USA in 1992. Six scholars from different parts of the world—all with expertise in science and mathematics education—were invited to participate in a 4-week full time working symposium on International Perspectives on Science, Mathematics, Engineering, and Technology Education. The project was designed by Dr. Kenneth J. Travers, then Director of the Division of Research, Evaluation and Dissemination, with the endorsement of Dr. Luther Williams, then Assistant Director, Directorate for Education and Human Resources. I was privileged to chair the group formed by Paul Black from England, Mohamed El-Tom from Sudan, Bienvenido Nebres from the Philippines, Tibor Nemetz from Hungary, and Michael Matthews from Australia. Our task was to investigate current programs of these four fields of study at all levels of schooling and to review ongoing reform initiatives in science and mathematics education in the world and their global implications for the twenty-first century (D'Ambrosio et al., 1993).

The symposium was unique in several ways. Probably the most innovative feature was its interdisciplinary and international nature. Probably the most relevant conclusion was the recognition of clear indications that in the curricular designs of the future, science, mathematics, engineering, and technology education will probably appear as a unified field of study.

Much of our work in the symposium dealt with the future. We tried to probe well into the twenty-first century—as far as the year 2061 when Halley's Comet will reappear—into a future when today's children will be in charge and coping with the effects of our decisions. Given the accelerated pace of change and the increased interdependence of nations, the decisions we make today seem somehow more critical than do those made by our predecessors generations ago. During the 4 weeks, we had the opportunity to meet with representatives of major professional societies, such as the American Association for the Advancement of Science, the National Science Teachers Association, the Mathematical Association of America, the National Council of Teachers of Mathematics and to interact with staff of the NSF.

It is relevant that Mathematics, as a science, has specificities. Mathematician Steve Kennedy (2003):

> Math is different from the other sciences. In a very real sense the problems, motivations and verification of mathematics come from inside the discipline itself, whereas the other sciences look to the world of phenomena for problems and affirmation. The chemist whose experiment yields a result within six decimal places of his theoretical prediction has good reason to feel pretty pleased with his theorizing. A mathematician rarely finds herself in such an empirically happy place vis-à-vis her theories. Usually a mathematician has only the cold reassurance of logic for comfort; the universe does not deign to validate our work except indirectly, when the work proves useful as a model in another science (Kennedy, 2003; p. 180).

In mathematics education, to approximate mathematics to the sciences is to show that mathematics is fully integrated with the scientific method, which is an essential component of multidisciplinary and interdisciplinary research. This integration is intrinsic to The Laboratory Method in mathematics education, as proposed in 1902 by Eliakim H. Moore, then President of the American Mathematical Society, in a remarkable article on the Foundations of Mathematics. Moore says:

> [...] the boy will be learning to make practical use in his scientific investigations—to be sure, in a naive and elementary way—of the finest mathematical tools which the centuries have forged; that under skilful guidance he will learn to be interested not merely in the achievements of the tools, but in the theory of the tools themselves, and that thus he will ultimately have a feeling towards his mathematics extremely different from that which is now met with only too frequently—a feeling that mathematics is indeed itself a fundamental reality of the domain of thought, and not merely a matter of symbols and arbitrary rules and conventions (Moore, 1903; p. 408).

Rather than proposing a shortcut, Moore proposes restoring mathematics education to the original roots of mathematics development in modernity. The advances proposed since the sixteenth century recognize mathematics as the main support of scientific inquiry.

1.3 A New Approach to Mathematics and Mathematics Education

A new mathematics depends, of course, on basic mathematics. But to what extent shall we insist on the basics? Thanks to the amazing technology available, it is possible to accelerate the acquisition of the basic mathematics that is necessary—only a small part of the necessary basics is in the usual programs, much of the program is irrelevant. It is important to step, rapidly, into new styles of mathematics. The basic mathematics includes mainly concepts, not techniques. Curricular development should focus on accelerating the teaching of the things that are effectively basic in traditional mathematics, which are concepts. Instead, much of the time and energy of teachers still goes into insisting on obsolete skills.

The difficulty is to bridge the gap between the internal advances of mathematics and their utilization.

Some examples of shortcuts for presenting advanced mathematics in a simple and contextual way are the proposals exposed in the books *Calculus Made Easy: Being a Very-Simplest Introduction to Those Beautiful Methods of Reckoning Which Are Generally Called by the Terrifying Names of the Differential Calculus and the Integral Calculus.*, first published in 1910, by Silvanus Thompson and generally repudiated by mathematicians, and *Lectures on Physics*, based on Richard Feynman's lectures from 1961 to 1963 (cited above), strongly criticized by his peers. Both authors were distinguished scientists, not mathematicians, but users of advanced mathematics. Both rapidly presented the concepts with adequate rigor for the purpose of their uses. Finding the equilibrium between accessible presentation and acceptable rigor is a major challenge. Regrettably, this is poorly done by mathematics educators.

A great challenge for mathematicians and mathematics educators is to recognize new ideas in mathematics and to develop methods for transmitting this to teachers. The need to face this challenge was clearly stated by David Hilbert in his memorable lecture in the International Congress of Mathematician of 1900: "A mathematical theory is not to be considered complete until you have made it so clear that you can explain it to the first man whom you meet on the street" (Hilbert, 1902; p. 438).

Of course, the same remark applies to children.

Children must be prepared for a future that we cannot envisage. Education, in this era of science and technology, challenges the established approaches "validated" by results in standardized tests. The goals of education go much beyond merely preparing for professional success. Education has a responsibility to build up saner attitudes toward the self, toward society, and toward nature. We are primarily faced with preparing teachers to assume a different attitude in their teachings, responding to the new challenges. Educators must be creative.

I believe the key problems in the preparation, both in pre- and in-service, of teachers of mathematics are related to inadequate visions of the purposes of education and of the role of mathematics teachers as educators. Prospective and already in-service teachers of mathematics should be always reflecting about the changes in education that are consequences of profound changes in society, particularly in the demographic scenario, in production, in information, in communication, and in the environment.

Here, I elaborate on the purposes of education and the responsibilities of teachers, particularly mathematics teachers, in building up a saner future. I identify two goals for societies to create and maintain establish educational systems:

- To promote citizenship, which prepares the individual to be integrated and productive in society and is achieved by transmitting values which are rooted in the past and show rights and responsibilities in society.
- To promote creativity, leading to progress and pointing to the future, which is achieved by helping people to fulfill their potentials and to rise to the highest of their capability.

The practice of education is in the present and in this concept, it must deal with the encounter of past and future. The major challenge to educators is to manage, in the process, the encounter of the past and of the future, that is, the transmission of values rooted in the past, which leads to citizenship, and the promotion of the new, for an uncertain future, which means creativity.

Responsible educators must guide their actions in meeting these two goals being aware that:

- To promote citizenship, a critical view of the present as the link of past and future is of major importance. We do not want educators to transmit a docile citizenship, in the sense of simply accepting rules and codes that violate human dignity. They must understand the necessity of insubordination when needed.
- To promote creativity, educators must discuss ethical issues and awareness of the consequences of their performance. We do not want those of our students who become bright scientists to design lethal weapons or those who become economists to create instruments that reinforce the practices of brutal capitalism or those who become public and private gestors and administrators to practice and allow inequity, arrogance, and bigotry. Educators must instill the feeling that in all our behavior we must be critical of the norms and rules to be followed and aware of the consequences of our actions.

I synthesize the goals that I hold important in education hence in mathematics education, as the transmission of values rooted in the past, which leads to citizenship, but not docile citizenship; and the promotion of the new, for an uncertain future, which means creativity, but not irresponsible creativity.

Didactics and pedagogy are strategies to manage cultural encounters in the present of the past, the "old," represented by parents and teachers, and of the future, the "new," represented by daughters and sons and students. Both home and schools are the encounter in the present of the past and of the future. I have appealed to the metaphor of the mythological god Janus in previous papers (D'Ambrosio, 2015). Discussed as generation gap by many sociologists, this may be the main factor of increasing violence in society. Parents and teachers are seen as heralds of the past, repeating traditional values, curricula and knowledge, and students, eager about fulfilling their wishes and hopes for the future, anxious to probe into the new and the unknown, are seen as unruly, capable of turbulent and even violent behavior. This is frequent in traditional schooling. The film The Wall: Another brick in the wall, by Pink Floyd, showing the appeal of a very violent behavior to a child, is shocking.[5]

[5] https://www.youtube.com/watch?v=FbeszMeyguQ&list=RDFbeszMeyguQ#t=6

From this and many other points of view, schools must change. In 2001, in a seminar at the Institute for Information Technology in Education, of UNESCO based in Moscow, Seymour Papert denounced the enormous amount of resources that are wasted in obsolete education:

> Using computers connected to the Internet students can obtain better and quicker access to sources of historical as well as scientific knowledge; they can explore economics as well as physics by making models and simulations; the rigor of mathematics can be extended to areas that were previously inaccessible. *But in the midst of these explosions of change the institution of School has remained as remarkably constant over time as it is across countries.* So why am I wasting time drawing attention to familiar facts and problems that are already being addressed? The answer is saddening: Although the problem is widely recognized, its depth is seldom appreciated. Most of those billions of dollars are being wasted (emphasis in original) (Papert, 2001; pp. 1–2).

Indeed, this waste means that much of the traditional content that exhausts current programs should be drastically changed. It is a big mistake to insist on mathematics curricula simply because they satisfy criteria of rigor and the internal organization of mathematics. Some defend that the satisfaction of such criteria are enough to justify content.

To prepare children to be proficient in obsolete mathematics may result in their marginality in the future, because they will possess outdated knowledge. Avoiding this anguish is an important feature of social justice. For me, social justice should be understood as an attempt to satisfy the basic needs for a good life: freedom and choice; health and bodily well-being; good social relations anchored on security, peace of mind, and respect for spiritual experience.

We must avoid giving students the illusion that by passing the tests and obtaining good grades they are prepared for the future. This illusion is fallacious and a denial of social justice. The inadequacy of tests is not new. More than 200 years ago, Évariste Galois clearly denounced the illusion of tests: "Are you quite happy to do well in the test? Do you believe you will be finally appointed as one of the two hundred geometers that will be admitted? You believe you are prepared: you are mistaken, this is what I will show you in a next letter" (Galois, 1831). He died before writing the next letter.

It is important to recognize that improving opportunities for employment is a real expectation that students and parents have of school. But preparation for the job market is indeed preparation for the capability of dealing with new challenges. Many careers exist that require different kinds of knowledge and experiences but remain unfilled because of the lack of able candidates. There is a need for change. But what to change and how to change? Ideally, the advances of research in mathematics education produce better-qualified teachers, ones capable of promoting innovative education. Regrettably, the focus on passing tests dominates school systems and it is reinforced by offering teachers rewards, such as salary increases, if their students are successful in the tests. School officials often support this practice, because they are rewarded with grants and other government subsidies. This reward system is a subtle form of corruption, which paves the way to explicit corruption, a flagrant violation of social justice.

Responsible governance should look carefully at the disequilibrium among preparation of graduates and the needs of the job market. Robert Reich, Professor of Public Policy of the University of California at Berkeley, extensively discussed this disequilibrium some years ago.

> Education for all, which is frequently given as a strategy for Social Justice, has many problems and the fact that more and more people are becoming educated, with emphases in science, technology and engineering, sounds like a good thing. It is, indeed, progress. But it is an illusion that "education for all" is the key to economic growth and prosperity and good jobs. We have to analyze the context in which this progress takes place and the fitness and quality of it. There is no point in preparing children for jobs that will probably be extinct when they reach adulthood (Reich, 1992).

Education for all leads to an extraordinary expansion of people going to school with the hope of finding good jobs. This increases the possibility of reducing social inequity. But there are reasons for caution, as pointed out by Viviane Forrester in her influential book *The Economic Horror*, 1999. The expansion can dilute the quality of graduates, giving space to less able individuals entering the system. It is frequent to see bright students poorly employed and disillusion of obtaining a degree. The ruthless and often fruitless fight for a permanent job is a fact all over the world. A cause may be the obsolescence of curricula, as repeatedly I discussed in this chapter. Many programs remain firmly attached to the traditional curricula, disregarding the disequilibrium between the preparation of graduates and the dynamics of the job market.

These remarks may be interpreted by many as suggesting a reduction of the importance of mathematical content. This interpretation is grossly mistaken. **We need more and better mathematical content**, but not the same content. I repeat that methodological innovation should be directed to making advanced mathematics attractive and teachable. "Compromising rigor, in benefit of generating interest and motivation, cannot be interpreted as conceptual errors, or as relaxing the importance of a serious mathematics in schools" (D'Ambrosio, 2012; p. 20).

1.4 Mathematics and Mathematics Education in a Changing Civilization

Mathematics is a fascinating cultural endeavor. It is seen as the imprint of rationality and, indeed, it is the dorsal spine of modern civilization. All the spectacular achievements of science and technology have their bases in mathematics. And the institutions of modern civilization—mainly economics, politics, management, and social order—are rooted in mathematics. It is no surprise that accomplished scholars are devoted to mathematics. A good number of successful citizens who did not accomplish well in mathematics in their school years, and sometimes even failed, put their trust in mathematics in the educational systems.

> Administrators, teachers, parents, students, and the population in general, see mathematics along with reading and writing as the principle subjects in schools. But society regards those who do well in mathematics as geniuses, and those who fail are stigmatized. There is a lack of recognition that there are different interests, different creativity, and different talents among different individuals, and particularly among different children. Mathematics acts a selector on intellectual elites. These elites too often pursue the same patterns of society, impregnated with arrogance, inequity, and bigotry, which is a clear violation of social justice (D'Ambrosio, 2012; p. 21).

When looking at mathematics education, we may identify two positions:

1. To use education as a strategy for teaching mathematics (a position defended by alienated mathematicians).
2. To teach mathematics as a strategy for good education (a position defended by socially committed educators).

Here, I resort to another metaphor. Position 1 sees mathematics as the center of the intellectual universe. Everything, children and curricula, turns around mathematics, as a rigid center, cold and austere in the words of Bertrand Russell: "Mathematics … possesses not only truth, but supreme beauty—a beauty cold and austere, like that of sculpture." Metaphorically, this is a Ptolomaic position. Position 2 claims that curricula, particularly methodology, are movable around children, the source of energy for our planet. Metaphorically, this is a Copernican position.

The focus of our mission as educators resides in children and also young and elderly adults—in general, those who are the reason and the source of energy for educational action. In this Copernican view, the disciplines, which revolve around those being educated, are merely instruments in this action. Disciplines are, thus, in permanent reformulation, reflecting social and cultural contexts and the queries, wishes, and needs of those being educated.

We have to look into history and epistemology with a broader view. The denial and exclusion of the cultures of the periphery, so common in the colonial process, still prevails in modern society. The denial of knowledge that affects populations is of the same nature as the denial of knowledge to individuals, particularly children. To propose directions to counteract ingrained practices is the major challenge of educators, particularly mathematics educators. Large sectors of the population do not have access to full citizenship. Some do not even have access to the basic needs for survival. This limited access is the situation in most of the world, and it occurs even in the most developed and richest nations. Further discussion about these matters is the objective of the Program Ethnomathematics. I will briefly give some fundamental ideas of the program.

The Program Ethnomathematics is a transdisciplinarian and transcultural research program on the evolution of behavior and knowledge in the human species. It benefits from results and methodology research in paleoanthropology, anthropology and neurocognition, mythology, history, sociology (communitarian life, politics, economics, education), and cultural studies in general. The central question is how the human species developed, in its evolution in the most diverse natural and sociocultural environments, strategies to cope with the ample environment. To face this challenge, research must be transcultural and transdisciplinarian and must borrow results and methodology of different fields. It is basic for the Program Ethnomathematics to explain and understand the strategies developed by the human species to deal with reality in its broadest sense. The ensemble of these strategies is a complex system of knowledge and behavior, focused in the satisfaction of the pulsations of survival and transcendence, characteristic of the human species.

This research addresses the emergence and evolution of behavior and knowledge since hominization through homo sapiens sapiens. As proposed by Jean Piaget, Lev Vygotsky, and many other researchers of human and animal behavior, this research is greatly aided and receives hints and ideas, thanks to the observation and analyses of the behavior and knowledge construction of children and adults, from birth to death. Particularly relevant in this line of research is the project of Alison Gopnik to study the evolution of observation and of their interpretations from birth to 5 years (Gopnik, 2009). We can have insights about the evolution of the species in analyzing the evolution of each individual human being and of their social interactions.

The Program Ethnomathematics initially contemplated the History and Philosophy of Mathematics and its pedagogical implications (D'Ambrosio, 1992). Soon it became clear that it is impossible to discuss history, philosophy, and pedagogy of mathematics isolated from other disciplines and from other forms of knowledge, in general.

Although I have written extensively about this, it is important to clarify that the word ethnomathematics is the result of an etymological exercise. I was looking for a simpler sentence to express "the ways, the arts and the techniques developed by the humans in specific natural and sociocultural environments to understand, to explain, to deal with daily needs and to satisfy will" when, browsing into etymological dictionaries, I found three Greek roots that approximately express what I was looking for:

techné (or *tics*) means the ways, arts and techniques developed by the humans.
ethno means specific natural and sociocultural environments.
mathema means to understand, to explain, to deal with daily needs and will.

Thus I composed the short phrase "*tics of mathema in distinct ethnos.*" Clearly, this led to an even simpler expression, the composed word *tics* + *mathema* + *ethno*. Next step was obvious: the word ethnomathematics. Although the word had been used before, mainly by ethnographers, the way I conceived it was a shortening of the sentence "the ways, the arts and the techniques developed by the humans in specific natural and sociocultural environments to understand, to explain, to deal with daily needs and to satisfy will."

This calls for a warning: do not use ethnomathematics in the sense of ethno+mathematics and even less as ethnic-mathematics, as it is used by ethnographers, anthropologists, and even mathematics educators concerned with cultural issues. Those interested on detailed discussion should look into the many publications on the theme (D'Ambrosio, 2006).

A new world order is urgently needed. Our hopes for the future depend on learning—critically—the lessons of the past. When we look at the history of mathematics since the early mathematical manifestations of man (and woman), we recognize the development of techniques to compare, to classify and to organize, to measure and to count, to infer and to conclude, much before mathematics is formalized. We also recognize mathematical ideas in the confluence of various modes of understanding, such as the religions, the arts, the techniques, and the sciences. We must assume a *transdisciplinarian* posture, and we need to look at all development and modes of understanding in different cultural environments, in different traditions—that is, we must assume a transcultural posture. This new posture may restore to mathematics its characteristic of being the most universal mode of thought and may allow it to face the most universal problem facing humanity, which is survival with dignity.

The enormous changes in society, particularly due to demographic dynamics, have raised the exclusion of large sectors of the population, both in developed and undeveloped nations, to unbearable levels. The exclusion of countries from the benefits of progress and advancement is unsustainable. Any explanation for the current perverse concept of civilization requires a deep reflection on colonialism. This reflection should not aim at blaming one group or another and should not be an attempt to redo the past. Rather, it is the moment to understand the past as a first step to move into the future. Because mathematics has everything to do with the State of the World, its autonomy in the curriculum, and its central role as the dominating discipline and as an educational sphere in itself, should be reconsidered. Paraphrasing Gromov (1998), we shall need for this the creation of a new breed of mathematical teachers, able to mediate between mathematics and the other disciplines (Gromov, 1998). But current curricula, in all levels of education, look like a selection of non-overlapping sets. Each discipline has its own domain. As a result, there is a lack of perception among teachers of the relation of mathematics to a broader vision of the world and of society. Curriculum is the strategy for the educational action. Educational action should offer three instruments that, together, provide what is essential for citizenship in a world moving swiftly toward a planetary civilization. These instruments are the communicative instruments, the analytic/symbolic instruments, and the technological instruments. They constitute the modern trivium, which I call respectively **literacy**, **matheracy**, and **technoracy**. This trivium is a proposal for a curriculum based on developing a broad perception of the complexity of the world and of society and providing the instruments to deal with such complexity.

Literacy is the critical capability of processing information, such as the use of written and spoken language, of signs and gestures, of codes and numbers (D'Ambrosio, 2012). Nowadays, reading must also include the competency of numeracy, of interpretation of graphs and tables, and of the other means of informing the individual. Reading even includes understanding the condensed language of codes. These competencies have much more to do with screens and keys than with pencil and paper.

Matheracy is the critical capability of inferring, proposing hypotheses, and drawing conclusions from data. It is a first step toward an intellectual posture, which is almost completely absent in our school systems. Matheracy is closer to the way mathematics was present both in classical Greece and in indigenous cultures (D'Ambrosio, 2012). The concern goes much beyond counting and measuring. Matheracy proposes a deep reflection about humans and society and aims at explaining and understanding reality. It is, indeed, symbolic analysis. This is the central idea behind the origins of mathematics. This competency should not be restricted to an elite, as it has been in the past. It is not the result of appropriation of skills but is instead acquired through competency to analyze.

Technoracy is the critical familiarity with technology. Of course, the operative aspects of it are, in most cases, inaccessible to the lay individual. But the basic ideas behind technological devices, their possibilities and dangers, and the morality supporting the use of technology, are essential issues to be raised among children at a very early age. As a historian, my resource is the critical perception of past and of future as a guide for action in the present, and history shows us that ethics and values are intimately related to technological progress (D'Ambrosio, 2012). Proficiency in mathematics means much more than counting, measuring, sorting, comparing, and solving problems aimed at drilling. Regrettably, even conceding that problem solving, modeling, and projects are practiced in some mathematics classrooms, the main importance is usually given to developing skills, particularly in the manipulation of numbers and operations. But problems and situations present in daily life are new and unexpected. Students should be prepared to tackle the new. The three instruments together, which obviously include reading, writing, and basic mathematics, constitute what is essential for citizenship in a world moving swiftly toward a planetary civilization.

1.5 Concluding Remarks

Civilization, as well as life of all the animal species, is threatened. There will not be, as we are told in the Epic of Gilgamesh or in the biblical episode of Noah, a privileged group of humans that will survive. I understand the threat to the species as the broadest violation of social justice. I tried to avoid, in this concluding chapter, commenting on or reinforcing the proposals of the previous chapters. All are written with extreme competence, presenting improvements of mathematics education aimed at social justice, something essential for citizenship. My objective in writing this concluding chapter was to bring to the attention of mathematics educators the need to give their thoughtful and serious consideration to a broader concept of social justice, focusing on the State of the World and the real threat to civilization. Paraphrasing Bertrand Russell and Albert Einstein in the 1955 Pugwash Manifesto,[6] a New Thinking is needed to achieve social justice, meaning equilibrium and safety,

[6] https://pugwash.org/1955/07/09/statement-manifesto/

in a world menaced by exhaustion of resources, which leads to war and fear. Mathematicians and mathematics educators have powerful means of developing new concepts and techniques to cope with the major threats to the survival of civilization (D'Ambrosio, 2012).

References

D'Ambrosio, U. (1976). Report of the B-3 section on 'overall goals and objectives for mathematics teaching (why do we teach mathematics?)'. In H.Athen, & H. Kunle (Eds.), *Proceedings of the Third International Congress on Mathematical Education* (pp. 221–230, 224). Karlsruhe, Germany:Zentralblatt für Didaktik der Mathematik.

D'Ambrosio, U. (1992). Ethnomathematics: A research program on the history and pedagogy of mathematics with pedagogical implications. *Notices of the American Mathematical Society, 39*(10), 1183–1185.

D'Ambrosio, U. (1993). Diversity and equity: An ethical behavior. In J. R. Becker, & B. J. Pence (Eds.), *Proceedings of the 15th Annual Meeting of the North American Chapter of the International Group for the Psychology of Mathematics Education* (pp. 31–37). San José, CA: The Center for Mathematics and Computer Science, San José State University. (ERIC No. ED372917); p.31.

D'Ambrosio, U. (2006). *Ethnomathematics. Link between traditions and modernity*. Rotterdam: Sense Publishers.

D'Ambrosio, U. (2012). Mathematicians, mathematics educators and the state of the world. *Journal of Research in Mathematics Education, 1*(1), 5–28. https://doi.org/10.4471/redimat.2012.01

D'Ambrosio U. Two realities, the academia and the exterior: We have to think in a new way. In Beatriz Silva D'Ambrosio e Celi Espasandin Lopes (Editors),Creative insibordination in Brazilian mathematics education research, Raleigh, NC: Lulu Press, Inc. , 2015; pp.41–52.

D'Ambrosio, U., & Almeida, M. C. (2017). Ethnomathematics and the emergence of mathematics. In *The nature and development of mathematics: Cross disciplinary perspectives on cognition, learning and culture* (to appear).

D'Ambrosio, U., Willians, L., Travers, K. J., Black, P., El-Tom, M., Matthews, M., Nebres, B., Nemetz, T. (1993). Science, mathematics, engineering, and technology education for the 21st century. In *Summer Symposium on Educating for Citizenship in the 21st Century, Final Report, July 1992*. National Science Foundation, Directorate for Education and Human Resources, Division of Research, Evaluation and Dissemination, Washington, DC, 1993 [ED 373 993SE 054 763 ; NSF-94-67].

Feynman, R. P., Leighton, R. B., & Sands, M. (1963). *The Feynman lectures on physics* (p. 3). Reading: Addison-Wesley Publishing Company.

Galois, E. (1831, January). Letter published in the *Gazette des Ecoles: Journal de l'Instruction Publique*, de l'Université, des Séminaires, numéro 110, 2è année.

Gopnik, A. (2009). *The philosophical baby: What children's minds tell us about truth, love, and the meaning of life*. New York: Farrar, Straus and Giroux.

Gromov, M. (1998). Possible trends in mathematics in the coming decades. *Notices of the American Mathematical Society, 45*(7), 846–847.

Gromov, M. (2010). Interview by Martin Raussen and Christian Skau. *Notices of the American Mathematical Society, 57*(3), 391–403 p. 401.

Hilbert, D. (1902, July). Mathematical problems. Lecture delivered before the International Congress of Mathematicians at Paris in 1900. *Bulletin of the American Mathematical Society*. p. 438.

ICSU. (2010). *Earth system science for global sustainability: The grand challenges*. Paris: International Council for Science (p. 20). Retrieved from http://www.baltex-research.eu/supplementary/BSSG26/03_ICSU_Grand_Challenges_Nov2010.pdf

Kennedy, S. (2003). Who wants to be a millionaire? *American Scientist, 91*(2), 180.

Moore, E. H. (1903). On the foundations of mathematics. *Science, XVII*(428), 401–416.

Paige, G. D. (2002). *Nonkilling global political science* (p. 30). Bloomington, IN: Xlibris.

Paiva, M. F. (2013). *A Matemática no Ensino Fundamental II: Utilizando conceitos da Astronomia como ferramenta no processo de ensino e aprendizagem*. Dissertation, UNIBAN/Universidade Bandeirantes Anhanguera de São Paulo, Brasil.

Papert, S. (2001, January–March). Knowledge for the Knowledge Society: A Russia-oriented perspective on technology and school. *IITE Newsletter* (Unesco Institute for Information Technologies in Education, pp. 1–2).

Pugwash Manifesto. (1955). *Pugwash conferences on science and world affairs*. Retrieved from https://pugwash.org/1955/07/09/statement-manifesto/

Rees, M. (2013, March 8). *Science* editorial, 339.

Reich, R. B. (1992). *The work of nations: Preparing ourselves for 21st century capitalism*. New York: Vintage.

Stubhaug, A. (2000). *The Mathematician Sophus Lie* (p. 143). New York: Springer.

Chapter 2
1, 2: What Did We Do? 3, 4: Let's Learn Some More! Research on Early Schooling Mathematics Education in Brazil

Rute Borba, Gilda Guimarães, Edda Curi, and Cristiano Muniz

Abstract In this chapter, we describe and analyse studies conducted by Brazilian researchers interested in investigating the processes of teaching and learning mathematics in the early years of schooling. To this end, we discuss the research carried out by Working Group 01, WG01, of the Brazilian Society of Mathematics Education, SBEM, focused on the study of mathematics education in kindergarten and elementary school. We consider the interests of WG01 members in relation to the research themes chosen and the theoretical-methodological approaches adopted to investigate the complex phenomena of mathematical learning. The studies developed by this group, which involve students and teachers, have focused on issues that still receive little attention; on the early-education classes; on content introduced as part of recent curriculum changes in Brazil; and on public policies that aim to improve Brazilian mathematics education. The chapter also highlights the innovations associated with the research conducted by the group, as well as aspects that future research should consider. We hope to contribute to a better understanding on how teaching and learning of mathematics occur, and how it becomes a part of the early years of schooling in Brazil and other parts of the world.

2.1 Introduction

This chapter discusses trends in the research presented by Working Group 01 (WG01) on mathematics in kindergarten and elementary school education, which is a part of the Brazilian Society of Mathematics Education (SBEM—Sociedade Brasileira de Educação Matemática). Our analysis focuses on the studies presented in the VI International Seminar for Research in Mathematics Education (Seminário

R. Borba (✉) · G. Guimarães
Universidade Federal de Pernambuco, Recife, Brazil

E. Curi
Universidade Cruzeiro do Sul, São Paulo, Brazil

C. Muniz
Universidade de Brasília, Brasília, Brazil

A. J. Ribeiro et al. (eds.), *Mathematics Education in Brazil*,
https://doi.org/10.1007/978-3-319-93455-6_2

Internacional de Pesquisa em Educação Matemática—SIPEM), which was held in 2015. However, the chapter also considers studies presented in previous editions of the seminar, and other Brazilian studies in which researchers describe and analyse investigations in Brazil focused on the teaching and learning of mathematics in kindergarten and elementary school. We bring as reference works by authors from recent national and international scientific events, such as the XII ENEM (Encontro Nacional de Educação Matemática) and the ICME 13 (International Congress on Mathematical Education), both of which occurred in 2016, to discuss and complement the production of WG01.

In Brazil, early childhood education corresponds to the age range 0- to 5-year-olds, and to the 1st to 5th years of elementary school, intended for children from 6 to 10 years old. Children up to 3 years old attend nurseries and 4- and 5-year-olds attend kindergarten. Within these first years of schooling, it is important to also recognise the existence of the Youth and Adult Education mode (EJA—Educação de Jovens e Adultos), aimed at young people and adults who did not have access to schooling during their childhood years, experienced little success repeatedly, and/or dropped out of school.

The preparation of teachers for these levels of education is provided through teacher education in pedagogy courses, which usually include one or two disciplines related to the didactics of mathematics teaching, with a workload of only 60–80 h each. This workload is not enough for deepening mathematical skills that should be developed to provide the future teacher with the necessary tools to teach this area of knowledge. It also means that opportunities to discuss curricula and the organisation of mathematics teaching are limited. This is a strong indication that in-service education programmes are needed to complete the processes initiated in the pedagogy courses.

WG01 was founded with the aim of discussing and disseminating research on teaching and learning of mathematics in kindergarten and elementary school, and organising them according to the following four mathematical areas described in the National Curriculum Parameters (BRASIL, 1997): *numbers and operations*, *magnitudes and measurements*, *space and shape* (geometry) and *data analysis* (statistics, probability and combinatorics). Since its inception, the research developed has focused, in general, on the following areas: (a) children's and adults' knowledge of mathematical concepts in early schooling, developed inside and outside of the classroom; (b) resources for teaching mathematics in kindergarten and elementary school; (c) initial and in-service education of teachers of early childhood and elementary education and their practices; and (d) the inclusion of students with and without special difficulties associated with learning mathematics in these years of schooling.

The second part of this chapter, through an analysis of research trends and a discussion regarding the specific objects investigated, brings the main objectives and the theoretical frameworks that support the studies, describes the research methods adopted and summarises the main results. It also situates the research concerning mathematics education of the early years of schooling developed in Brazil in the international arena. The studies are presented in categories to highlight the foci of the research works and the results Brazilian students present in mathematical learning during early-years education.

Thus, the title of the chapter is justified insofar as it indicates what researchers of WG01 have investigated, the main results of the research presented at SIPEM, the future trends of research that are needed to consolidate further understandings of learning at the beginning of schooling and, consequently, what teaching processes might be the most effective. The chapter also considers contributions to initial and in-service teacher education programmes, and the practices of teachers who teach mathematics in kindergarten and elementary school.

2.2 Overview of WG01 Investigations at SIPEM

From SIPEM I, held in 2000, to SIPEM V, held in 2012, 76 studies were presented, 37 of which focused on the teaching and learning of mathematics in elementary school, 31 focused on teacher education and 8 focused on mathematics teaching and learning resources in early schooling. This distribution shows that the research is more student-centred, although teachers and teaching and learning resources have also been targets of investigations.

During SIPEM I, WG01 concentrated exclusively on the first 5 years of elementary schooling, but, at the end of the event, it was decided that studies on kindergarten education (4–5 years old) would be incorporated into WG01 research, therefore, recognising that in kindergarten, prior to the 5 years of elementary schooling, mathematical learning occurs, and that this aspect ought to be part of WG01 investigations.

In line with this decision, since SIPEM II work focused on kindergarten education has been presented. One study (Selva, 2003) discussed the use of bar charts, and investigated the feasibility of involving very young children in the study of statistical concepts, although there is little evidence that they are taught at this level of education. The results of this study align with the framework recently proposed by Kinnear and Clark (2016), stating that children from 5 years of age can already present statistical notions, such as the use of fundamental aspects of data handling and the use of their data observations to make decisions.

Elia and Mulligan (2016), in a retrospective study presented at ICME-13, state that, internationally, there has been an increase in research focused on early childhood education in recent years, although they also point to the wide variation amongst countries in the ages eligible for kindergarten and elementary school. Studies concerning the training of teachers for work in early childhood education have also increased internationally, but an issue that permeates the concern of national and international researchers is what kind of mathematics should be addressed in early childhood education. In some countries, such as Germany, according to Gasteiger and Benz (2016), it is not common to have explicit mathematics classes in kindergarten. This is also not usual in Brazil. In this sense, it is necessary to broaden the discussion about teaching at this level of education.

Studies focusing on the role of teacher education in teaching and learning mathematics in the early years of schooling have increased during the five editions of the SIPEM. Much of this research addressed mathematical content that teachers regard as difficult, such as rational numbers.

Research on textbook analysis was first presented at SIPEM III in 2006. These studies reinforced the interest of WG01 members in investigating resources for teaching and learning. Since SIPEM III the group has been concerned with these textbook-analysis studies because of the impact that this resource has on Brazilian education.

For the first time, at IV SIPEM, in 2009, WG01 was involved in discussing difference and inclusion. A study addressing the teaching of natural numbers to deaf children motivated the discussions (Nogueira & Silva, 2009). At the next SIPEM, the group debated the issue again. At the SIPEM VI, a WG specifically focused on inclusion and differences was established, and some of WG01 members chose to migrate to present studies in this new working group (WG13).

In the first five editions of SIPEM, different methodologies were used in the studies, including: participant observation, experimental research, document analysis, phenomenological approaches, didactic engineering and case studies, and a variety of theoretical frameworks and constructs informed the activities of data collection and analysis.

Considering jointly the research into the teaching and learning of kindergarten and elementary school-aged students, and the teaching and learning resources in early schooling, 30 out of the studies presented in the first five SIPEM were related to *numbers and operations*, 5 investigated *geometry* content, and 10 addressed statistics and combinatorics (*data analysis*).

In studies involving *numbers and operations*, the main references were Gérard Vergnaud, Raymond Duval, Jean Piaget, Delia Lerner and Patricia Sadovsky, whilst for the studies involving teacher education, Lee Shulman, Donald Schön, Deborah Ball and João Pedro da Ponte were often cited. These references have had an important influence on Brazilian studies, particularly in research carried out by members of WG01, and other researchers interested in the teaching and learning of mathematics in the early years of schooling. In this area, additive and multiplicative conceptual fields, including natural and rational numbers, were widely investigated.

Data analysis was also a content area that consistently attracted attention, with studies on statistical and combinatorial thinking. *Geometry* was studied more in the SIPEM I, and *magnitudes and measurements* were not the focus of research from editions SIPEM I to V. Other topics, such as mental calculation, assessment, inclusion and algebraic thinking, recognised as important issues in elementary education, were not addressed in depth within WG01, although these have been investigated within the Brazilian mathematics education community. An example is the recent research by Silva (2016), presented at the ENEM XII, who investigated the flexibility of mental calculation by children of the fourth year in multiplication and division operations.

The choice of themes, on the one hand, favoured teaching focuses, such as statistics, probability and combinatorics (i.e. *data analysis*), that have become part of the areas of mathematics in the early schooling years relatively recently, following the publication of the Parâmetros Curriculares Nacionais (BRASIL, 1997). On the other hand, there was little or no emphasis on other areas such as *geometry* and *magnitudes and measurements*. The trend to reduce interest in geometry seems to be global. As Clements and Sarama (2016) observe, this theme has played a limited role in mathematical research related to the early years of schooling and practice in the classroom, because of the lack of emphasis on curriculum and little teacher knowledge in this area. Regarding the lack of research on magnitudes and measurement, Chambris and Dougherty (2016) have noted that, besides our little knowledge on concepts and skills related to measurement, in several countries the possible connections between magnitudes and measurement and other areas of mathematics are not being taught.

2.3 SIPEM VI: Foci of Research in Mathematics Education in the First Years of Early Schooling

A diversity of research foci and methods were presented in the 15 studies that formed the basis of discussions in WG01 during SIPEM VI in 2015. This diversity can be observed in the tables below (Tables 2.1 and 2.2).

Table 2.1 shows mathematics areas and school levels of the 15 studies presented at the seminar, and four research foci are identified: students' learning of mathematical content inside and outside the classroom; teacher education; public policies related to teaching mathematics at the beginning of schooling; and the dialogic relationship between cognitive and emotional processes. Some of these issues had not been previously debated in WG01. In particular, it is interesting to note that the public policies implemented by the Brazilian government to improve the quality of teaching and learning in basic education were discussed in five papers, whilst one contribution considered a second new theme: students' emotional relations with mathematics and their learning of the discipline.

Regarding the foci of the research, students' learning and teacher education continue to be emphasised in WG01 investigations. This is not surprising, because investigations about students and teachers are always of interest to researchers in mathematics education, whether in Brazil or other countries.

The keen interest in public policies may have been motivated by the implementation and expansion in recent years of national and state government programmes for teacher education, the analysis of textbooks and large-scale assessments, which are aimed at teachers and students from the beginning of schooling. The analysis of public policies has become, in a very relevant way, the focus of investigations for researchers in mathematics education in Brazil, and this was particularly evident amongst the researchers who participated in WG01 during SIPEM VI.

A possible explanation for the limited number of studies focusing on emotional processes in WG01 may be the fact that researchers in this area choose to discuss their work in the working group designed specifically for the study of cognitive and linguistic processes (WG09).

Table 2.1 Foci of research in mathematics education in the early years of schooling in WG01

Focus of research	Author(s)	Area (content)	School levels
Student learning	Azerêdo (2015)	Numbers and operations (multiplicative fields)	Elementary school
	Marques and Silva (2015)	Magnitudes and measures	Elementary school
	Scucuglia and Rodrigues (2015)	Numbers and operations (digital mathematical performances)	Elementary school
	Silva and Borba (2015)	Data analysis (graphs and tables interpretation)	Elementary school
	Silva, Campos, Canova and Pinheiro (2015)	Numbers and operations (fractions in quotient situations)	Elementary school
	Tôrres and Muniz (2015)	Numbers and operations (numeric registration)	EJA (youth and adult education)
Teacher education	Alencar (2015)	Numbers and operations (multiplicative fields)	Elementary school
	Guimarães and Oliveira (2015)	Data analysis (statistics—classification)	Kindergarten and elementary school
	Vieira and Lobo da Costa (2015)	Geometry (use of technology)	Elementary school
Public policies	Bellemain (2015)	Magnitudes and measures (length)	Elementary school
	Curi (2015)	Numbers and operations and geometry (spatial relations and spatial geometric figures)	Elementary school
	Pacheco and Pires (2015)	Geometry (teacher's relations with curriculum material)	Elementary school
	Pinto Júnior, Tatagiba, Alarcão, Pereira and Souto (2015)	Numbers and operations, geometry, magnitudes and measures and data analysis	Elementary school
	Vieira and Nasser (2015)	No specific content	Elementary school
Cognitive and emotional processes	Medeiros (2015)	No specific content	Elementary school

In relation to the mathematical areas involved, Table 2.1 indicates that *numbers and operations* remains the mathematical topic that the members of WG01 most investigated, with seven studies directed at this area. Furthermore, in addition to natural numbers, there are studies on rational numbers, as both the additive and the multiplicative conceptual fields are investigated. Four studies on *geometry* and three on *data analysis* were also presented and, in this edition of SIPEM, three studies involved *magnitudes and measurements*. It is interesting to note that studies focusing on geometry previously addressed diagnostic studies of students and in the SIPEM VI involved documentary research and studies with teachers in seeking to under-

Table 2.2 Foci, authors, titles, theoretical and methodological references of the research presented at the SIPEM VI

Focus	Author(s)	Title	Theoretical Reference(s)	Method
Student learning	Azerêdo (2015)	Registros semióticos do campo multiplicativo: um instrumento para o ensino nos anos iniciais	Duval (2009, 2011); Nunes and Bryant (1997); Vergnaud (1991, 2009)	Intervention study (teaching)
	Marques and Silva (2015)	Alfabetização na perspectiva da racionalidade aberta	Galvão and Di Pierro (2007); Pérez (2008)	Diagnostic study
	Silva and Borba (2015)	Desempenho de estudantes em itens sobre tratamento da informação na Provinha Brasil de Matemática	Guimarães and Oliveira (2011)	Documental research and diagnostic study
	Silva, Campos, Canova and Pinheiro (2015)	Equivalência em situação quociente: uma análise de dois estudos	Nunes and Bryant (1997, 2009); Streefland (1984, 1997)	Intervention study (teaching)
	Scucuglia and Rodrigues (2015)	A produção de performances matemáticas digitais nos anos iniciais do ensino	Borba and Villarreal (2005); Gardner (1993)	Intervention (teaching)
	Tôrres and Muniz (2015)	Identificação e análise de conhecimentos numéricos de jovens e adultos, em explicações orais e escritas	Brousseau (2008); Freire (1987); Schoenfeld (1992); Vergnaud (1990)	Diagnostic study
	Alencar (2015)	Os referenciais teóricos norteadores de pesquisas sobre a formação contínua de professores dos anos iniciais no campo conceitual multiplicativo	Shulman (1986); Vergnaud (1990)	Literature review
Teacher education	Guimarães and Oliveira (2015)	Relação entre saber classificar e o domínio da linguagem oral como determinantes na explicação oral de professores dos anos iniciais	Ortiz (1994); Piaget and Inhelder (1983); Proulx, Bednarz, and Kieran (2006); Schneuwly and Dolz (1997)	Diagnostic study
Teacher education	Vieira and Lobo da Costa (2015)	Perspectivas de mudanças no ensino de geometria com o uso de tecnologia digital: experiências em grupo de estudos de professores dos anos iniciais	Leontiev (2004); Mishra and Koehler (2006); Murphy and Lick (1998); Shulman (1986)	Action-research

(continued)

Table 2.2 (continued)

Focus	Author(s)	Title	Theoretical Reference(s)	Method
Public policy	Bellemain (2015)	Ensinar comprimento no ciclo de alfabetização	Chevallard (1985, 2002); Douady and Glorian (1989)	Documental research
	Curi (2015)	Orientações curriculares, livros didáticos, Prova Brasil de matemática do 5° ano e práticas de sala de aula: resultados de uma pesquisa longitudinal	Lerner and Sadovsky (1996); Piaget and Inhelder (1993); Sacristán (2000); Van Hiele (1986); Vergnaud (1996)	Documental research
	Pacheco and Pires (2015)	Espaço e forma nos anos iniciais e o uso de materiais curriculares	Brown (2009); Crowley (1994); Parzysz (1988); Piaget and Inhelder (1993)	Documental research and diagnostic study
	Pinto Jr., Tatagiba, Alarcão, Pereira and Souto (2015)	Matrizes de Matemática do Saeb: um estudo comparativo entre 3° e 5° anos do Ensino Fundamental	Neidorf, Binkley, Gattis, and Nohara (2006)	Documental research
	Vieira and Nasser (2015)	PNAIC no Estado do Rio de Janeiro: investigando as práticas dos formadores numa perspectiva interdisciplinar	Fazenda (1992, 2003); Japiassú (1976); Santomé (1998)	Case study
Cognitive and emotional processes	Medeiros (2015)	Satisfação e Conhecimento Matemático: uma pesquisa sobre afetos com alunos dos anos iniciais do ensino fundamental em uma escola pública do Distrito Federal, Brasil	Gómez Chacón (2003); González Rey (2005); Morin (2011); Vygotsky (2007)	Participant observation

stand difficulties with teaching in that area. Studies that reference magnitudes and measurements directly addressed the knowledge of students and discussed public policies related to this area.

It is noteworthy that the researchers of WG01 have turned their attention to different areas and mathematical content and contributed to discussions regarding teaching and learning different concepts of mathematics in early schooling. However, WG01 researchers have not studied intensively some specific content taught in the early years, such as algebra.

There may be a need for further investigation into the impact that algebra learning from the early school years can have on the development of more sophisticated algebraic thinking, as indicated by Kieran, Pang, Schifter, and Ng (2016). Another path of research, influenced by Carraher and Schliemann (2015), is noted by

Teixeira, Magina, and Merlini (2016) who sought to investigate the understanding of algebraic structures in multiplication problem situations.

As for the levels of schooling addressed in WG01 research during the SIPEM VI, 14 studies were concerned with the regular initial years (from 1st to 5th grades of elementary school), either from the perspective of students or teachers working at this level. One study also addressed the kindergarten level and one focused on Youth and Adult Education (EJA). This was the first study regarding EJA presented in WG01 in the six editions of SIPEM.

As observed in relation to the previous SIPEM, there is a continuing need for studies concerning the kindergarten years, considering that mathematical learning starts early and it is possible, in appropriate ways (in a playful way, for example), to treat mathematical concepts with very young children. Further studies with teachers and students of the EJA (Youth and Adult Education) are also necessary, since this is a frequent teaching modality in Brazil, as in the country there are still many young people and adults with little schooling. To understand better how students with little schooling develop mathematical knowledge inside and outside the school is, therefore, a very present challenge still in Brazilian reality.

2.4 SIPEM VI: Theoretical and Methodological Bases of Research in Mathematics Education in the Early Years

The 15 papers presented at SIPEM VI are shown in Table 2.2, highlighting the theoretical references that guided the studies and the methods used.

A variety of theoretical references supported the research. Some theoretical references are common to several studies; others are particular, in accordance with the subject of investigation. The references sometimes are specific to the field of mathematics education, and in other studies, the foundation draws from scholars in other areas such as psychology, sociology, language/linguistics and education.

Classical theories by scholars such as Jean Piaget referring, in particular, to the classification of collections of elements (Piaget & Inhelder, 1983) and with respect to geometry (Piaget & Inhelder, 1993) were used. Other classical theories referenced came from the work by Lev Vygotsky, discussing the historical-cultural perspective of learning (Vygotsky, 2007), and the assumptions of Leontiev (2004) about the process of involvement in activities. The impact of classical theories (constructivist, socio-interactional and historical-cultural) is, thus, evident in mathematics education of the early years of schooling.

In SIPEM VI, Gérard Vergnaud, with his theory of conceptual fields, continues as a reference in many of the studies: with regard to conceptualisation (Vergnaud, 1990) and the classification of additive and multiplicative problems (Vergnaud, 1991, 1996, 2009). Other references of a psychological nature include the studies of Terezinha Nunes and Peter Bryant, in particular, in the classification of multiplicative problems and in the discussion of equivalence in quotient situation (Nunes & Bryant, 1997, 2009). One of the studies referred to multiple intelligences (Gardner, 1993). In this sense, the strong influence of psychology in the Brazilian studies of early school years is observed, especially in the consideration of the cognitive elements of mathematical conceptualisation.

From a psychological and linguistic perspective, Raymond Duval also continues to be a reference for studies on the role of semiotic representations in students' learning of mathematics (Duval, 2009, 2011). Other research outside SIPEM also uses the theory of register of semiotic representation as a reference in studies of the early years of schooling. Such research includes, for example, investigation by Borba, Azevedo, and Bittar (2016), presented at ICME 13, who analysed the contribution in textbooks of various symbolic representations and conversions between them in the development of combinatorial reasoning.

Other approaches based on linguistic studies, such as the work by Schneuwly and Dolz (1997), are referenced with respect to verbal and nonverbal elements of spoken language, as is Oritz's work (1994) on the need for communicative competence of the teacher; and Proulx et al. (2006) are referenced with respect to teachers' explanations as a regulatory function of learning and of students' development of knowledge.

Scholars focused on mathematics didactics were also used as the basis of WG01 studies discussed during SIPEM VI. Guy Brousseau and his theory of didactic situations served as a reference regarding the notion of the didactic contract (Brousseau, 2008). The anthropological theory of the didactic (Chevallard, 1985, 2002) and the argument of Douady and Glorian (1989) on the need to articulate three domains (i.e. geometric, of magnitudes and of measurements) were the theoretical basis of a study on magnitudes and measurements. The teaching approaches of Lerner and Sadovsky (1996) and Saiz (2006) were also referenced. These cited references indicate the importance of the didactics of mathematics in WG01 studies.

From the area of educational research, Lee Shulman and his position regarding teachers' content and pedagogical knowledge (Shulman, 1986) continued to be a theoretical reference used in the studies considering the practices of teacher, as was the case in previous editions of SIPEM. The work of Mishra and Koehler (2006) was used as a theoretical basis concerning technological content knowledge. Some studies cite Paulo Freire, in particular as regards the recovery of extracurricular knowledge (Freire, 1987); Gimeno Sacristan, as regards curricula (Sacristán, 2000); Fazenda (1992, 2003), Japiassú (1976) and Santomé (1998), who are cited with regard to interdisciplinary learning; Murphy and Lick (1998), to support the creation of a study group; Morin (2012), considering knowledge as a complex construct; Almeida (2010), discussing learning by culture in line with school knowledge; Pérez (2008), comprising literacy as a resignified production; and illiteracy concepts mentioned by Galvão and Di Pierro (2007). Neidorf et al. (2006) are referenced regarding the analysis of testing. It is observed that the breadth of educational theorists used in SIPEM involves researchers not specifically concerned with the learning of mathematical knowledge.

Many mathematics educators were also used as references for the studies in SIPEM VI. Alan Schoenfeld's work is cited as related to metacognition processes (Schoenfeld, 1992) and Tony Brown's studies of teachers' relations with curriculum materials are also mentioned (Brown, 2009). Borba and Villarreal (2005) were used in a study with the concept of *human-with-media* interaction, and geared to specific mathematical content; other authors were referenced concerning the teaching and learning of rational numbers (Streefland, 1984, 1997) and the theorising of geometric thought (Parzysz, 1988; Van Hiele, 1986). The discussions of feelings and emotions in mathematical learning referred to Gómez Chacón (2003), González Rey (2005) and Morin (2011).

In short, the various frameworks of WG01 studies presented at SIPEM VI tend to have their roots in classical theories (i.e. constructivist, socio interactionist and historical-cultural). These studies are strongly influenced by psychology, specifically concerning the relations with cognition and, in particular, the analysis of the role of symbolic representations in the teaching and learning of mathematics. They are also influenced by the didactics of mathematics and studies of education of a broader nature. We see, therefore, these diverse references pointing to the variety of perspectives and approaches that are possible in research in mathematics education in the early years of schooling.

Table 2.2 also indicates a great variety of methods used in the research discussed in WG01 relating to the early years of schooling. This collective of studies shows that there is a need for more intervention studies (i.e. teaching children enrolled in kindergarten and elementary school).

2.5 Main Results of the Studies of WG01 at SIPEM VI

2.5.1 Elementary School Students' Learning

With respect to this research focus, six studies were presented and discussed in WG01 during the SIPEM VI, including three diagnostic studies with data from tests and observations (one involving documental research and diagnostic study) and three intervention studies (i.e. research that involves teaching). Five of the studies were conducted with elementary school children and one with EJA students. There was clearly great interest amongst WG01 researchers in conducting diagnostic studies directly with children, young people and adults.

In the sixth edition of SIPEM, like in the previous editions, the most frequently investigated mathematical area was *numbers and operations*, but there was also interest in the study of *data analysis*. Investigations addressed most recently included curriculum content, such as interpreting graphs and tables, and difficult content, such as multiplicative structures.

One of the studies (Azerêdo, 2015) aimed to understand the role of treatment and conversion of semiotic registers (as discussed by Duval, 2009, 2011) of 105 3rd, 4th and 5th grade students in activities involving multiplication and division, in proportion and distribution problems and a conversion activity in the direction text-drawing. The results signalled considerable difficulty in solving problems and the usage of a variety of semiotic registers, with the use of formal algorithm evidenced only by 5th grade students. When the children converted text into drawing, the results indicated an increase in the success rate, which points to the mediation role played by drawings.

Another study referred to school and extracurricular knowledge of students based on narratives and observations (Marques & Silva, 2015). The research consisted of a study of the relations between mathematical knowledge developed within and outside school, discussing cultural learning, as indicated by Almeida (2010). The study was conducted with 13 children aged 7 and 8 years old in a multigrade[1] class at a riverside community in the Amazon. For data collection, classes were observed, as were recreational activities and the journeys between school and children's homes on the school boat. Most of the time, children related *learning* only to the classroom, not recognising their learning beyond the school environment. When reporting their experiences of the harvesting and marketing of *açaí* (a fruit from northern Brazil), children showed knowledge of magnitudes and measurements, when comparing *açaí* tree heights and when demonstrating their knowledge of the supply-and-demand market system using the example of the price variation of a "rasa" (a measure of capacity) of *açaí*.

This study verifies how important it is to have more studies focused on mathematical knowledge that children and adults develop in ways that go beyond what is learnt in school, reinforcing the well-known work developed by Carraher, Carraher, and Schliemann (1985). WG01 researchers recognise strongly that much mathematics is learned out of school and discuss the role of school in mathematical development.

A third diagnostic study (Silva & Borba, 2015) regarded children's data analysis in *Provinha Brasil* (INEP, 2011), also known as Children's Literacy Assessment. It is a diagnostic evaluation that aims to investigate the development of literacy skills of children enrolled in the second year of elementary education in Brazilian state schools. *Provinha Brasil* is applied twice a year (at the beginning and the end of a school year) to have a more accurate diagnosis of children's learning in reading skills and mathematics. Adherence to this assessment is optional, and the application is at the discretion of each education department of the Brazilian states, with the implementation date of *Provinha Brasil* following a decision of each state's school system.

In addition to observing the performance of 40 children from a state school of Rio Grande do Sul in each of the 18 data-analysis items of various editions of the *Provinha Brasil*, the researchers analysed documents from the National Institute of

[1] Multigrade classes are common in Brazil, especially in rural areas where there is a shortage of teachers, students or resources. In these classes, students of different ages and levels of schooling are taught by the same teacher.

Studies and Research Anísio Teixeira (INEP).[2] The authors problematise how national evaluation questions can relate to the performance of children in interpreting graphs and tables items. The results indicate that the mode of presenting the items, the contexts involved and the numerical values influence the success and the children's interpretation ability, corroborating with the studies of Guimarães and Oliveira (2011). Usual beliefs of many mathematics educators, such as improved performance in evaluation due to the use of images or that familiar contexts facilitate reasoning, however, were not confirmed in the study. Regarding large-scale assessments and items related to graphs and tables in *Provinha Brasil*, the findings suggest that they should cover wider items and not only those familiar to children, since the images shown in statements or familiar with the subject does not always favour a proper assessment of children's performance. A wider cover of items can enable children to understand the functionality of graphs as a tool to communicate data, as indicated by Leavy and Sloane (2016).

Another investigation with students was presented with a focus on multiplicative structures (Silva, Campos, Canova, & Pinheiro, 2015), particularly on how students learn about equivalence between fractions in quotient situations. Research by Nunes and Bryant (1997), Nunes et al. (2004), and Streefland (1984, 1997) were used as theoretical frameworks for discussing the teaching and learning of rational numbers. Two studies were carried out: one with 163 5th-grade students and another with 20 students aged between 10 and 11 years old, who responded to diagnostic instruments and learnt about quotient situations. It was found that, initially, a very low percentage of students (less than 2%) showed understanding and could justify the equivalence between fractions. However, it was observed that some students, after exploring the quotient meaning, presented correct representations and understandings of equivalence between fractions. Nunes (2016), in a research review presented at ICME 13, defends that quotient situations bring opportunities for the students to relate with everyday situations, to use quantitative reasoning and to use representational tools that help them mathematise situations. It was noted that understanding the meaning of fractions can be extended if, during teaching, teachers use ratios to help students make sense of the symbols and how to manipulate them. Similar results were obtained by Utimura and Curi (2016) in a study, presented at ENEM XII, with 4th grade teachers and students learning rational numbers. The authors noted that meanings underlying representations came to be understood by students through teacher mediations and student interactions.

In another study with children (Scucuglia & Rodrigues, 2015), the role that music and film can play in mathematical learning was examined in the context of digital mathematics performances (DMP). The study took as one of its theoretical bases the notion of *human-with-media* interaction, as discussed by Borba and Villarreal (2005). In the study, teaching sessions were conducted with four groups

[2]The National Institute of Educational Studies Anísio Teixeira (Instituto Nacional de Estudos e Pesquisas Educacionais—INEP) is under the Ministry of Education (MEC) and aims to support the formulation and implementation of public policies for the area of education in Brazil, promoting studies, research and periodic reviews of the Brazilian educational system.

of students (47 in total) in which they produced digital mathematical performances from video recordings. One way to get students to reflect from recordings can be by producing DMPs, thereby relating meaning, emotion and mathematical communication. However, besides good artistic quality (visual and musical, for example), in the production of a DMP it is necessary for the mathematical meanings involved to be fully met, that is, mathematics should be kept in first plan.

A single study with youth and adult students was presented at the SIPEM VI at WG01 (Tôrres & Muniz, 2015). This study aimed to identify and analyse the production of mathematical knowledge of young people and adults in varying degrees of oral and written formalisation, of cognitive and meta-cognitive knowledge of the decimal numerical system, focusing on the impossibility of determining accurately the existing boundaries between students' school and extracurricular math skills. Just as with children, it was observed that the production of school mathematical skills by young people and by adults was affected by the locus and didactic contract (from the perspective of Brousseau, 2008) established between educator and students. Even outside of the classroom, it was observed that the rules of the didactic contract are present in the minds of students and are revealed as they perform mathematical tasks, even in the absence of the teacher. Also, students explain what they know or do not know using self-regulation arising from their beliefs and intuitions about mathematics.

2.5.2 *The Training of Teachers Who Teach Mathematics in Elementary Education*

Focus on teacher training was a trend observed over the six editions of SIPEM. Nearly 40% of the research studies presented in WG01 focused on this theme. This corroborates the evidence presented by Lee and Lin (2016) indicating the ever-growing volume of international research in recent years regarding the education of elementary school teachers. However, contrary to the international trend of the privileging of studies with teachers in initial teacher education courses, WG01 has contemplated studies almost equally about initial and continuing teacher training.

As shown in Tables 2.1 and 2.2, three of the studies presented in 2015 focused on teacher training. All three studies involved the elementary school level; the first study was a literature review, the second was a diagnostic study and the third was an action-research project. The absence of studies with kindergarten children or EJA students indicates a need for research directed towards these areas, because they have specific characteristics that are paramount for the training of elementary school teachers.

The topics that were treated in these studies were, once again, *numbers and operations* (more specifically, multiplicative structures) and *data analysis* (regarding element classification), which would indicate a greater concern with teacher training in the most recently inserted area in the Brazilian curriculum (i.e. data analysis) and contents that are considered difficult (such as multiplication, division and rational numbers). There is a consensus around the idea that they are difficult topics, in particular, how difficult it is to understand rational numbers and, especially, how to teach this topic, which has been discussed in several studies, as noted by Tirosh (2000).

The first of these studies (Alencar, 2015) presents literature review of theoretical frameworks that guide research towards continuing education of early-school teachers on the multiplicative conceptual field (as described by Vergnaud, 1990). This literature review was based on the databank of dissertations and theses of CAPES.[3] It was found that the focus of the investigations centres on theoretical issues (Shulman, 1986, amongst others), discussing teacher knowledge, teacher professional development, and reflective practice essential for an effective teacher training that enables mathematical developments by elementary school students.

The second project, a diagnostic study, involved future teachers (Guimarães & Oliveira, 2015) and aimed to investigate the construction of criteria for conducting classifications and students' oral explanations about the categorisations made by colleagues, verifying the influence of initial education in the ability to classify and explain categorisations made. To do this, classification activities by 113 students from three different nationalities (Brazilian, Canadian and Spanish) were observed. Analyses of the results indicated that the participants had difficulties in creating classification criteria regarding *exhaustiveness* and *exclusiveness*. Other studies also observed difficulties with classification (Estrella & Olfos, 2016; Kinnear & Clark, 2016), showing that classification is not an act as simple as sometimes thought. Learning to collect and organise data is a fruitful activity for students from early childhood education so that they can develop statistical reasoning (Kafoussi, 2016; Vidal-Szabó, Estrella, & Morales, 2016). Added to this, an oral explanation in mathematics can facilitate learning by linking the domain of oral discourse and the mathematical concept domain. The domain of only one of these aspects is insufficient for ensuring a quality explanation. It is essential that teachers be aware that their oral discourse may provide a greater or lesser quality of student learning and that it can lack a domain and articulation between ordinary language, mathematics language and the mathematical concepts.

Another study (Vieira & Lobo da Costa, 2015) sought to analyse processes associated with the appropriation of digital technology in a study group composed of elementary school teachers, through the use of software using 2- and 3-dimensional geometric figures. This was an action-research project aimed at encouraging teachers to mobilise technological pedagogical knowledge, technological knowledge of content (as indicated by Mishra & Koehler, 2006) and specific knowledge content in the discussion of geometry concepts. It was observed that the teachers understand the nature of the possibilities and limitations of software applications and envision how these are related to utility in their teaching practices.

[3] The Coordination of Higher Education Personnel Training (Coordenação de Aperfeiçoamento de Pessoal de Ensino Superior—CAPES) is a foundation linked to Brazil's Ministry of Education (MEC) and is engaged in the expansion and consolidation of post-graduate studies (i.e., for Masters and Doctorate degree programmes) in all Brazilian states.

2.5.3 Relations of Public Policies with Mathematics in the First Years of Elementary School

In SIPEM VI, one of the strongest research focuses was on public policy, with five studies addressing elementary schooling public policies. Again, research on kindergarten and adult education were not presented and public policies aimed at younger children or young people and adults in early schooling were not discussed.

Three of the studies involved documentary research, one articulated documentary research with a diagnostic study of teachers and students and the other was a case study on the National Pact for Literacy at the Right Age (Pacto Nacional pela Alfabetização na Idade Certa—PNAIC; BRASIL, 2014), described below. The four areas specified in the National Curriculum Parameters (PCN) (i.e. *numbers and operations*, *geometry*, *magnitudes and measurements* and *data analysis*) were involved in the studies, highlighting the breadth of research focused on public policy, using varied content and mathematical concepts from the analysis of documents and observations of teachers and students.

The first of these studies (Bellemain, 2015) aimed to analyse nationwide curriculum guidance documents and books approved by the public policy textbook evaluation programme (Programa Nacional do Livro Didático—PNLD) for the first 3 years of schooling (literacy cycle), regarding the teaching of concepts related to length.

The National Textbook Programme (PNLD) aims to subsidise the pedagogical work of teachers by distributing collections of textbooks to students of basic education. These collections are evaluated by researchers and teachers of different educational levels and described in a guide for teachers. Each year, the Ministry of Education of Brazil (MEC) acquires and distributes books contained in the guides to all students at a school level from elementary, middle and high school education.

In this first study related to public policies, relations were observed between what is prescribed in the curriculum guidance documents and the curriculum presented in textbooks. It found a positive variety of types of activities proposed in the books, but the emphasis on teaching length lies in the use of units and conventional instruments, which differs from what is preconised in the curriculum guidelines. It was also noted that content treating aspects of length is strongly focused on strengthening numbers and numerical operations (contrary to the recommendations by Douady & Glorian, 1989, on the need to articulate geometric, magnitude and the measurement domains) and focusing on numbers and operations bring little benefit to teachers on how to explore connections with extracurricular practices.

The importance of investigations of educational resources such as textbooks is emphasised and has been a constant concern for WG01 members. Research in this area has pointed to improvements in the quality of textbooks in Brazil and indicates further work that still is needed. Public policies related to these improvements have been in place in Brazil since the mid-1990s, in contrast to other countries, in which government actions in this direction have been implemented only recently. In Ireland, for example, it was only in 2011 that the National Council for Curriculum and Assessment decided to intervene in the development of textbooks (O'Keeffe, 2014), and in other countries, such as the Czech Republic, according to Moraová (2014), many books that are currently used were published in the early 1990s and have become outdated over time.

The second study regarding the focus upon public policies (Curi, 2015) aimed to analyse curriculum guidelines, textbooks and external evaluations that public policy bodies have developed to assist teachers in their mathematics teaching practices in the classroom. The study was motivated by the difficulties teachers experience in accessing and understanding official documents, in particular with regard to *numbers and operations* and *geometry*. It involved an analysis of curriculum documents from the Secretary of Education of São Paulo that report on results of the SAEB (an evaluation system described below), collections of textbooks, testimonials and teacher reports.

The System of Basic Education Assessment (SAEB—Sistema de Avaliação da Educação Básica) consists of two processes: The National Basic Education Assessment (ANEB—Avaliação Nacional da Educação Básica) and the National School Performance Assessment (ANRESC—Avaliação Nacional do Rendimento Escolar). ANEB is carried out by sampling in each Brazilian state and focuses on the efforts of educational systems, and receives the name SAEB in its dissemination. ANRESC is more extensive and detailed than ANEB and focuses on each Brazilian school, receiving the name *Prova Brasil* in its dissemination.

In this second study, it is evident that there is consistency between the prescribed curriculum documents and the assessment instruments proposed by public bodies and aimed at improving the teaching of mathematics. However, in the *presented curriculum* (textbooks) and the *curriculum-in-action* (experienced in the classroom), using the terminology proposed by Sacristán, (2000), there are major differences between the *prescribed* and the *evaluated curricula*. In the *presented curriculum*, gaps were observed, for example, in addressing characteristics of the decimal number system (as also indicated by Lerner & Sadovsky, 1996), an absence of some of the meanings associated with numerical operations, and, in relation to geometry, books do not contain different representations of space—contrary to recommendations made by Saiz (2006). Moreover, the greatest difficulties are the actions of the teachers, who, despite presenting planning objectives consistent with the curriculum documents, reveal a practice that resembles their experiences as students of basic education.

The third study that focused on public policies (Pacheco & Pires, 2015) problematised the fact that education departments of different Brazilian regions have produced and implemented curriculum materials without proper monitoring and without drawing on the results of scientific research. The objective was, therefore, to analyse how these materials are interpreted and used in the classroom by teachers in the early school years. The study focused specifically on material related to geometry. An analysis was made of the curriculum materials associated with the project "Mathematics Education in the Early Years", in the light of the model of the teacher's relations with curriculum material, proposed by Brown (2009). Observations of two teachers' classes in which the material was used (in the first and fifth year of elementary education in state schools in São Paulo) were undertaken and semi-structured interviews with teachers were also carried out. It was noted that curricular materials can have greater impact on teaching practice than the prescribed curriculum. It was also observed that teachers use the materials in different ways (reproduce, adapt or create situations), reflecting their views and knowledge. Similarly, Cortês and Muniz (2016) analysed implicit curriculum concepts in the practices of early schooling teachers and observed that public policy teacher education has partially influenced the practices; the prescribed curriculum has had little reach in practice, and learning assessments usually occur at the end of the process, making curricular reorganisations impossible.

The fourth study focused on public policies (Pinto Junior, Tatagiba, Alarcão, Pereira, & Souto, 2015) observed convergences and divergences between 3rd and 5th grade reference matrices of the SAEB (INEP, 2010) described above. The analysis of documents in which the results associated with SAEB are presented (Brasil, 1997, 2002, 2012, 2013) demonstrated that the assessments present convergences between 3rd and 5th grade reference matrices, both in their dimensions of content and cognitive dimension, although the 3rd grade reference was constructed 12 years after the 5th grade reference, and by another group of experts. Advances in skills that are measured by the national assessment system were also observed, except in the area of *data analysis*, which is probably justified by more recent discussions in skills to be developed in statistics education.

The fifth study, focused on the analysis of public policies (Vieira & Nasser, 2015), is a case study that aimed to analyse the challenge of developing an interdisciplinary perspective upon mathematics and Portuguese in the teacher education programme, National Pact for Literacy at the Right Age (PNAIC).

PNAIC is a national public policy that aimed at ensuring a high quality of literacy education during the first 3 years of schooling. This programme, conducted in all Brazilian states in 2014, led experts in the areas of language and mathematics to develop and implement interdisciplinary in-service education processes.

This fifth study applied questionnaires with 14 Portuguese and 13 mathematics teachers who are PNAIC group members of the state of Rio de Janeiro. The results indicated the need for breaking paradigms, such as the teaching of isolated disciplines (see Fazenda, 1992, 2003; Japiassú, 1976; Santomé, 1998, who point towards an interdisciplinary search for completeness, but considering also the particularities of the disciplines) deepening concepts; valuing collective decisions and exchange of knowledge between areas; providing different perspectives and more productive discussions. In a similar direction, Souza, Souza, and Passos (2016) found contributions of the language area, in children's stories suggested in PNAIC notebooks for the mathematical literacy of children of the first 3 years of schooling.

2.5.4 *Dialogues Between Cognition and Emotion in Mathematics Learning*

One study presented at SIPEM VI (Medeiros, 2015) focused on affection and the satisfaction of students in the early years of elementary education concerning mathematical knowledge, emphasising that mathematics learning involves cognition and affection (Gómez Chacón, 2003). The paper represents an area little explored in WG01. The investigation was characterised as participant observation within a 4th-grade class (children 8–10 years) in a state school. The study noted that for a better acquisition of mathematical knowledge by the students, it is also necessary to consider the satisfaction of students' learning processes. Similarly, in another study (Muniz, 2016) presented at ENEM XII, a WG01 researcher sought to understand the production of subjective senses in the history of mathematics learning of children at risk, i.e. those that have their fundamental rights violated or threatened by injury, whether by act or omission of the society or the state; or missing, omission or abuse of parents or guardians. The study enabled a schemes analysis and revealed elements of the complex learning processes of these children.

2.6 Final Thoughts

This chapter has aimed to show the approaches adopted in WG01 to investigate the development of mathematical knowledge of students in and out of school, initial and continuing teacher education, a variety of resources for the teaching of mathematics and an inclusive perspective of mathematics education. Some topics have been intensely investigated, such as the mathematical conceptual development of children and teacher training. Others have been the subject of a good number of investigations, such as teaching resources (textbooks and new technologies). Some topics require further research and debate, especially kindergarten education, teaching and learning of young people and adults in early education, the development of mathematical knowledge in extracurricular environments and the inclusion of all children in mathematical learning. Some of the topics have been the research targets of other working groups of the Brazilian Society of Mathematics Education; nonetheless, it is considered important that they are also targets of specific discussions by the research group aimed at studies on teaching and learning of mathematics in the first years of schooling.

The body of studies developed by WG01 has involved the following different mathematical areas: most of the studies related to *numbers and operations*; a good number of studies focused on *geometry* and *statistics* and *combinatorics* (but no study focused on *probability*), and a few studies addressed magnitudes and measurements. Other Brazilian studies regarding probability in the early years (such as Batista & Borba, 2016) have been conducted, although these studies have not been presented at WG01. Therefore, the debate regarding different concepts and mathematical areas are valued by Brazilian researchers of the initial years of schooling, in order to provide a balanced weight to diverse mathematical knowledge in the classroom.

Different problematisations have generated distinct themes and diverse objects of investigations, with different research goals from diverse theoretical frameworks, which are largely established in the international literature. These references are from the area of mathematics education, in conjunction with other areas of knowledge, such as education, psychology, linguistics and sociology.

In WG01 studies, several investigative methods were used (mostly diagnostic and very few intervention projects): case studies, participant observation and documental research. This variety of perspectives may contribute to better investigate processes associated with teaching and learning, and enable both directly and indirectly the understanding of practices and uses of mathematics. It is noteworthy, however, that few of the investigations aimed to analyse specifically teaching and learning in the classroom, and this is one of the focuses that needs more research.

Several specific concepts have been the focus of WG01 investigation with emphasis on contents that are considered difficult, both in the understanding of children and of teachers. For example, *multiplicative structures* have been the subject of research from different perspectives—the student, the teacher, the curriculum and teaching resources. More recent topics of the curriculum proposals, such as statistics and combinatorics, have also been the subject of constant research. Algebraic thinking, however, is a topic that still needs to be the focus of research, since recent curriculum proposals have noted the need to work with algebraic concepts from the beginning of schooling. This should boost investigations into practices in the classroom and teacher education to work on the development of algebraic thinking.

The focus of research in recent areas of national and state curricula and in some contents rarely worked in class seem to have been some of the strongest contributions of WG01 of the Brazilian Society of Mathematics Education, including: (a) the investigation on how school students can develop different modes of reasoning, such as geometric, statistical and combinatorial, amongst other ways of thinking, and (b) establishing the possibility of treating more complex concepts with children from kindergarten and elementary school.

Amongst the consensuses resulting from the various studies conducted by WG01 is a vision of mathematics articulated with other areas of education and supported in meaningful contexts for students, including daily extracurricular activities. Beliefs, emotions and desires are also aspects to be considered.

Also, in many studies conducted by the researchers of the group, it is recognised that symbolic representations play a very important role in mathematics, and in explanation, argumentation and communication in teaching and learning processes. These are essential tools for mathematical knowledge and are not only accessories to this understanding and, therefore, require close attention in the monitoring of teaching and learning processes, mostly in the first years of schooling.

In particular, regarding government action, it is noteworthy that several WG01 studies examine public policy regarding teacher training, courseware production and large-scale assessment of students' mathematical performances. It is necessary to investigate further teachers' understanding in relation to these policies and their implementation in the teaching of mathematics, to understand better how the prescribed curriculum, based on research, can be approached in the classroom. Assessment processes also require special attention to examine the knowledge acquired and in development by the students; and serve to support the work of teachers in the classroom, allowing reflections and improvements in the quality of mathematics teaching in early school years. Thus, WG01 studies on the learning of students have sought to allow teachers a better understanding of the processes of teaching and learning, including the evaluation processes conducted locally and those that result from public large-scale evaluation policies.

The investigations of Brazilian public policies, such as the evaluation of textbooks, continuing teachers' education and assessment of student learning, amongst others, point to some of the advances that these policies have provided for Brazilian mathematics education, preparing teachers for practice in the classroom, providing high quality educational resources and checking advances in the mathematical knowledge of students in early schooling. They also show that there is still much to be done.

2.7 Articles Presented by WG01 at the SIPEM VI

Alencar, E. (2015). Os Referenciais teóricos norteadores de pesquisas sobre a formação contínua de professores dos anos iniciais no campo conceitual multiplicativo. *Anais eletrônicos do Seminário Internacional de Pesquisa em Educação Matemática* (SIPEM VI), Pirenópolis-Brasil.

Azerêdo, M. (2015). Registros semióticos do campo multiplicativo: um instrumento para o ensino nos anos iniciais. *Anais eletrônicos do Seminário Internacional de Pesquisa em Educação Matemática* (SIPEM VI), Pirenópolis-Brasil.

Bellemain, P. (2015). Ensinar comprimento no ciclo de alfabetização. *Anais eletrônicos do Seminário Internacional de Pesquisa em Educação Matemática* (SIPEM VI), Pirenópolis-Brasil.

Curi, E. (2015). Orientações curriculares, livros didáticos, Prova Brasil de matemática do 5° ano e práticas de sala de aula: resultados de uma pesquisa longitudinal. *Anais eletrônicos do Seminário Internacional de Pesquisa em Educação Matemática* (SIPEM VI), Pirenópolis-Brasil.

Guimarães, G., & Oliveira, I. (2015). Relação entre saber classificar e o domínio da linguagem oral como determinantes na explicação oral de professores dos anos iniciais. *Anais eletrônicos do Seminário Internacional de Pesquisa em Educação Matemática* (SIPEM VI), Pirenópolis-Brasil.

Marques, V. & Silva, C.A. (2015). Alfabetização na perspectiva da racionalidade aberta. *Anais eletrônicos do Seminário Internacional de Pesquisa em Educação Matemática* (SIPEM VI), Pirenópolis-Brasil.

Medeiros, A. (2015). Satisfação e Conhecimento Matemático: uma pesquisa sobre afetos com alunos dos anos iniciais do ensino fundamental em uma escola pública do Distrito Federal, Brasil. *Anais eletrônicos do Seminário Internacional de Pesquisa em Educação Matemática* (SIPEM VI), Pirenópolis-Brasil.

Pacheco, D. & Pires, C.M.C. (2015). Espaço e Forma nos anos iniciais e o uso de materiais curriculares. *Anais eletrônicos do Seminário Internacional de Pesquisa em Educação Matemática* (SIPEM VI), Pirenópolis-Brasil.

Pinto Junior, W., Tatagiba, A., Alarcão, C., Pereira, C. & Souto, W. (2015). Matrizes de matemática do SAEB: um estudo comparativo entre 3° E 5° anos do Ensino Fundamental. *Anais eletrônicos do Seminário Internacional de Pesquisa em Educação Matemática* (SIPEM VI), Pirenópolis- Brasil.

Scucuglia, R. & Rodrigues, A. (2015). A produção de performances matemáticas digitais nos anos iniciais do ensino. *Anais eletrônicos do Seminário Internacional de Pesquisa em Educação Matemática* (SIPEM VI), Pirenópolis- Brasil.

Silva, A. F. G., Campos, T. M. M., Canova, R. F., & Pinheiro, M. G. C. (2015). Equivalência em situação quociente: uma análise de dois estudos. *Anais eletrônicos do Seminário Internacional de Pesquisa em Educação Matemática* (SIPEM VI), Pirenópolis-Brasil.

Silva, J.A. & Borba, R. (2015). Desempenho de estudantes em itens sobre tratamento da informação na Provinha Brasil de Matemática. *Anais eletrônicos do Seminário Internacional de Pesquisa em Educação Matemática* (SIPEM VI), Pirenópolis-Brasil.

Torres, P. & Muniz, C. (2015). Identificação e análise de conhecimentos numéricos de jovens e adultos, em explicações orais e escritas. *Anais eletrônicos do Seminário Internacional de Pesquisa em Educação Matemática* (SIPEM VI), Pirenópolis-Brasil.

Vieira, E.R. & Lobo da Costa, N.M. (2015). Perspectivas de mudanças no ensino de geometria com o uso de tecnologia digital: experiências em um grupo de estudos de professores dos anos iniciais. *Anais eletrônicos do Seminário Internacional de Pesquisa em Educação Matemática* (SIPEM VI), Pirenópolis-Brasil.

Vieira, E.R. & Nasser, L. (2015). PNAIC no Estado do Rio de Janeiro: investigando as práticas dos formadores numa perspectiva interdisciplinar. *Anais eletrônicos do Seminário Internacional de Pesquisa em Educação Matemática* (SIPEM VI), Pirenópolis-Brasil.

References

Alencar, E. S. (2015). A divisão no campo conceitual multiplicativo: Comparações entre pesquisas. *Anais XII Encontro Nacional de Educação Matemática – Enem*, São Paulo, Brasil.

Almeida, M. C. (2010). *Complexidade, saberes científicos, saberes da tradição*. São Paulo: Ed. Livraria da Física (Col. Contextos da ciência).

Batista, R., & Borba, R. (2016). No jogo é a moeda que diz, não é a gente que quer não: O que dizem crianças sobre a probabilidade. *Revista Vidya, 36*(2).

Borba, R., Azevedo, J., & Bittar, M. (2016). Brazilian Primary School textbooks: Symbolic representations in combinatorial situations. *Paper presented at the 13th International Congress on Mathematics Education*, TSG 12, Hamburg, Germany, June 24–31.

Borba, M. C., & Villarreal, M. E. (2005). *Humans-with-media and the reorganization of mathematical thinking: Information and communication technologies, modelling, experimentation and visualization* (Vol. 39). New York: Springer.

BRASIL. (1997). *Ministério da Educação. Secretaria da Educação Básica. Parâmetros Curriculares Nacionais – Matemática* (Vol. 3). Brasília: MEC.

BRASIL. (2002). Ministério da Educação. Secretaria da Educação Básica. *Saeb 2001: Relatório Nacional*. Brasília: Inep, 303p.

BRASIL. (2012). *Ministério da Educação. Elementos Conceituais e Metodológicos para Definição dos Direitos de Aprendizagem e Desenvolvimento do Ciclo de Alfabetização (1°, 2° e 3° anos) do Ensino Fundamental*. Brasília: MEC.

BRASIL. (2013). *Avaliação Nacional da Alfabetização (ANA): Documento básico*. Brasília: INEP.

BRASIL. (2014). *Ministério da Educação. Secretaria da Educação Básica. Pacto Nacional pela Alfabetização na Idade Certa – Alfabetização Matemática*. Brasília: MEC.

Brousseau, G. (2008). *Introdução ao estudo das situações didáticas: Conteúdos e métodos de ensino*. São Paulo: Ática.

Brown, M. W. (2009). The teacher-tool relationship: Theorizing the design and use of curriculum materials. In J. T. Remillard, B. A. Herbel-Eisenmann, & G. M. Lloyd (Eds.), *Mathematics teachers at work: Connecting curriculum materials and classroom instruction* (Studies in Mathematical Thinking and Learning Series, A. Schoenfeld, Ed.). New York: Routledge.

Carraher, T. N., Carraher, D. W., & Schliemann, A. D. (1985). Mathematics in the streets and in schools. *British Journal of Developmental Psychology, 3*(1), 21–29.

Carraher, D., & Schliemann, A. (2015). Powerful ideas in elementary school mathematics. In L. English & D. Kirshner (Eds.), *Handbook of international research in mathematics education*. New York: Routledge.

Chambris, C., & Dougherty, B. (2016). *Abstract of TSG 9 Teaching and learning of measures (focus on primary education)*. Retrieved from http://www.icme13.org/topic_study_groups

Chevallard, Y. (1985). *La transposition didactique: Du savoir savant au savoir enseigné*. Grenoble: La Pensée Sauvage.

Chevallard, Y. (2002). Organiser l'étude. Structures & fonctions. In J.-L. Dorier, M. Artaud, M. Artigue, R. Berthelot, & R. Floris (Eds.), *Actes de la 11ª École d'été de Didactique des Mathématiques* (pp. 3–22). Grenoble: La Pensée Sauvage.

Clements, D. H., & Sarama, J. (2016). Young children's conceptualization and learning of geometric figures. *Paper presented in13th International Congress on Mathematics Education*, TSG 12, Hamburg, Germany, June 24–31.

Cortês, S. A., & Muniz, C. A. (2016). Considerações sobre a organização e o desenvolvimento curricular pelo professor e sua relação com o processo de ensino e aprendizagem de matemática nos anos iniciais. *Anais XII Encontro Nacional de Educação Matemática – Enem*, São Paulo, Brasil.

Crowley, L. M. (1994). O modelo de Van Hiele no pensamento geométrico. In M. Lindquist, & A. Shulte (Org.), *Aprendendo e Ensinando Geometria*. São Paulo: Atual.

Douady, R., & Glorian, M. J. P. (1989). Un processus d'apprentissage du concept d'aire de surface plane. *Educational Studies in Mathematics, 20*(4), 387–424.

Duval, R. (2009). *Semiósis e pensamento humano: registro semiótico e aprendizagens intelectuais*. (Trad. Lênio Fernandes Levy e Marisa Rosâni Abreu da Silveira). São Paulo: Editora Livraria da Física.

Duval, R. (2011). *Ver e Ensinar a Matemática de outra Forma: entrar no modo matemático de pensar: Os registros de representação semiótica* (Org. Tânia M. M. Campos; trad. Marlene Alves Dias. 1. ed.). São Paulo: PROEM.

Elia, I., & Mulligan, J. (2016). *Abstract of TSG 1 early childhood mathematics education (up to age7)*. Retrieved from http://www.icme13.org/topic_study_groups

Estrella, S., & Olfos, R. (2016). Early statistics: Graphical representations and transnumeration in 3rd grade students. *Paper presented in13th International Congress on Mathematics Education*, TSG 15, Hamburg, Germany, June 24–31.
Fazenda, I. C. (1992). *A Integração e Interdisciplinaridade no ensino brasileiro: Efetividade ou ideologia?* São Paulo: Loyola.
Fazenda, I. C. (2003). *Interdisciplinaridade: Qual o sentido?* São Paulo: Paulus.
Freire, P. (1987). *Pedagogia do oprimido*. Rio de Janeiro: Paz e Terra.
Galvão, A. M., & Di Pierro, M. A. (2007). *Preconceito contra o analfabeto*. São Paulo: Cortez.
Gardner, H. (1993). *Inteligências Múltiplas: A Teoria na Prática*. Porto Alegre: Artes Médicas.
Gasteiger, H., & Benz, C. (2016). Professional knowledge for early mathematics education. *Paper presented in13th International Congress on Mathematics Education*, TSG 45, Hamburg, Germany, June 24–31.
Gómez Chacón, I. (2003). *Matemática emocional: Os afetos na aprendizagem matemática*. Porto Alegre: Artmed.
González Rey, F. (2005). *Sujeito e subjetividade*. São Paulo: Thomson.
Guimarães, G., & Oliveira, P. (2011). Educação Estatística e Avaliações em Larga Escala: uma análise de itens e descritores. In *CIAEM - Conferência Interamericana de Educação Matemática*. Recife: UFPE.
INEP. (2010). Instituto Nacional de Estudos e Pesquisas Educacionais. In *Matriz de Referência para Avaliação da Alfabetização Matemática Inicial*. Brasília: MEC.
INEP. (2011). Instituto Nacional de Estudos e Pesquisas Educacionais. *Provinha Brasil de Matemática*. Brasília: MEC.
Japiassú, H. (1976). *Interdisciplinaridade e Patologia do saber*. Rio de Janeiro: Imago.
Kafoussi, S. (2016). Teaching statistics in primary school: Collecting and organizing data. *Paper presented in13th International Congress on Mathematics Education*, TSG 15, Hamburg, Germany, June 24–31.
Kieran, C., Pang, J., Schifter, D., & Ng, S. F. (2016). Early algebra, research into its nature, its learning, its teaching topical diagnosis studies. In *13th International Congress on Mathematics Education*, Springer Open.
Kinnear, V. A., & Clark, J. A. (2016). Young children's abductive reasoning about data. *Paper presented in13th 13th International Congress on Mathematics Education*, TSG 15, Hamburg, Germany, June 24–31.
Leavy, A. M., & Sloane, F. (2016). Prospective primary teachers' understandings of graphs. *Paper presented in13th International Congress on Mathematics Education*, TSG 15, Hamburg, Germany, June 24–31.
Lee, H. F., & Lin, P. J. (2016). A review of number and arithmetic in teacher education. In *8th International Conference on Technology and Mathematics Education and Workshop of Mathematics Teaching*. Taiwan, May 16–17 (pp. 105–113).
Leontiev, A. (2004). *O desenvolvimento do psiquismo* (2nd ed.). São Paulo: Centauro.
Lerner, D., & Sadovsky, P. (1996). O sistema de numeração: Um problema didático. In C. Parra & I. Saiz (Eds.), *Didática da Matemática: Reflexões psicopedagógicas*. Porto Alegre: Artmed.
Mishra, P., & Koehler, M. J. (2006). Technological pedagogical content knowledge: A framework for teacher knowledge. *Teachers College Record, 108*(6), 1017–1054.
Moraová, H. (2014). Pre-service and in-service teachers' preference when selecting mathematics textbooks. In *Proceedings of the International Conference on Mathematics Textbook Research and Development*, UK (pp. 351–357).
Morin, E. (2011). *O método 3: O conhecimento do conhecimento* (4nd ed.). Trad. Juremir Machado da Silva. Porto Alegre: Sulina.
Morin, E. (2012). *Os sete saberes necessários à educação do futuro*. São Paulo: Cortez.
Muniz, C. A. (2016). As crianças que calculavam: O ser matemático como sujeito produtor de sentidos subjetivos na aprendizagem. *Anais XII Encontro Nacional de Educação Matemática – Enem*, São Paulo, Brasil.

Murphy, C., & Lick, D. (1998). *Whole faculty study groups: A powerful way to change schools and enhance learning*. California: Corwin.

Neidorf, T. S., Binkley, M., Gattis, K., & Nohara, D. (2006). *Comparing Mathematics Content in the National Assessment of Educational Progress (NAEP), Trends in International Mathematics and Science Study (TIMSS), and Program for International Student Assessment (PISA) 2003 Assessments (NCES 2006-029). U.S. Department of Education*. Washington, DC: National Center for Education Statistics.

Nogueira, C., & Silva, M. C. (2009). A escrita numérica de crianças surdas fluentes em libra. *Anais do IV Seminário Internacional de Pesquisa em Educação*. Taguatinga, Brasil: Matemática – SIPEM.

Nunes, T. (2016). Teaching and learning about whole numbers in primary school. Topical Diagnosis studies. In *13th International Congress on Mathematics Education*, Springer Open.

Nunes, T., & Bryant, P. (1997). *Crianças fazendo matemática*. Porto Alegre: Artes Médicas.

Nunes, T., Bryant, P., Pretzlik, U., Evans, D., Wade, J., & Bell, D. (2004, January 28–31). *Vergnaud's definition of concepts as a framework for research and teaching*. Paper presented at the annual meeting for the Association pour la Recherche sur le Développement des Compétences, Paris.

Nunes, T., & Bryant, P. (2009). *Key understandings in mathematics learning, paper 3: Understanding rational numbers and intensive quantities*. London: Nuffield Foundation.

O'Keeffe, L. (2014). Mathematics textbook analysis; supporting the implementation of a new mathematics curriculum. In *Proceedings of the International Conference on Mathematics Textbook Research and Development,* UK (pp. 365–371).

Ortiz, E. (1994). El perfeccionamiento del estilo comunicativo del maestro para su labor pedagógica. *Lenguaje y Textos, 8* [Universidad de la Coruña].

Parzysz, B. (1988). "Knowing" vs "seeing": Problems of the plane representation of space geometry figures. *Educational Studies in Mathematics, 19*(1), 79–92.

Pérez, C. L. V. (2008). Alfabetização: Um conceito em movimento. In R. L. Garcia & E. Zaccur (Orgs.). *Alfabetização: Reflexões sobre saberes docentes e saberes discentes*. São Paulo: Cortez.

Piaget, J., & Inhelder, B. (1983). *Gênese das Estruturas Lógicas Elementares* (3nd ed.). Rio de Janeiro: Zahar Editores.

Piaget, J., & Inhelder, B. (1993). *A representação do espaço na criança*. Porto Alegre: Artmed.

Proulx, J., Bednarz, N., & Kieran, C. (2006). Caractéristiques des explications orales en classe de mathématiques: construction d'un cadre d'analyse pour rendre compte de la pratique des futurs enseignants et futures eiseignants de mathématiques du secondaire. *Canadian Journal of Science, Mathématics and Technology Education, 6*(3).

Sacristán, J. G. (2000). *O currículo: Uma reflexão sobre a prática*. Porto Alegre: ArtMed.

Saiz, I. E. (2006). A direita...de quem? Localização espacial na educação infantil e séries iniciais. In M. Panizza (Ed.), *Ensinar matemática na educação infantil e séries iniciais: Análise e propostas* (pp. 143–166). Porto Alegre: Artmed.

Santomé, J. T. (1998). *Globalização e interdisciplinaridade: O currículo integrado*. Porto Alegre: Artmed.

Schneuwly, B., & Dolz, J. (1997). Les genres scolaires. Des pratiques langagières aux objets d'enseignement. *Repères, 15*, 27–40.

Schoenfeld, A. H. (1992). Learning to think mathematically: Problem solving, metacognition, and sense making in mathematics. In D. Grouws (Ed.), *Handbook of research on mathematics teaching and learning* (pp. 334–370). New York: MacMillan.

Selva, A. (2003). Trabalhando com manipulativos e gráficos de barras: A resolução de problemas aditivos por crianças de 6 a 8 anos. In *Anais do II Seminário Internacional de pesquisa em educação Matemática*. Santos, Brasil: SIPEM.

Shulman, L. S. (1986). Those who understand: Knowledge growth in teaching. *Educational Researcher, 15*(2), 4–14.

Silva, A. F. G., Campos, T. M. M., Canova, R. F., & Pinheiro, M. G. C. (2015). Equivalência em Situação Quociente: uma análise de dois estudos. In: *VI SIPEM - Seminário Internacional de Pesquisa em Educação Matemática*. Pirenópolis, Brasil: VI SIPEM.

Silva, C. C. R. (2016). Flexibilidade de cálculo mental nas operações de multiplicação e divisão. *Anais XII Encontro Nacional de Educação Matemática – Enem*, São Paulo-Brasil.

Souza, T. F., Souza, A. P. G., & Passos, C. L. B. (2016). Programa de formação de professores Pacto Nacional Pela Alfabetização na Idade certa: um olhar para as práticas de leituras e escritas nas aulas de matemática. *Anais XII Encontro Nacional de Educação Matemática – Enem*, São Paulo, Brasil.

Streefland, L. (1984). Search for the roots of ratio: Some thoughts on the long term learning process (Towards... A theory) Part 1: Reflections on a teaching experiment. *Educational Studies in Mathematics, 15*, 327–348.

Streefland, L. (1997). Charming fractions or fractions being charmed? In T. Nunes & P. Bryant (Eds.), *Learning and teaching mathematics. An international perspective* (pp. 347–372). Hove, UK: Psychology Press.

Teixeira, C., Magina, S., & Merlini, V. (2016). Introdução do raciocínio funcional para estudantes do 5° ano do Ensino fundamental. *Anais XII Encontro Nacional de Educação Matemática – Enem*, São Paulo, Brasil.

Tirosh, D. (2000). Enhancing prospective teachers' knowledge of children's conceptions: The case of division of fractions. *Journal for Research in Mathematics Education, 31*(1), 5–25.

Utimura, G. Z., & Curi, E. (2016). Algumas revelações de alunos e professoras 4° ano do Ensino Fundamental sobre o significado parte-todo dos números racionais. *Anais XII Encontro Nacional de Educação Matemática – Enem*, São Paulo, Brasil.

Van Hiele, P. (1986). *Structure and insight*. Orlando: Academic Press.

Vergnaud, G. (1990). La théorie des champs conceptuels. *Recherches en Didactique des Mathématiques, 10*(23), 133–169.

Vergnaud, G. (1991). *El Niño, Las Matemáticas y La Realidad: Problemas de La enseñanza de Las Matemáticas en La Escuela Primaria*. México: Trillas.

Vergnaud, G. (1996). A Teoria dos Campos Conceituais. In J. Brun (Ed.), *Didáctica das Matemáticas* (pp. 155–191). Lisboa: Instituto Piaget.

Vergnaud, G. (2009). *A criança, a matemática e a realidade: problemas do ensino da matemática na escola elementar*. Tradução de Maria Lúcia Faria Moro. Curitiba: Ed. da UFPR.

Vidal-Szabó, P.; Estrella, S., & Morales, S. (2016). Early statistic: Situations, invariants and representations for the treatment of data tables of kindergarten to grade 2. *Paper presented in13th International Congress on Mathematics Education*, TSG 15, Hamburg, Germany, June 24–31.

Vygotsky, L. (2007). *A formação social da mente*. São Paulo: Martins Fontes.

Chapter 3
Research on Mathematics Education in Middle School in Brazil

Claudia Lisete Oliveira Groenwald, Carmen Teresa Kaiber, and Silvia Dias Alcântara Machado

Abstract This chapter discusses research carried out by the Working Group Mathematics Education in Middle School (6th, 7th, 8th, and 9th grades), WG02, of the Brazilian Society of Mathematics Education, SBEM. The objective of WG02 is to collect, map, and present research on mathematics education in the middle school, to disseminate the results obtained to the community of mathematics teachers and educators, and to identify the main requirements concerning teaching and learning mathematics in these school grades. Amongst the investigations examined were studies addressing public policies as well as curricular requirements and methodological approaches associated with these school grades. Teachers' thoughts and beliefs and the influence of their conceptual frameworks on the development of their work as teachers were also examined.

3.1 Introduction: The Formation and Articulation of WG2

The Working Group *Mathematics Education in Middle School* (WG2) is formed by researchers that investigate the mathematics teaching and learning process in middle school in Brazil (6th, 7th, 8th, and 9th grades, when students are between 11 and 14 years old) and in the country's youth and adult literacy programme (EJA). Researchers seek their theoretical references in the light of learning theories, in didactics, in the main trends of mathematics education and in curricular theories (Groenwald & da Silva, 2013).

Along with WG2, several other working groups were formed during the first edition of the International Seminar of Research on Mathematics Education (SIPEM I, Seminário Internacional de Pesquisa em Educação Matemática), held in the year 2000. One of these groups, *Mathematics Education in High School* (WG3), has had a close relationship with WG2, meeting together as one group from the second to

C. L. O. Groenwald (✉) · C. T. Kaiber
Universidade Luterana do Brasil, Canoas, Brazil

S. D. A. Machado
Pontifícia Universidade Católica de São Paulo, São Paulo, Brazil

A. J. Ribeiro et al. (eds.), *Mathematics Education in Brazil*,
https://doi.org/10.1007/978-3-319-93455-6_3

the fifth edition of the event, between 2003 and 2012, while still maintaining their distinct identities and memberships. During SIPEM V, the members of both groups decided that the WG2 and WG3 should work independently again, starting from SIPEM VI, in 2015. But the result was not as expected, and the two groups have decided to work together again as a single group in the future.

In order to better understand the scope of WG2, we clarify a few aspects of the Brazilian education system, which includes not only elementary and college education, but also mathematics teacher education.

In Brazil, elementary school is organised as three levels, namely kindergarten, with three grades and students aged between 3 and 5; elementary school, with nine grades and students aged between 6 and 14 (elementary school is subdivided in elementary School I, from the 1st to the 5th grades and is equivalent to the elementary, or grade school in the US education system, and elementary school II, from the 6th to the 9th grades and is equivalent to middle school in the US education system); and high school, with three grades and students between 15 and 17 years of age. Also, the country has a youth and adult literacy programme (Educação de Jovens e Adultos—EJA), offered to students over 18 who did not finish school grades at the expected age.

In relation to teaching qualifications, mathematics teachers obtain a university degree in Mathematics Teacher Education lasting 3200 h. A mathematics teacher thus qualified may teach in middle school, high school, and EJA.[1]

It should be noted that the teacher qualified to teach in early childhood education and elementary education I is the professional formed by undergraduate courses in early childhood education or in higher education courses with a teaching qualification, being a multipurpose teacher (who assumes all the subjects in class).

Therefore, the organisational structure of elementary and high school, as well as of teacher education means that many teachers teach in both education levels and also that researchers often develop studies addressing these two education levels. These factors, taken together, contributed to the decision to merge WG2 and WG3 in the future.

In this chapter however, we concentrate upon the research trajectory of WG2, beginning by reflecting upon the editions of SIPEM prior to SIPEM VI, before moving on to an analysis of the works presented in the last edition of the event. We also discuss the direction of future efforts by WG2.

3.2 Investigation Trends of WG2: A Research Record

Since SIPEM I, in 2000, WG2 has concentrated upon studies carried out from different perspectives. According to Pires, Lopes, and Fainguelernt (2000), the studies presented in that first edition of SIPEM were grouped under three areas: (1) didactic

[1]In Brazil, a mathematics teacher qualifies to teach k11 to k17 in the North American education system.

and methodological approaches, which discuss elements related to the preparation of curriculum suggestions, the use of didactic and methodological resources, and work associated with (the learning of) specific areas of mathematical content; (2) teacher thinking and the influence of conceptual frameworks in a teacher's work, especially concerning students, learning, motivation, and students' interests and behaviours; (3) and curriculum issues, official curriculum programmes, and the influence they have on mathematics teaching and learning.

The discussions and reflections developed by members of WG2 in SIPEM I illustrated a concern with the process of incorporating the curriculum guidelines emerging at the time, indicating that, despite the absence of significant criticism of the notions embedded in these guidelines, these did not manifest in the classroom. Besides, the poor performance of students in mathematics, expressed by the high retention rates in the discipline and, mainly, the results obtained by middle school students in mathematics performance assessments carried out in the country by the National System of Elementary School Evaluation (SAEB) were highlighted as indicators of the condition of teaching of the discipline.

Therefore, the report issued by the organisers of WG2 (Pires et al., 2000), along with the studies published in the proceedings of the first SIPEM, showed that, although these studies were carried out from a variety of theoretical and methodological perspectives, they were articulated around discussions about curricula and curriculum guidelines.

In the II SIPEM, one of the concerns of the members of WG2 was related to the implementation of public policies regarding the teaching of mathematics in Brazil (Pires, 2003). The study also underscores the fact that the decisions made about mathematics teaching were historically characterised by questionable procedures. The analyses of the school reform efforts called 'Francisco de Campos', in 1931, and 'Gustavo Capanema', in 1942 showed, for instance, that curricular innovations in mathematics as proposed by Euclides Roxo in the 'Francisco de Campos' reform were challenged by the following reform effort, but based on poor arguments. An example was the unification of the mathematics topics of algebra, arithmetic, and geometry in one single discipline, so as to address the three of them from an unified approach, and the guideline that established that the teaching of deductive geometry should be preceded by a practical approach to geometry. These ideas gained strength subsequently. Though the 'Francisco de Campos' reform included the extension of the concept curriculum beyond a mere list of contents to be taught, to include a discussion of didactic directions, this important landmark did not materialise in the following reform effort (Pires, 2003).

Pires (2003) also underscores that, besides the lack of actions towards curricula implementation, follow-up and evaluation of innovations proposed simply were not implemented, which does not afford appropriate assessments about what was done, whether correctly or not. One of the consequences thereof was the existence of prescribed curricula (as detailed in official publications), and of actual curricula (as developed and taught by teachers in the classroom). Apart from that, consistent data to implement the required changes and to invest in what brings about good results were also lacking.

Maranhão and Machado (2009) emphasised the fact that the investigations discussed and that sparked the interest of WG2 during SIPEM II and III addressed methodological, historical, social, cultural, linguistic, psychological, pedagogical, curricular, and didactic aspects (considering the tripartite structure student-teacher-mathematical knowledge, or segments of this structure) in middle school. The authors also drew attention to the fact that these aspects characterised the interests of WG2 about the topics of the studies that were presented in the two editions of SIPEM.

Another point of relevance that led to research interests in the following SIPEM (IV and V) was the implementation of the National Curricular Standards (PCN) (Brasil, 1998) by the Education Ministry of Brazil and, in the same period, the adoption of the National Curriculum Guidelines for Teacher Education introduced by the National Education Council of Brazil. These initiatives were supported by a Federal Government Act (Act 9394), from December 20, 1996, which defined the competence of the Federal Government, in cooperation with States, the Federal District, and Municipalities of Brazil to establish directives to curricula so as to guarantee uniform elementary education across the country.

The question of centralization or decentralisation of curricula was hence a predominant issue in the discussions within WG2, underscoring the fact that decentralisation, despite affording the advantage of including regional aspects in curricula, also brings about problems, since by assigning curriculum development to states and municipalities, regional inequalities would emerge. Economically and socially developed regions have the best conditions to prepare updated curriculum projects, while less developed areas do not enjoy such comparatively favourable structures to incorporate research advancements into specific knowledge areas or into didactic and pedagogical fields, which highlights inequalities and makes the access to quality education more difficult.

Finally, in the latest edition of SIPEM, in 2015, the discussions that emerged from the presentation of the works led to reflections that addressed the importance and the concepts around the organisation of a National Common Curriculum Basis (BNCC) currently under deliberation by the Education Ministry of Brazil. At the event, the possible consequences of BNCC on the elementary school curriculum and on the results of third-party evaluation initiatives like the Elementary School Evaluation System (SAEB) and the Program for International Student Assessment (PISA) were discussed. In this topic, WG2 plays an essential role, holding detailed discussions spanning all segments of school and academia so that implementation efforts by BNCC reach the schools across the country, influencing the actual curriculum adopted at this teaching level, and affecting the quality of the teaching and learning process. In the following section, we present a more detailed analysis of the research trends evident in the most recent seminar.

3.3 Current Research Trends in WG2

This section describes the research presented and discussed in SIPEM VI. Seven studies were the results of PhD theses, while three were MSc dissertations. This emphasises the growing interest in graduate studies by teachers and other professionals in mathematics education in Brazil, also observed based on the increasing number of *stricto* sensu courses offered by the universities in the country.

Concerning the theoretical framework that guided the studies presented, several authors and trends stand out, especially those addressing the development of the teaching and learning process of elementary school contents, the curriculum as a field of research, didactics in French mathematics, and the social-cultural theory. It should be underlined that digital technologies were also discussed, in two studies, in line with a quite strong research trend in mathematics education. This small number of studies focused on digital technologies (20% of the total number of studies) may be partly associated with the fact that a work group currently investigates mathematics education technologies specifically.

The research methodologies adopted in the studies presented in WG2, SIPEM VI, are qualitative in nature. The category of qualitative research that was most often present was the case study (40%). The development and application of activities, adopted in 90% of the studies, was the qualitative research category that was adopted mostly in outstanding case studies.

Concerning the themes that have been included recurrently in the discussions held by WG2, in addition to contemporary issues in Brazilian education, the studies presented were grouped under the categories Didactic and Methodological Approaches, Curriculum and Official Programmes, and Teacher Thinking about Teaching and Learning (Table 3.1). In addition to this characterisation, the WG2 understood the importance of underlining the theme or content, research focus, the participating subjects, and the research that analysed curricular documentation in elementary school.

Regarding the theme Didactic and Methodological Approaches, which is a trending topic in a significant number of investigations discussed in all meetings, five studies were presented, four of which were focused on students and one that included the analysis of teachers' practice in the use of geometric construction techniques.

The theme Curriculum and Curricular Documentation was addressed in four studies. Three of these looked into the analysis of curricular documentation such as teaching programmes, school programmes, and the National Curricular Standards (PCN). One investigation introduced a curricular proposal for elementary school II focused on the competencies that are important for a student as he or she concludes this education level to become an engaged citizen.

Concerning the theme Teachers' Thinking, only one study was presented. Barbosa (2015) discussed the teaching process around the fundamental theorem of arithmetic (FTA) by students of the 6th grade of middle school considering the

Table 3.1 Research presented by WG2 in SIPEM VI

Title/authors	Theme/content	Participants
Didactic and methodological approaches		
An Electronic Didactic Sequence for Decimal Numbers for the 6th Grade in Junior High School—Fiuza and Groenwald (2015)	Decimal numbers	6th grade students—Middle school
Linear Equations: a Recovery of Contents Based on an Electronic Didactic Sequence—Lemos and Kaiber (2015)	Linear equations	7th grade students—Middle school
The Technological-Theoretical Discourse and its Contribution to the Development of Techniques to Solve Geometric Construction Problems—de Jesus (2015)	Geometric construction techniques	Math teachers—Middle school and high school
Patterns in Teaching and Learning Mathematics—dos Santos (2015)	Patterns	8th and 9th grade students—Middle school
Embodied or Symbolic? A Journey through the Concept Image Developed by a Student—Freire and Lima (2015)	Rational numbers	6th grade students—Middle school
Curriculum and official programmes		
Curriculum Based on Competencies for Junior High Schools of the 15th CRE—Zanoello and Groenwald (2015).	Curriculum based on competencies	Math teachers—Middle school
Institutional Relationships in Brazilian Official Documents about Linear Polynomial Equations—Barbosa and Lima (2015)	Linear equations	–
Irrational Numbers in Elementary Education: Curriculum Documents and Students' Knowledge—Rezende and Nogueira (2015)	Irrational numbers	Brazilian and French students—Middle school and high school
The Teaching Object in Algebra: Research, Curriculum Programs, and the Teachers' Voice—Panossian and Moura (2015).	Teaching of algebra	Math teachers
Teacher thinking		
Learning the Fundamental Theorem of Arithmetic by Students of the 6th Grade of Junior High School—Barbosa (2015)	Fundamental theorem of arithmetic	Teachers and students—Middle school

Source: SIPEM VI annals

analysis of the answers given by a group of teachers taking a continuous formation programme and by training teachers of mathematics. However, we deem it important to underscore this theme, in view of the fact that it was also addressed in the research carried out by Panossian and Moura (2015), who collected data from teachers' written notes and recorded statements about the constitution of algebra as a teaching object. Another investigation that covered the same subject was that by Zanoello and Groenwald (2015), which discussed the directions towards a curriculum proposal based on the teachers' views in three categories, namely teachers' concepts of teaching, the teachers' work in classrooms, and the participation of teachers in events and continuous formation programmes. The relevance of this theme, for WG2, lies in the fact that the system of ideas and the concepts held by teachers to this education level direct the actions in the classroom.

In addition to the classification based on themes, other categories surfaced as relevant in the investigations discussed in WG2, namely elementary school contents, analysis of curricular documentation in elementary school, and participants of studies.

In the category covering the elementary school contents addressed in the studies, four investigations were about arithmetic (more specifically decimal numbers, rational numbers, and irrational numbers, as well as the fundamental theorem of arithmetic), four were about algebra, two were about linear equations, and only one was about geometry.

The investigation about decimal numbers by Fiuza and Groenwald (2015) was characterised by a methodological proposal that integrated the main content with the transversal theme Work and Consumption for 6th grade students in middle schools using communication and information technologies as a didactic tool. The content was organised as an electronic didactic sequence, and the objective was to assess the potential of this sequence, which was implemented (developed, applied, and evaluated) using computerised system as teaching strategy to study the concepts of the theme analysed. According to the authoresses of the study, the didactic sequence was structured so as to enable students to learn in their own learning pace. For the teacher, the aim was to look into the difficulties experienced by students, which represents a way to collect information for an individualised activity planning aiming to help students solve these problems.

The study presented by Freire and Lima (2015) discussed the concepts of rational numbers as fractions, and analysed the solution of one problem about the subconstruct 'quotient' before and after the content about fractions was taught. As a diagnostic investigation, the two solutions obtained by a 6th grader of a middle school are analysed as a means to identify the image of the concept invoked to solve the problem proposed.

The investigation about irrational numbers by Rezende and Nogueira (2015), which is also diagnostic in character, analysed the knowledge about irrational numbers in middle school and high schools in Brazil and the equivalent grades in France. The study was based on the solution of nine activities during individual interviews. Based on the theory of conceptual fields introduced by Vergnaud, the results indicated that the knowledge about irrational numbers acquired by French students matches the knowledge Brazilian students acquired. The authoresses also concluded that it was not possible to find any significant differences between the students from both countries.

The research presented by Barbosa (2015) discussed the learning process around the fundamental theorem of arithmetic by 6th graders in middle school based on the analysis of the answers given by a group of mathematics teachers and training teachers. The questions were: (1) What strategies are appropriate for students to understand the concept of prime number? (2) How to promote the comprehension of the decomposition of numbers in prime factors and its use as a means to simplify calculations? (3) What are the arguments adopted by students during the signification process cited previously? (4) What procedures do they adopt? (5) What are the most common mistakes? (6) What is the cause of these mistakes? These questions were answered considering the results of a previous study presented by Barbosa (2005) and tasks proposed by the same author, based on the theory of conceptual fields taught to 6th grade students of a middle school. The answers revealed that the learning process of the TFA by these students is not linear, with continuities and ruptures, since the students make mistakes. Considering the situations that the students experience, they produce generalisations, which was perceived as a set of ideas manifested by these students' teachers.

The work presented by dos Santos (2015), *Patterns in Teaching and Learning Mathematics* aimed to identify the strategies used by a group of 8th and 9th grade students of a middle school to solve tasks about patterns, regularities, and generalisations focused on tasks about arithmetic and geometric patterns. The results showed that the students experienced difficulties to represent the general member, and could not reach a mathematical expression, but produced one out loud. Also, students thought and seemed to have rhetorical mathematical knowledge, taking support essentially from discursive arguments, as opposed to mathematical language. Concerning the work with images, the knowledge demonstrated about patterns was intuitive.

The research presented by Lemos and Kaiber (2015) investigated the development and application of a didactic sequence about linear equations to recover concepts and procedures about that content. The proposal focused on the work with students with difficulties to learn the topic and was developed with 21 7th graders of a middle school. The results showed that most difficulties are about conceptual questions, which affected the development of tasks such as maintaining equality using the distributive property, use of the additive and multiplicative principles, and the transition from the natural language to the algebraic language, which influenced the performance of the students in solving problems.

The study presented by Barbosa and Lima (2015) discussed the teaching of algebra in Brazil concerning linear equations and the analysis of teaching programmes in the country on nation-wide and state-wide scales as a preliminary model of the mathematical praxes around the solution of linear equations based on tasks, techniques, and technologies. These were used to characterise, analyse, and compare the programmes considering PCN data and teaching programmes implemented in 15 Brazilian states. The authors underscore the fact that official documents analysed implicitly show that the teaching of linear equations is a tool to solve social problems, indicating that the metaphor about scales is mentioned in five regional documents as didactic resource, and indicate that it should not be used strictly for manipulation purposes.

The research presented by Panossian and Moura (2015) looked into the constitution of the object of teaching algebra based on the fundamentals of the historical-cultural theory. The authors took, as basic premise, the idea that the object of algebra as seen by mathematics researchers differs from that seen by the researchers on mathematics teaching. Diagnostic in character, the study unveiled a panorama of research about the process of teaching and learning algebra and analysed curriculum proposals based on data obtained from teachers' written and voice records, which are essential elements in the teaching of algebra, as acknowledged by the researchers, namely the role of symbolism, the variable, the relationships between magnitudes and the process of generalisation, even though there are different conceptions about each of them and the approach they are given in teaching. The authors emphasise that the different conceptions about the nature of algebra and of teaching algebra are also revealed by curricular programmes and that, based on the historical-cultural theory, it is possible to see that the relationship between generalisation as a cognitive skill of a student to be developed and/or a process to be carried out over mathematical objects is necessarily dialectical. The cognitive skills of a student in generalising, according to the authors, is developed while the student conducts generalisation processes. At the same time, the generalisation processes develop to the measure the student has the cognitive skills required.

The work presented by de Jesus (2015) was the only one addressing geometry. The objective was to investigate how technological and theoretical contents about the mathematical object bisection and the technique used to construct it were made explicit as constitutive elements of the techniques used to solve geometric construction tasks. The tasks were focused on the construction of the definition of bisection and the explanations of the properties that surface from this definition, from which the teachers, in a continuous education course, had to justify the involved geometric constructions. The techniques developed also were shown to be a useful tool to solve new tasks, enlarging their scope of applications.

Four were the studies in the category Curricular Documentation of Elementary School. One of them had the objective to present a curriculum proposal for the high school grades focused on the skills that are important and necessary for a student at this level to become a citizen (Zanoello & Groenwald, 2015). The analysis of curricular documentation such as PCN and the mathematics curriculum adopted in the region the research was carried out, together with the opinions of teachers about the possibility and relevance of the proposal of a curriculum based on competencies for mathematics were considered the results of the research.

The three other studies in this category focused on the analysis of curricular documentation such as teaching programmes, school programmes, and PCN concerning the knowledge presented by students or by teachers' opinions. These three studies covered specific themes and contents, such as the object of teaching algebra (Panossian & Moura, 2015), the institutional relationships in official Brazilian documentation about linear equations (Barbosa & Lima, 2015), and irrational numbers from the perspective of curricular documentation and students' knowledge about the topic (Rezende & Nogueira, 2015), contrasting with the general character of a competency-based curriculum.

The PCN, which is a paper published by the Ministry of Education of Brazil contains guidelines for teaching systems to optimise their curricula and teaching programmes. Although the PCN is not an official programme, it was used in the analysis of the four studies in this category.

Despite having distinct objectives and methodologies, the four investigations about curricula and curricular documentation were discussed throughout the work of WG2 in SIPEM VI and extended beyond the discussions on mathematics education: the issue of Brazilian education public policies. Therefore, WG2 reached a consensus about the fact that the discussion on curriculum and curricular proposals requires a more comprehensive discussion concerning the way Brazilian education policies are conducted. The main condition to be met for the development of education as a whole and of mathematics education, in particular, is the implementation of public policies that guarantee student populations access to quality education, helping them remain at school, and the participation of society in the definition, implementation, management, and evaluation of these policies. WG2 understands that it has to play a role also in this aspect.

In the category of participating subjects in research, considering the ten investigations presented in WG2, five had students as participants, four had teachers, and one addressed the institutional relationships in official Brazilian documentation on a specific content, linear polynomial equations (Barbosa & Lima, 2015).

The investigations concerning students, their interests, motivations, learning, and cognitive development show that research places the student at the centre of the education process, which is partly confirmed by the fact that three out of the four investigations whose participants were students addressed tasks in the classroom aiming at investigating aspects of knowledge acquisition, looking into the difficulties regarding learning in terms of specific educational proposals (Fiuza & Groenwald, 2015; Lemos & Kaiber, 2015), and analysing the cognitive development of students in the work with a given content, rational numbers (Freire & Lima, 2015). The discussions in WG2 on the topic confirm the view researchers have of the importance of research on learning and student development. However, this question was not discussed alone, since it was considered in view of the understanding that the teacher, the student, and knowledge contents maintain intrinsic relationships that influence one another and that, even when one investigation focuses specifically on one of these elements, the others are always present.

This understanding directed the discussions about the investigations that had teachers as participants. Only the research presented by Barbosa (2015) covered the view or the understanding teachers have of a specific question. The study discussed the learning process of FTA from the standpoint of teachers and training teachers. The other three studies also considered the opinions of teachers about this topic of research.

The work presented by de Jesus (2015) looked into the contributions of technological-theoretical works in the development of techniques to solve geometric construction tasks based on the analysis of the opinions mathematics teachers have on this question. The investigation presented by Zanoello and Groenwald (2015) aimed to lay the foundations for a curriculum based on competencies for middle school, considering also the concepts teachers have about it in the region the study was carried out. Similarly, the research conducted by Panossian and Moura (2015), which looked into the teaching of algebra based on curricular programmes and research, also considered teachers' opinions.

Therefore, the understanding about the importance of teacher thinking and the influence of its conceptual framework on the development of a teacher's work, already addressed in SIPEM I, in 2001, actualises itself and persists as the focus of investigations in WG2, indicating the pathway for future research.

3.4 Future Investigations in WG2

The data, analyses, and reflections presented in this chapter, by focusing on the investigations concerning middle school in Brazil, indicate that the studies discussed in WG2 offer important contributions about the transformation of approaches to mathematics contents at school and the mathematics education formation programmes.

We understand that WG2 has the social commitment of discussing public policies that influence the curriculum of middle school and of considering the problems and the requirements associated with this education level, which may spark questions and investigations. By addressing the classroom environment, WG2 investigates the didactical and methodological approaches to teach contents that form the curriculum, analysing when and how to develop them and the possibilities concerning school evaluation, both internal and external to the school environment.

The discussions and reflections made by WG2 indicate that the future requirements about research tend to remain focused on the themes mentioned above, though the group is always attentive to new requirements that may appear from emerging problems concerning this school level.

References

Barbosa, G. S. (2005). *Construção dos Conceitos de Múltiplo e Divisor à Luz da Psicologia de Vygotsky*. Dissertation, Universidade Santa Úrsula.

Barbosa, G. S. (2015). A Aprendizagem do Teorema Fundamental da Aritmética por Alunos do 6° Ano do Ensino Fundamental. In *Anais do VI SIPEM*, Goiás, Brazil.

Barbosa, J. T., & Lima, A. P. A. B. (2015). As relações Institucionais em Documentos Oficias Brasileiros sobre Equações Polinomiais do Primeiro Grau. In *Anais do VI SIPEM*, Goiás, Brazil.

Brasil (Secretaria de Educação Fundamental). (1998). *Parâmetros curriculares nacionais: matemática: Introdução aos parâmetros curriculares nacionais*. Brasília: MEC/SEF.

de Jesus, G. B. (2015). O discurso tecnológico-teórico e sua contribuição na emergência de técnicas para resolver tarefas de construção geométrica. In *Anais do VI SIPEM*, Goiás, Brazil.

dos Santos, L. G. (2015). Padrões no Ensino e na Aprendizagem Matemática. In *Anais do VI SIPEM*, SBEM, Goiás.

Fiuza, R. P., & Groenwald, C. L. O. (2015). Sequência didática eletrônica com a temática Números Decimais para o 6° ano do Ensino Fundamental. In *Anais do VI SIPEM*, Goiás, Brazil.

Freire, P. C., & Lima, R. N. (2015). Corporificada ou Simbólica? Uma Jornada pela Imagem de Conceito de um Aluno. In *Anais do VI SIPEM*, Goiás, Brazil.

Groenwald, C. L. O., & da Silva, M. A. (2013). *Educação Matemática; contribuições para as séries finais do Ensino Fundamental e Médio*. Canoas: ULBRA.

Lemos, A. L., & Kaiber, C. T. (2015). Equações de 1° Grau: uma recuperação de conteúdos a partir de uma Sequência Didática Eletrônica. In *Anais do VI SIPEM*, Goiás, Brazil.

Maranhão, C., & Machado, S. (2009). *Educação Matemática nos Anos Finais do Ensino Fundamental e no Ensino Médio*. São Paulo: MUSA.

Panossian, M. L., & Moura, M. O. (2015). O objeto de ensino da álgebra: pesquisas, programas curriculares e a fala dos professores. In *Anais do VI SIPEM*, Goiás, Brazil.

Pires, C. C. (2003). Educação Matemática nas séries finais do Ensino fundamental: As pesquisas focalizadas pelo GT2. *Livro de resumos do II* SIPEM: SBEM.

Pires, C. M. C., Lopes, A. J., & Fainguelernt, E. K. (2000). Educação Matemática nas séries finais do Ensino Fundamental. Livro de Resumos do I SIPEM: SBEM.

Rezende, V., & Nogueira, C. I. (2015). Números Irracionais na Educação Básica: Documentos Curriculares e Conhecimentos dos Alunos. In *Anais do VI SIPEM*, Goiás, Brazil.

Zanoello, S. F., & Groenwald, C. L. O. (2015). Currículo por Competências para os anos finais do Ensino Fundamental da 15ª CRE. In *Anais do VI SIPEM*, Goiás, Brazil.

Chapter 4
Brazilian Mathematics Education in High School

Célia Maria Carolino Pires, Elenilton Vieira Godoy, Marcio Antonio da Silva, and Vinício de Macedo Santos

Abstract The aim of this chapter is to share with the national and the international academic community an overview of Brazilian mathematics education research, based on the papers approved and presented at the International Seminars on Research in Mathematics Education (SIPEM) by the Working Group Mathematics Education in High School (WG03). Aimed originally at creating a space that would gather research addressing this level of basic schooling in Brazil (students from 15 to 17 years old), in this chapter the group resumes its trajectory. To present and analyze the results of the research studies, we organized the themes in three axes, namely: Teaching and Learning; Curriculum and Curricular Materials; Youth and Adult Education (EJA), Special Education, Professional Training and Teacher Training. We concluded that the studies associated with the Teaching-Learning axis were responsible for more than 60% of the presentations in these events. However, in the last three editions, the studies associated with Curriculum and Curricular Materials and with EJA, Special Education, Professional and Teacher Training axes divided the attention, especially in the last SIPEM, where presentations involving the Curriculum theme gained more space. We conclude the chapter by suggesting that the groups involved in research about high school investigate some paramount questions the debates have missed so far.

C. M. C. Pires
Universidade Cruzeiro do Sul, São Paulo, SP, Brazil

E. V. Godoy (✉)
Federal University of Paraná, Curitiba, PR, Brazil
e-mail: evgodoy@terra.com.br

M. A. da Silva
Federal University of Mato Grosso do Sul, Campo Grande, MS, Brazil

V. de M. Santos
University of São Paulo, São Paulo, SP, Brazil
e-mail: vms@usp.br

A. J. Ribeiro et al. (eds.), *Mathematics Education in Brazil*,
https://doi.org/10.1007/978-3-319-93455-6_4

4.1 Introduction

High school education is an expression that identifies one of the stages of the basic education in Brazil. It is subsequent to elementary school and previous to higher education. The current Guidelines and Bases of National Education—Law number 9.394/96 (Lei de Diretrizes e Bases da Educação Nacional—Lei 9.394/96) establishes two levels of school education: basic education (elementary and high school) and higher education, and also it sets the modes: youth and adult education program, professional or technical education and special education for elementary or high school level of basic education and it still establishes the distance education mode for the basic and higher levels.

The inclusion of high school in the basic education and its progressively mandatory characteristic shows the recognition of the political and social importance it owns. Young people[1] (from 15 to 17 years old) were attracted by a growing demand on schooling and the need to compete in the narrow labor market as well as socialize the population in a new logic of the labor world. The high school education represents a challenge to the public policies. Indeed, the high school education has never had a very clear identity apart from the stepping-stone to university or the professional education. Since mid-1990s, the Brazilian public high school has been expanded in a meaningful way. However, the federal government recently put its mandatory premise through the Constitutional Amendment number 59/2009, which broadens the mandatory school from 6- to 17-year-old people, following a trend of the growing number of students who finish the major stage of basic education, answering the international pressures. High school expansion policies answer not only the grass-roots aspirations for more schooling, but also the needs of placing the country in the international scenery according to development criteria which privileges the education and the children and younger level of schooling.

Concerning the teaching quality offered by the schools in this level of schooling, there are great controversies and critics related to the high school former students development. In this aspect, the controversial issue involves both the diverse interests that formulate and manage the educational policies in different federal bodies (in federal, state, and municipal scope) and the effectiveness of the same policies. The curricular debate is extremely intense when it refers to this final stage of basic education in Brazil, including the subjects' organization model offered and its developed contents and methodologies.

Since the Law of the Guidelines and Bases of National Education (LDBEN 9394/96), the high school education is considered the final stage of the basic education in Brazil. As a result of the LDBEN 9394/96, official documents were drawn up for different teaching levels, in which, for the high school education, the 1999 National Curriculum Parameters (PCNEM) stand out. Aiming to educate a citizen

---

[1]According to the 2015 Educational Census (BRASIL, 2015), there are about 8,300,189 students enrolled in the regular high school and 1,308,786 in the youth and adult education program.

that would be prepared for the labor market and also for the continuation of the studies in the more technological and multidisciplinary contemporary society, the PCNEM proposed a teaching and learning process that privileged the development of skills and construction of cognitive competence.

Thus, the document shifted from the schooling subject contents as an ending in themselves. In doing so, the curriculum-organizing axis became "contextualization and interdisciplinarity."

> The central criterion is the contextualization and interdisciplinarity, that means, it is the subject potential that allows connections between a diversity of mathematical concepts and different ways of mathematical thinking or even the cultural relevance of the subject concerning its applications inside or outside mathematics and its historical importance in the development of sciences itself. (Brasil, 1999a, p. 87–88).

Changes occurred in different teaching levels of Brazilian education, from the regulation point of view analyzed in this text, have had a great impact on expanding research on sciences and mathematics, teaching, and mathematics education, and regarding the latter, especially for mathematics, it owns a status of empowered and legitimated knowledge acting as a mandatory subject for all the students of high school.

Therefore, we consider important to analyze those shifts, the formulation of educational policies for high school education and its impact on the research field of mathematics education, from a research mapping presented on the six editions of the Research International Seminar on Mathematical Education—SIPEM in the Mathematics Education in High School Group. To begin with, we gathered the subjects in four axes: Curriculum and Curricular Materials and Youth and Adult Education Program (EJA), Special Education, Professional and Teacher Education.

4.2 Research Classification and Features According to Each Axis

The Teaching and Learning axis encompasses 60.5% of the presentations. Thirty four working papers were classified in this axis (category). Most of them are concentrated on algebra learning (ten papers), followed by the ones dedicated to geometry learning (nine papers). The others are shown in the following chart:

Chart 4.1 Teaching and Learning Research

Axis	SIPEM						**TOTAL**
	I	II	III	IV	V	VI	
Geometry learning	1	1	3	3	1		9
Arithmetic learning			3		1	1	5
Algebra learning	2	2	1	1	2	2	10
Calculus learning			1		1		2
Trigonometry learning			1				1

Information treatment learning			1				1
Methodological resources			1			1	2
Contextualization and Interdisciplinarity			1	1			2
Proofs and Demonstrations			1				1
Reading, writing, and learning				1			1
TOTAL	3	3	13	6	5	4	34

The Curriculum and Curricular Materials axis encompasses 27% of papers with very different subjects, but with some emphasis on curricular materials, prescribed and practiced curriculum, mathematics purpose on curriculum, and theoretical conceptions and methodologies. In the chart below, the other subjects are presented:

Chart 4.2 Curriculum and Curricular Materials Research

Axis	SIPEM						**TOTAL**
	I	II	III	IV	V	VI	
Curricular material			1			2	3
Compared studies						1	1
Prescribed and practiced curriculum				1		2	3
Educational reforms			1				1
Purposes	1				1		2
Competence			1				1
Theoretical and epistemological concepts				1	1	1	3
TOTAL	1	0	3	2	2	6	14

In the Youth and Adult Education Program (EJA), Professional Education and Special Education axis, 9% of the papers are concentrated mostly in Youth and Adult Education Program (EJA), as it is shown in the following chart:

Chart 4.3 Youth and Adult Education Program (EJA), Professional and Special Education Research

Axis	SIPEM						TOTAL
	I	II	III	IV	V	VI	
Youth and adult education				2	1		3
Special education			1				1
Professional education				1			1
TOTAL	0	0	1	3	1	0	5

Finally, in the Teacher Training axis, there are 3.5% of the papers, in which one of the articles refers to conceptions underlying high school teacher training, and another one approaches teacher training practices, as shown in the following chart:

Chart 4.4 Teacher Training Research

Axis	SIPEM						TOTAL
	I	II	III	IV	V	VI	
Conceptions		1					1
Practices					1		1
TOTAL	0	1	0	0	1	0	2

After the first categorization, we aimed at classifying the papers, starting from their titles in the Teaching and Learning, Curriculum and Curricular Materials, Youth and Adults Education, Professional Education, Special Education, and Teacher Training Research axes. Thus, for each axis we prepared the summary reading of the presented papers in the six events of SIPEM and created four subcategories as follows: (a) aims; (b) theoretical framework; (c) methodology; and (d) results. Finally, we identified the key elements envisioned in each of the four subcategories we designed to be described.

4.2.1 Teaching-Learning Categorization Axis

In the Teaching and Learning axis, 34 papers were classified, divided into seven categories as presented in Table 4.1.

Table 4.1 shows that the majority (55.5%) of the presented working papers in the Teaching and Learning axis is about algebra learning or geometry learning. The paper in trigonometry learning category, by Silva and Neto (2006), represents only 3% of the total, the same percentage as for information treatment learning, which, since II SIPEM has turned into a specific working group. It is interesting to notice that there are more papers associated with calculus learning, by Rezende (2006) and Silva (2012), than with trigonometry learning. Therefore, this needs a deeper investigation, since trigonometry is a broadly explored subject during the 3 years of high school, and teachers understand it as a difficult subject to work with, according to research done by Godoy and Pires (2003).

Table 4.1 Teaching and Learning Axis

Categories	Quantity	Percentage (%)
Algebra learning	10	29.0
Arithmetic learning	5	15.0
Calculus learning	2	6.0
Geometry learning	9	26.5
Information treatment learning	1	3.0
Trigonometry learning	1	3.0
Methodological resources	6	17.5
TOTAL	34	100

Table 4.2 Aims subcategory

Keyword	Quantity	Percentage (%)
Student	15	44.0
Content	14	41.0
Teacher	5	15.0
TOTAL	34	100

After classifying the papers into seven categories in Table 4.1, we analyzed each of the 34 papers based in the following subcategories: (a) aims; (b) theoretical framework; (c) methodology; and (d) results, intending to map and characterize the papers in WG03 in the Teaching and Learning axis.

4.2.1.1 The Aims as a Subcategory

The immersion in each paper from the abovementioned subcategories tried to identify one word or a key expression that would characterize them. As aims, we identified the keywords "student," "content," and "teacher"; that is, in the papers in this axis, the aims are related to one of these three words. It is worth highlighting that we chose the keyword that characterized mostly the paper, since, in a way, every paper included at least two of these three keywords. Table 4.2 presents the results found.

Table 4.2 shows a low percentage of papers with the teacher as a main aim. It is possible that such percentage is due especially to the existence of the Working Group "Mathematics Teacher Training" (GT 07).

For this reason, the papers classified in the algebra learning category, written by Moretti (2000), Santos and Mattos (2006), Groenwald and Becher (2009), Júnior (2003), and Lucas and Gualandi (2015), had mathematical content as main aim; finally, the study taken by Rosenbaum (2012) focused on a teacher, from a discussion about how to align constructivist learning perspectives to the teaching planning of trigonometric functions.

The papers classified under the geometry learning category, by Tinoco and Nasser (2000), Pirola, Quintiliano, and Proença (2003), Proença and Pirola (2006, 2009), had the student as main aim; Pietropaolo (2006), Silva (2006), Bezerra (2009), and Teles and Bellemain (2009) highlighted the mathematical content as main aim; lastly, Oliveira et al. (2012) pointed out the teacher, starting from the learning object (OA) draft involving the conical studies.

The papers classified into the methodological resources category, by Franco and Moretti (2006), Pereira, Mattos, and Nascimento (2006), and Nishio and Leal (2009), highlighted the mathematical content as main aim; Ninow and Kaiber (2015) and Oliveira and Lopes (2009) papers emphasized the teacher; Pirola et al. (2006) focused on the students, by investigating their development in solving problems with assertions that presented expendable data.

The papers that fell within arithmetic learning, by Xavier and Lutaif (2006) and Silva and Perovano (2012), highlighted the student; Fonseca (2006) and Campos and Rodrigues (2006) emphasized the mathematical content; Pinto and Laudares (2015) focussed on the teacher from the presentation about objects learning (AO) directed to high school professional courses in the electronics field.

In the papers within calculus learning category, Rezende (2006) and Silva (2012) enhance as main objective the mathematical content; Silva and Neto's paper (2006) classified on the trigonometry learning category focussed on students through verifying their knowledge about definitions, the relationships, and the problems involving the trigonometric ratios in the right triangle; Pacheco and Medeiros's paper (2006), classified in information treatment learning, also pointed out the students through analyzing their performance when approaching problems characterized as verbal problems within the scope of the introductory combinatorial analysis.

Finally, among the aims associated to the keyword "student" we highlight: to verify the mathematical knowledge; to verify the argumentation capacity; to analyze the performance; to identify the previous knowledge; to map the competences and abilities; to diagnose obstacles; to create contextualized and interdisciplinary problem-situations. Concerning the keyword "content," the following objectives are highlighted: to submit new approaches with or without the help of new technologies; to apply practice content; to analyze the approach given by the material; to present the content stemming from the problem-solving. It is important to point out that one objective which stood out for the others was what we called "content by content"; in other words, the study exclusively aimed for the mathematical content. Finally, regarding the keyword "teacher," the objectives found were: the presentation of learning objects; to discuss constructivist perspectives; to investigate the use of different reading and writing strategies in mathematics teaching in high school; to investigate the viability of developing working projects.

4.2.1.2 The Theoretical Framework as a Subcategory

In this subcategory, the variety and the quality of theoretical framework presented in the papers inhibited the building of key elements. With no intention of pointing out the main theoretical framework of each paper analyzed by Working Group 03, and taking into account our mapping that, except for a few cases, occurred while reading the abstracts, we will try to gather the papers that share similar theoretical framework.

In this regard, the papers classified in the algebra-learning category, by Moretti (2000), Alonso and Moraes (2003), and Santos and Mattos (2006), present, among other foundations, the socio-interactionist Vygotsky perspective (1989). Lima (2000), Júnior (2003), Lucas and Gualandi (2015), and Gonçalves, Lima, and Karrer (2015) present, in their theoretical framework, the French didactics authors, with emphasis on Douady (1984), with his object-tool dialectics, and Duval and Egret (1993) with the semiotics representation registers. Groenwald and Becher (2009)'s paper is based on authors who deal with algebra and the algebraic thought, especially Usiskin (1995) and Kaput (2005). Yet, Ribeiro and Júnior (2012) took as a basis the meaningful learning theory by Ausubel and Novak (*apud* Moreira, 2001) and the studies on previous knowledge developed by Pozo (2000). Lastly, Rosenbaum's paper (2012) was based on the hypothetical trajectories of learning (THA), by Martin Simon (1995).

In the papers classified into the geometry learning category, by Bezerra (2009) and Teles and Bellemain (2009), there is prevalence of French didactics authors with a predominance of Artigue (1988), who deals with methodology on didactics engineering, and Vergnaud (1990), with the conceptual field theory. Yet, Proença and Pirola (2006, 2009) use Klausmeier and Goodwin's idea (1977), which takes up the conceptual theory; Silva (2006) emphasizes Van Hiele's model (1984) about the development levels of geometrical thought. Pietropaolo (2006) and Oliveira et al.'s papers (2012) outline Balacheff (1988), who handles with the analysis of difficulties of basic education students in the proof learning process; Healy and Hoyles (2000) deal with the studies on students' proof conceptions; and Duval (1991) discusses about the reasoning structure and the demonstration learning. Tinoco and Nasser (2000) based their study on Hoyles (1997), who deals with tests involving ability of argumentation, and on Dreyfus and Hadas (1994), who discuss the students difficulties concerning Euclid's axioms. Finally, Pirola et al. (2003) is based in authors that look after the abandonment of the geometry teaching, highlighting Perez (1995), Fainguelernt (1995), Biembengut and Silva (1995), and Pirola (1995).

Each of the papers classified in methodological resources category presented different theoretical framework, as already expected due to the very nature of the category. Therefore, Franco and Moretti (2006) based their work on Duval's semiotics representation register (1993). Pereira et al. (2006) relied on Delors (2003) who deals with the four pillars of knowledge: to learn to know, to learn to do, to learn to live, to learn to be. Ninow and Kaiber (2015) based their work on Hernandez's working projects (1998). Oliveira and Lopes (2009) took as a basis Vygotsky's historical-cultural approach (1998, 2001), and the studies about mathematical language and the use of reading and writing in mathematics classes from the point of view of Corrêa (2005), Fonseca and Cardoso (2005), Malta (2003), Mesquita (2001), Pimm (1999), Santos (2005), and Smole and Diniz (2001). Pirola et al. (2006) underpins with the theory about the learning development by Krutetskii (1976).

The papers classified in arithmetic learning category, by Xavier and Lutaif (2006), Silva and Perovano (2012), and Campos and Rodrigues (2006), present, among others, theoretical framework about the French didactics, highlighting Vergnaud (1990) and the conceptual field theory, and Bachelard (1996) and Brousseau (1997a, 1997b, 2007), and the obstacles studies. Fonseca (2006) is based on authors who study the students difficulties with real numbers concept, emphasizing Baldino (1994), Pinto and Tall (1996), Igliori and Silva (2001), and Moreira and David (2005), and on authors who deal with the number conceptualization, such as Otte (2001, 2003), Conway and Guy (1999), and Conway (2001). Pinto and Laudares (2015) are supported by Wiley (2000), who studies about learning objects, and by Oliveira (2001) and Masetto (2013), who discuss about computing education.

Resende's paper (2006), classified in calculus learning category, presents, as theoretical framework, authors who deal with failure in calculus teaching, among them Barufi (1999), Bean (2004), and Fernandes and Healy (2006), and the calculus basic concepts of learning difficulties with particular reference to Tall and Vinner (1981), Tall (1990, 1993), and Tall and Rasslan (2002). Yet Silva's paper (2012) is based on Doll Jr.'s studies (1997) about the criteria for content selection and organization. Silva and Neto's paper (2006), classified within the trigonometry learning category, is sustained by the National Curricular Parameters (Brasil, 1999b), and in studies which involve trigonometry with emphasis on Nacarato et al. (2005), Sobrinho, Lima, and Thomaz Neto (2004), Briguenti (1998), and Manrique and Bianchini (2005). Finally, Pacheco and Medeiros's paper (2006) classified in information treatment learning category is based on authors who study problems (Echeverria & Munício, 1994; Saviani, 1980; Medeiros, 1987), problem-solving (Grossinkle, 1968; Lester, 1982), and the problem-solving in failure (Pérez, Torregrosa, & Pérez, 1988).

4.2.1.3 Methodology as Another Subcategory

In the "methodology" subcategory, all the papers analyzed belong to a qualitative research approach. The qualitative research mode that appeared in most papers (70%) was field research. Among the tools for data collection associated with field research, the most prominent ones were tests, with 21%, and preparation and application of activities, with 29%. The other modes of qualitative research were: bibliographic (15%), experimental (9%), theoretical rehearse (3%), and phenomenological (3%) modes.

4.2.1.4 Results as Another Subcategory

In the "results" subcategory, it was not possible to characterize each paper from its word or key expression, since, when emphasizing a cause, the paper naturally presented a consequence to such a cause. Table 4.3 presents the results associated with the "results" subcategory.

Table 4.3 Results Subcategory

Keyword	Quantity	Percentage (%)
Finding	28	52.0
Cause	9	16.50
Consequence	9	16.50
Proposal	8	15.0
TOTAL	54	100

Stemming from these results, we can notice that the papers presented in WG03, "Teaching and Learning" axis determine much more than propose. From our perception, the papers need to determine and indicate the causes and the possible consequences. They should provide actions to settle the difficulties the students presented with the definitions and with the solving of problem situations, their passivity to accept without enquiring the mathematical statements, and put forward actions so that students enjoy geometry; and that mathematics teaching should include more than exclude.

The results analysis of the 34 abstracts of the papers classified in the Teaching and Learning axis was done starting with the key subjects "student," "content," and "teacher."

With regard to the key subject "student," the following results are emphasized: (a) concerning the findings: mathematics teaching promotes exclusion and refusal of students (Xavier and Lutaif, 2006); the students present difficulties (Silva & Neto, 2006; Silva & Perovano, 2012); they present shortcomings (Tinoco & Nasser, 2000); there is a lack of or little knowledge (Pirola et al., 2003; Proença and Pirola, 2006); and the application of mathematical knowledge to the students reality make them fail to notice mathematics as monster (Santos & Mattos, 2006). (b) concerning the causes and consequences: the choice for the strategy in using the formula caused a primary, that is, conceptual, mistake (Pacheco & Medeiros, 2006); the lack of strengthening of argumentation activities (or even the justification of the given responses) by the teacher make the students accept the mathematical statements as true and with no enquiry (Tinoco & Nasser, 2000); the fact that the student does not enjoy geometry is related, among other factors, to the methodology used by the teacher (Pirola et al., 2003). (c) concerning the proposals: there is a need of more studies to be held regarding the subjective aspects which have marked the relation of the subject with the mathematical knowledge (Xavier and Lutaif, 2006); reduction in repetition rates at school may occur by means of inter- and transdisciplinary projects stemming from the projects pedagogy proposed by Hernandéz (2005) (Santos & Mattos, 2006); the characterization of the students' knowledge previous to each new educative experience is an important tool for the accomplishment of pedagogical interventions that are more efficient and that generate significant learning (Ribeiro and Júnior, 2012).

Concerning the "content" key subject, we highlight the following results: (a) regarding the findings: there is a few research and projects about the proof and demonstrations in the basic education mathematics teaching (Pietropaolo, 2006); the geometrical board is hardly ever used by the students to try to solve a problem; the real need of the use of mathematics turned up into a great interest by the students (lima, 2000); the role played by different forms of representation of the same mathematical objective in the teaching and learning process is evident (Franco & Moretti, 2006). (b) Regarding the causes and consequences: students' thinking is not stimulated and investigation activities are not provided, so students solve problems that are only application of the content developed (Silva, 2006). (c) Concerning the proposals: the counterexample is a teaching tool that may be exploited in the classroom under the perspective prescribed by Duval and Egret (1993) (Franco &

Moretti, 2006); all the numerical data suggested by the diagram need to be used (Teles & Bellemain, 2009) to discuss structural changes in the current mathematics high school education curriculum at the light of research guiding such changes (Silva, 2012).

Lastly, according to the key subject "teacher," we emphasize the following results: (a) concerning the findings: the student becomes more interested when learning is more effective (Pinto & Laudares, 2015); the use of research contributes to the teaching organization (Rosenbaum, 2012). (b) Regarding the causes and consequences: the results suggest that the process may turn the relationship between teacher and student more interactive and more effective for the construction of the mathematical knowledge (Oliveira & Lopes, 2009); the teaching sequence is not enough to guarantee that the constructivist approach is adopted by the teacher, because the teacher's action plays a key role on the construction of the students' knowledge (Rosenbaum, 2012). (c) With regard to the proposals, the use of working projects is viable in the classroom (Ninow & Kaiber, 2015).

4.2.2 Curriculum and Curricular Materials Categorization Axis

In the curriculum and curricular materials axis, 14 papers were classified in seven categories, according to Table 4.4.

Table 4.4 shows that the sum of the papers in the "curricular material," "prescribed and practiced curriculum," and "theoretical and epistemological conceptions" categories represent nearly two-thirds of the presented papers in the High School Education Working Group (WG03) and classified in "Curriculum and Curricular Materials" axis.

After classifying the papers in the seven categories presented in Table 4.4, we analyzed each of the 14 papers taking into account the following subcategories: (a) aims; (b) theoretical framework; (c) methodology; and (d) results.

Table 4.4 First Categorization Curriculum and Curricular Materials

Categories	Quantity	Percentage (%)
Curricular material	3	21.5
Compared studies	1	7.0
Prescribed and practiced curriculum	3	21.5
Educational reforms	1	7.0
Purposes	2	14.5
Competence	1	7.0
Theoretical and epistemological conceptions	3	21.5
TOTAL	14	100

Table 4.5 Second Categorization Curriculum and Curricular Materials axis

Categories	Quantity	Percentage (%)
Prescribed curriculum	7	50.0
Prescribed and practiced curriculum	3	21.5
Curriculum presented to the teachers	3	21.5
Other	1	7.0
TOTAL	14	100

Due to both the small quantity of papers in this category and the peculiarity of the axis, we chose to carry out a new classification, sharing the papers that emphasize: (a) prescribed curriculum; (b) prescribed and practiced curriculum; (c) curriculum presented to the teachers (Table 4.5). With this new classification, we came to the following distribution:

Just one paper was not classified among the mentioned modes. Written by Prof. Célia Maria Carolino Pires, Ph.D. the paper presents reflexions and arguments referred to the need to establish a working group about curricular organization and development in mathematics among the existing ones as part of SIPEM. All the other papers were classified in three categories. Half of the productions made reference to curriculum prescribed. We chose to analyze the aims, procedures, and results of these papers in each of these categories. Subsequently, we are going to present the features of each group.

4.2.2.1 Prescribed Curriculum as a Subcategory

When referring to prescribed curriculum, in Sacristán and Pérez-Gómez's conception (1998), we think the curriculum as part of the political and administrative decisions. However, it is all about how the educational public policies define what and how it will be taught.

In this subcategory, seven papers were about subjects related to curricular prescriptions. The aims or research problems of these productions were to: (a) characterize singularities and peculiarities of prescribed curriculum of Latin American countries subjects, starting with their historical, social, and economic constraints; (b) present high school education curricular proposal; (c) reflect on the high school education nature and about the education necessary for the student of this teaching level; (d) investigate, reflect about and analyze critically the official curricular production for the high school education in Brazil, trying to identify and discuss meanings, limits and possibilities of a mathematics curriculum that has as its assumption to answer the objective, socioeconomic, and cultural necessities of the students; (e) analyze the document entitled "Curriculum Guidelines for High School Education," published in 2006, which was the most recent national instruction issued when the paper was published, in order to search for relevant features that constitute epistemological and organizational antithesis: the dichotomies static knowledge versus dynamic knowledge, and linear organization versus network organization; (f) locate and discuss the places reserved for the educational subject over the twentieth century in different theoretical perspectives about the curriculum that was object of bibliographical studies; (g) discover how the trajectory of mathematics teaching in Brazilian high school education and its trends could be outlined from the works written by Professor D'Ambrosio about the subject in the period before the modern mathematics movement had set off in Brazil.

Due to the variety of goals, it seems difficult to articulate them with each other, but we can infer that the motivations for this research are connected with problematizing, with references to critic or contemporary theoreticians and to the description of curricular prescriptions, using compared studies. Those analyses are carried out by theoretical essays, literature reviews, and documental analysis, among other procedures.

4.2.2.2 Prescribed and Practiced Curriculum as Another Subcategory

In the subcategory "prescribed and practiced curriculum," there are papers analyzing the relations between the curricular process dimensions. Although these dimensions should be enmeshed, and suffering the action of one another, what stems from the three papers that compose this subcategory is that, in general, the prescribed curriculum is the main factor of the teacher practice in the classroom, constituting what is known as practiced curriculum or curriculum in action, the re-elaboration, in teacher practice, reflected on the activities and on the teacher's plan.

The aims of those three papers were: (a) to present research on the basic education curricular innovations that had been developed since 2000 in research group coordinated by the author; (b) to investigate possible factors that led to meaning production by two high school mathematics teachers when engaging in a teaching proposal developed at the light of critical mathematics education (EMC) and hypothetical learning trajectory (THA) of statistical measures; (c) to describe mathematics teaching in the basic education schools in Angola, particularly in middle school in the province of Cabinda, comparing the guidelines existing on the official documents of Angola's Ministry of Education and the teachers pedagogical practice observed in the classroom and identified in the interviews.

Two of these papers analyze, although through distinct references, the meanings or impacts of the hypothetical learning trajectory (THA) on high school education teachers practice. This framework (THA), developed by Martin Simon, was widely used in research related to curriculum in the group. The investigations that used this contribution to understand the use teachers make of the curricular prescriptions or how the curricular prescriptions affect those teachers' practices were recurrent.

The authors of these papers use interviews, documental analysis, and video analysis, among others, as methodological procedures.

Among the conclusions, it is also noteworthy the difference between prescriptions and practices; how much the prescribed curriculum withdraws the teachers from actions they would like to carry out and, at the same time, how much the teachers worry about following it at risk, although they hardly ever get to do it.

4.2.2.3 Curriculum Presented to Teachers as Another Subcategory

In the subcategory "curriculum presented to teachers" or planned for the teachers, making a kind of translation of the prescriptions for the classroom practices, three papers analyzed the same type of material: the textbooks.

The aims or research questions and purposes of these papers were: (a) is the object of knowledge, linear diophantine equations (EDL), considered a teaching object in textbooks? (b) how are the problems of research on mathematics textbooks configured? What conceptual framework can guide research on those mathematics education materials? (c) to investigate the role that the curricular materials can play when mediating learning and changes in teachers' practice, besides bringing some ideas for curricular materials analysis aiming at contributing for the debate on mathematics curriculum.

By applying a variety of theoretical references and methodologies, the researchers contributed to specific content analysis, textbooks analysis proposals, and general enquiries about the research involving these materials in mathematics education.

Among the results, the concern with the lack of attention paid to official curricular proposals and textbooks about the subject EDL was emphasized; however, we can expand this result to other possible investigations with specific subjects. Would the results be different? Another paper indicates a considerable number of research works that focussed on answering the descriptive questions. A reduced number of research works had as topic the identification and the comprehension of the relations between textbooks, or some of their features, and other variables. Finally, one of the papers presented the textbooks analysis proposals, showing that a set of research works had already studied the relation of causality existing between determined factors and textbooks, and states that it is necessary to respect the teacher's role and understand how an agent who interprets this material needs to adapt it to real situations in the classroom.

4.2.3 EJA, Special Education, Professional and Teacher Education Categorization Axis

Among the presented papers in GT03, in several seminars held, there is a set of seven papers about aspects correlated to high school education subject, but they are also bound to other subjects that are concerned with different working groups. Five of them refer to Youth and adult education program (EJA) (Lozada, Lozada, & Rozal, 2009; Mattos & Andrade, 2009; Pompeu, 2012) to professional teaching

(Nascimento & Nascimento, 2009) and to special education (Fernandes and Healy, 2006) in secondary school, and two others, Godoy and Pires (2003), and Maciel (2012) refer to the dimensions of teacher education, also in high school. If, on one hand, it is difficult to classify the main thematic focus of these papers due to the ambivalence and thematic dispersion, on the other hand there are elements in these papers that point to the multiple features of high school education in Brazil at the same time that they deal with aspects taken into account in the production by other working groups.

Therefore, there is a crossing and juxtaposition between levels and modalities that explain the fact that the mentioned papers do not need to be exclusively bonded to a single thematic group or research axis. Once providing that there is no correspondence among each subject or sub-subject to a single working group, and that there is not in the structure of the seminar groups as many groups as the subjects searched, it seems to prevail among the researchers the trend to link a paper to a specific group, the one which is closer to their interest of investigation from the given emphasis, however considering, the identification with the thematic axis that is the main object of the research. This flexibility does not entail thematic losses, conversely, it promotes the exploitation of a subject about different dimensions that the research about it comprises.

It is a coverage, a certain thematic dispersion that promotes the productive research field development.

Therefore, understanding that the research works approach high school education issues articulating several topics, it is important to highlight some of them by their emphasis on outlining the coverage of the investigation on high school education. Seven of the papers focussed mainly on: (1) public policies (one paper); (2) knowledge and teacher education (three papers); (3) mathematics learning and competence (two papers); (4) inclusion (one paper). In some instances, the same paper presents more than one emphasis.

4.2.3.1 Public Policies

The investigation on public policies for education constituted an important pathway, inter alia, to map research in the mathematics education area. Production policies and curricular implementation, teacher training official programs, the evaluation system in large scale adopted by the country in local, regional, national or international scope, special needs education programs, textbook program, etc. are some examples.

Among the papers on the axis Youth and Adult Education Program (EJA), Special Education, Professional and Teacher Training, we highlight the article presented by Matos and Andrade (A Experiência de Ensinar Matemática no PROEJA: seus limites e possibilidades/The experience of teaching mathematics in PROEJA: limits and possibilities) approaching (Youth and Adult Integration Program of Technical and Professional High School Education—PROEJA) student-worker-oriented public policies, but which favors the teacher practice. It is a paper held in a specific pole, where the program is developed, showing the potential of a public politic to promote the most diversified possibilities of interdisciplinary teacher practices around the country.

4.2.3.2 Knowledge and Teacher Training

The teacher practice issue and the initial or continued education process are the main interest of researchers in the mathematics education area because teacher training is an end activity for the researchers in institutions, although their main research object might be another other than teacher training. Furthermore, the dense production motivates very much the research in the area with significant advances on the comprehension and theoretical foundation about the nature and types of knowledge that are constituted in teaching, and that imply directly on the formulation of curricular guidelines, teacher training modes and programs. In the set of papers presented , we can highlight the ones by Nascimento (Matemática- ferramenta interdisciplinar na construção da aprendizagem na Educação Agrícola/ Mathematics-interdisciplinary tool in the construction of learning in agricultural education), Godoy and Pires (Professor do Ensino Médio e suas concepções/High school teacher and their concepts) and Maciel (Formação continuada e práticas de professores de Matemática: relações e perspectivas/Continuing education and practices of mathematics teachers: relations and perspectives). Nascimento's paper presents the mode of the so-called interdisciplinary pedagogical practices in the agricultural education as a pathway to motivate students' interest in learning mathematics. Yet, Godoy and Pires' paper investigates what the teachers think about the contextualization and interdisciplinary assumptions adopted as curriculum structuring axis, in case of mathematics in high school education in the country. Maciel's paper, in turn, is a comparative nature study analyzing pedagogical practices and experiences of continuing education in public and private schools.

4.2.3.3 Mathematics Learning and Competence

The learning of specific mathematics content and factors that interfere with this learning, as well as the relations of the students with the mathematical knowledge, are important and very privileged aspects of mathematics teaching studies; such studies refer, on one hand, to the students achievement, to the development of mathematical competence, and their difficulties, and, on the other hand, to the methods and teaching strategies. The most prominent set of papers were by Lozada et al. (Contribuições para o numeramento em turmas de Educação de Jovens e Adultos/ Contributions for number learning in groups of youth and adult education) and Pompeu (A experiência escolar de alunos Jovens e adultos e sua relação com a Matemática/Youth and adult students' school experience and how they deal with mathematics). The former authors explore mathematics modelling as a way to deal with students' difficulties, and the latter tries to analyze the relation of young and adult students with mathematics, bearing in mind their experience in lived situations inside and outside the school context.

4.2.3.4 Inclusion

The research agenda began to turn its attention to the issue of differences concerning socio, economical, cultural, ethno-racial conditions, and the ones related to people who have different kinds of disabilities, as far as such difference received special attention of inclusive policies guided by the idea of treatment and practices in mathematics education. The work presented by Ahmad et al. (O processo de inclusão de alunos cegos nas aulas de Matemática: as vozes dos atores/The process of inclusion of blind students in mathematics classes: the actors' voices) tries to investigate how the diversity and inclusion issue involving students without visual acuteness has been practiced in terms of what is postulated by legislation. Ultimately, the papers in the axis EJA, Special Education, Professional and Teachers Training address studies of bibliographical and documental nature, and studies held on *n* certain empirical field (classroom, group of students) or combining both ways. The studies assume predominantly a characteristic of experience reports. Although they discuss relevant aspects about teaching in this range, both the processes of change promoted with educational reforms over time and the teaching competence, it is possible to verify that the results of the studies, a still incipient theoretical production, demonstrate how challenging is to promote expansion and depth in the theoretical knowledge on the subject. Although some of these works are based on a theoretical framework that is also fundamental for several studies in mathematics education area inside and outside the country, such as Vygotsky, Lave & Wenger, Bishop, Charlot, Nóvoa, Sacristan, Pèrez-Gomez, among others, there is not, exactly, in the set of papers analyzed, a discussion that establishes strong relations with similar studies carried out in other countries. This may indicate that this happens because the research on Youth and Adults Education, Inclusive Education, and even the professional formation of the student and the high school teacher is still recent. Another possibility is that such research has characteristics of a particular social and educational reality and that, for this reason, the papers present an expressive reference to studies by Brazilian researchers.

4.3 Final Considerations

Throughout the six versions of SIPEM, the works associated with the Teaching-Learning axis accounted for more than 60% of presentations. However, in the last three SIPEMs, the works associated with Curriculum and Curricular Materials and EJA, Special Education, Professional and Teacher Training axes divided the attentions, especially for the last SIPEM, where the presentations involving Curriculum theme gained more space. There has not been, in our opinion, a apparent reason for the migration of works from one axis to another, however, something that could help us in this analysis would be the growth of groups that study the curriculum theme in the last versions of SIPEM, which eventually promoted the creation of the National Curriculum Forum, that is already in its fourth edition, and the publication of a special issue on mathematics curriculum, in the Bulletin of Mathematics Education (BOLEMA), in 2014.

Another factor that corroborates the decrease in works associated with the Teaching-Learning axis is the creation of Working Groups (WG) that present greater adherence to the results of researches that deal with high school mathematical contents, such as WG 06, "Mathematics Education: New Technologies and Distance Education." The use of new technologies in mathematics education is much emphasized in basic education; however, in the studies we analyzed from the Teaching-Learning axis, there are very few indications that technologies have been used, reinforcing our belief that works that approached high school mathematical contents migrated for WG 06. The same reasoning can be applied to works associated with the EJA, Special Education, Professional and Teacher Training axis, since that the working groups "Difference, Inclusion and Mathematics Education" (WG 13) and "Mathematics Teachers Training" (WG 7) were created throughout the six versions of SIPEM.

There are clamant questions to be investigated by the research groups involved in the subject of high school education: What objects of study should the lines of research that investigate the mathematics curriculum constitute? What dialogues need to be established between research on mathematics curricula and curricula theories in education? What should advances be in order to improve curricula presented in curricular materials such as textbooks, materials offered by education departments, videos, etc.? What suggestions should be offered to the Ministry of Education for the creation of continuous, systematic, transparent, and participative processes of curricular debate? What suggestions should be given to the initial and continuing training courses of mathematics teachers so that they can become more participatory actors of the curriculum development process? What suggestions should be given to SBEM in relation to the debate on the subject? What are unanswered questions in the field of mathematics curriculum research? Why are there no lines of research in the graduate programs in mathematics education, nor working groups in the events, exclusively focused on the curricular discussion in mathematics? Does research on curricula necessarily intersect with other areas, such as teacher training, technology, teaching and learning? What characterizes curriculum research and what differentiates it from other areas? What are the topics that could be addressed in the initial and continuing training courses of mathematics teachers so that they can learn more about curricular theories? What are the topics that could be addressed in the graduate courses in mathematics education so that future researchers can learn more about curricular theories? When we speak of research in curriculum what we understand by curriculum, beyond the etymological meanings?

These questions can be used as catalysts for new studies that will lead to the creation of new groups that will focus also on the other levels of education in Brazil.

4.4 Summary

This chapter aims to present to the national and international academic community an overview of Brazilian mathematics education research, from the analysis of the papers that were approved and presented in the events held in the scope of the Research International Seminar on Mathematical Education—SIPEM by the Mathematics Education in High School Group. It revisits its trajectory, which had as original purpose to create a space that gathered research directed to this basic educational level in Brazil, which deals with young students from 15 to 17 years old. To present and analyze the results presented by the group, we clustered the subjects in three axis as follows: Teaching and Learning; Curriculum and Curricular materials, and Teacher Education. We concluded that the research associated with the subject Teaching and Learning accounted for more than 60% of the presentations in these events. However, for the last three events, the research associated with Curriculum and Curricular Materials and Youth and Adult Education Program (EJA), Special Education, Professional and Teacher Training divided the attentions, especially during last SIPEM, where the presentations involving curriculum gained more space. We finished the chapter suggesting that there are clamant issues to be investigated by the research groups involved in the subject of high school education.

References

Alonso, E. P., & Moraes, M. S. S. (2003). Educação Matemática no ensino médio uma alternativa para o estudo de funções no ensino médio. In *II Seminário Internacional de Pesquisa em Educação Matemática (SIPEM)*. Santos, SP: Sociedade Brasileira de Educação Matemática (SBEM).

Artigue, M. (1988). *"Ingénierie Didactique". Recherches en Didactique des Mathématiques* (Vol. 9.3, pp. 281–308). Grenoble: La Pensée Sauvage-Éditions.

Bachelard, G. (1996). *O novo espírito científico: Contribuição para uma psicanálise do conhecimento*. Tradução Estrela dos Santos Abreu. Rio de Janeiro: Contraponto.

Balacheff, N. (1988). Aspects of proof in pupil's practice of school mathematics. In D. PIMM (Ed.), *Mathematics teachers and children* (pp. 216–235). London: Hodder and Stoughton.

Baldino, R. R. (1994). A ética de uma definição circular de número real. *Bolema-Boletim de Educação Matemática, 9*(10), 31–52.

Bean, D. W. (2004). *Aprendizagem pessoal e aprendizagem afastada: O caso do aluno de cálculo*. Tese de doutorado. Campinas, UNICAMP.

Bezerra, N. J. F. (2009) O GPS como instrumento didático auxiliar no processo de significação conceitual do ensino da geometria Analítica. In *IV Seminário Internacional de Pesquisa em Educação Matemática (SIPEM)*. Brasília, DF: Sociedade Brasileira de Educação Matemática (SBEM).

Brasil. (1999a). *Parâmetros Curriculares Nacionais (PCN): Ensino Médio: ciências da natureza, matemática e suas tecnologias*. Brasília: Ministério de Educação—Secretaria de Educação Média e Tecnológica.

Brasil. (1999b). Secretaria de Educação Média e Tecnológica. Parâmetros Curriculares Nacionais. Ensino Médio: Ciências da natureza e suas tecnologias. Brasília, DF.

Barufi, M. C. B. (1999). *A construção/negociação de significados no curso universitário inicial de Cálculo Diferencial e Integral. Tese de Doutorado*. São Paulo: FE-USP.

Biembengut, M. S., & Silva, V. C. (1995). Ornamentos X Criatividade. *A Educação Matemática em Revista, 3*, 39–44.

BRASIL. (2015). Retrieved June 24, 2016, from http://portal.mec.gov.br/index.php?option=com_docman&view=download&alias=17044-dados-censo-2015-11-02-materia&Itemid=30192.

Briguenti, M. J. L. (1998). *Alterando o ensino de Trigonometria em escolas de nível médio: A representação de algumas professoras*. Marília: Universidade Estadual Paulista. 339p. Tese (Doutorado em Educação).

Brousseau, G. (1997a). *Theory of didactical situations in Mathematics—Didactique des Mathématiques, 1970-1990*. Trad. Nicolas Balacheff, Martin Cooper; Rosamund Sutherland e Virginia Wafield. Londres, Kluwer Academic Publishers.

Brousseau, G. (1997b). *La théorie des situations didactiques*. Cours donné lors de l'attribution à Guy Brousseau du titre de Docteur Honoris Causa de l'Université de Montréal.

Brousseau, G. (2007). *Introdução ao estudo das situações didáticas: Conteúdos e métodos de ensino*. Tradução de Camila Bogéa. São Paulo: Ática.

Campos, T. M. M. de., & Rodrigues, W. R. (2006). Um olhar sobre aspectos do conceito de fração por meio de situações-problema. In *III Seminário Internacional de Pesquisa em Educação Matemática (SIPEM)*. Águas de Lindóia, SP: Sociedade Brasileira de Educação Matemática (SBEM).

Conway, J. H. (2001). In) (Ed.), *On numbers and games* (2nd ed.). Natick, MA: A K Peters.

Conway, J. H., & Guy, R K. (1999). *O livro dos números*. Tradução de José Sousa Pinto. Lisboa: Gradiva.

Corrêa, R. A. (2005). Linguagem matemática, meios de comunicação e Educação Matemática. In A. M. Nacarato & C. E. Lopes (Eds.), *Escritas e Leituras na Educação Matemática* (pp. 96–99). Belo Horizonte: Autêntica.

Delors, J. (2003). *Educação: Um tesouro a descobrir* (2nd ed.). São Paulo: Cortez.

Doll, W. E. Jr. (1997). *Currículo: Uma perspectiva pós-moderna*. Trad. Maria Adriana Veríssimo Veronese. Porto alegre: Artes Médicas.

Douady, R. (1984). *Jeux de cadre et dialectique outil objet dans l'enseignement des mathématiques*. Paris: IREM de l'Université de Paris VII.

Dreyfus, T., & Hadas, N. (1994). Euclides deve permanecer—E até ser ensinado. In M. M. Lindquist (Ed.), *Aprendendo e ensinando Geometria*. São Paulo: Atual.

Duval, R., & Egret, M. A. (1993). Introduction à la démonstration et apprentissage du raisonnement déductif. *Repères, 12*, 114–140.

Duval, R. (1991). Structure du raisonnement deductif et apprentissage de la demonstration. *Educational Studies in Mathematics, 22*(3), 233–261.

Echeverria, M. D. P. P. & Munício, J. I. (1994). P. Aprender a Resolver Problemas y Resolver Problemas para Aprender. In: Munício, J. I. P. (Org.) La Solución de Problemas. Madrid: Santillana, S.A., cap. 1, pp. 14-50.

Fainguelernt, E.K. (1995). *O Ensino de Geometria no 1° e 2° graus*. Revista da Sociedade Brasileira de Educação Matemática. São Paulo, ano III, n 4, p.45–53, 1 semestre.

Fernandes, S. H. A. A., & Healy, L. (2006). O processo de inclusão de alunos cegos nas aulas de Matemática: as vozes dos atores. In *III Seminário Internacional de Pesquisa em Educação Matemática (SIPEM)*. Águas de Lindóia, SP: Sociedade Brasileira de Educação Matemática (SBEM).

Fonseca, M. d. C. F. R., & Cardoso, C. d. A. (2005). Educação Matemática e Letramento: Textos para ensinar Matemática e Matemática para ler o texto. In A. M. Nacarato & C. E. Lopes (Eds.), *Escritas e Leituras na Educação Matemática* (pp. 66–70). Belo Horizonte: Autêntica.

Fonseca, R. F. da. (2006). A complementaridade entre os aspectos intensional e extensional do conceito de Número. In *III Seminário Internacional de Pesquisa em Educação Matemática (SIPEM)*. Águas de Lindóia, SP: Sociedade Brasileira de Educação Matemática (SBEM).

Franco, P. L., & Moretti, M. T. (2006). O contraexemplo no ensino da Matemática: uma análise com base nos registros de representação semiótica. In *III Seminário Internacional de Pesquisa em Educação Matemática (SIPEM)*. Águas de Lindóia, SP: Sociedade Brasileira de Educação Matemática (SBEM).

Godoy, E. V., & Pires, C. M. C. (2003). Professores do ensino médio e suas concepções. In *II Seminário Internacional de Pesquisa em Educação Matemática (SIPEM)*. Santos, SP: Sociedade Brasileira de Educação Matemática (SBEM).

Gonçalves, J. da S., Lima, R. N. de., & Karrer, M. (2015). Funções Composta e Inversa e os Registros de Representação Semiótica em Livros Didáticos do PNLD - 2015 do Ensino Médio. In *VI Seminário Internacional de Pesquisa em Educação Matemática (SIPEM)*. Pirenópolis, GO: Sociedade Brasileira de Educação Matemática (SBEM).

Groenwald, C. L. O., & Becher, E. L. (2009). Características do pensamento algébrico de estudantes do 1o ano do Ensino Médio. In *IV Seminário Internacional de Pesquisa em Educação Matemática (SIPEM)*. Brasília, DF: Sociedade Brasileira de Educação Matemática (SBEM).

Grossinkle, F. (1968). Mathematical Sentences and Problem Solving. In: Grossinkle, F. Teaching school mathematics. New York: Holt, Rinehart and Winston, cap. 17, pp. 302-313.

Healy, L., & Hoyles, C. (2000). A study of proof conceptions in algebra. *Journal for Research in Mathematics Education, 31*(4), 396.

Hernandéz, F. *Pedagogia de Projetos. Disponível em.* www.cdisp.org.br/pedagogico/projeto/. Acesso em 03/03/05.

Hernandéz, F., & Ventura, M. (1998). *A Organização do Currículo por Projetos de Trabalho* (5° ed.). Porto Alegre: Artes Médicas.

Hoyles, C. (1997). The curricular shaping of students' approaches to proof. *For the Learning of Mathematics, 17*(1), 7–16.

Igliori, S., & Silva, B. (2001). Concepções dos alunos sobre Números Reais. In J. B. Laudares & J. Lachini (Eds.), *Educação Matemática: A prática educativa sob o olhar de professores de Cálculo* (pp. 39–67). Belo Horizonte: Fumarc.

Júnior, A. T. (2003). Alguns problemas de programação linear e a importância dos registros de representação no processo ensino-aprendizagem de sistemas de inequações do 1° grau. In *II Seminário Internacional de Pesquisa em Educação Matemática (SIPEM)*. Santos, SP: Sociedade Brasileira de Educação Matemática (SBEM).

Kaput, J. (2000). *Teaching and learning a new algebra with understanding*. Documento retirado de http://www.simcalc.umassd.edu/downloads /KaputAlgUnd.pdf. em 21de Outubro de 2005.

Klausmeier, H. J., & Goodwin, W. (1977). *Manual de Psicologia Educacional: Aprendizagem e capacidades humanas* (M. C. T. A. Abreu, Trad.). São Paulo: Editora Harper & Row.

Krutetskii, V. A. (1976). *The psychology of mathematical abilities in schoolchildren*. Chicago: The University of Chicago Press.

Lester, F. K. (1982). Issues in teaching mathematical problem solving in the elementary grades. *School Science and Mathematics, 82*(2), 93–98.

Lima, R. N. de. (2000). Resolução de equações de terceiro grau através de cônicas. In *I Seminário Internacional de Pesquisa em Educação Matemática (SIPEM)*. Serra Negra, SP: Sociedade Brasileira de Educação Matemática (SBEM).

Lozada, A. de O., Lozada, C. de O., & Rozal, E. F. (2009). Contribuições para o numeramento em turmas de Educação de Jovens e Adultos. In *IV Seminário Internacional de Pesquisa em Educação Matemática (SIPEM)*. Brasília, DF. Sociedade Brasileira de Educação Matemática (SBEM).

Lucas, T. M., & Gualandi, J. H. (2015). As Diferenças Formas de Registro de Funções Exponenciais. In *VI Seminário Internacional de Pesquisa em Educação Matemática (SIPEM)*. Pirenópolis, GO: Sociedade Brasileira de Educação Matemática (SBEM).

Maciel, M. M. (2012). Formação continuada e práticas de professores de Matemática: relações e perspectivas. In *V Seminário Internacional de Pesquisa em Educação Matemática (SIPEM)*. Petrópolis, RJ: Sociedade Brasileira de Educação Matemática (SBEM).

Malta, I. (2003). *Linguagem, Leitura e Matemática*. Puc – Rio de Janeiro. 2003. Retirado de junho, 2000, partir de www.mat.puc-rio.br/preprints/pp200308.pdf.

Manrique, A. L., & Bianchini, B. L. (2005). Percepções de professores de matemática sobre trigonometria no projeto pró-ciências. In: *V Congresso ibero-americano de educação matemática.* Porto. 1CD-ROM.

Masetto, M. T. (2013). Mediação Pedagógica e Tecnologias de Informação e Comunicação. In: Moran, J. M.; Masetto, M. T.; Behrens, M. A. (Org.). Novas tecnologias e mediação pedagógica. 21 Campinas, SP: Papirus, p.11-72.

Mattos, J. R. L. de., & Andrade, L. O. M. (2009). A experiência de ensinar matemática do PROEJA: seus limites e possibilidades. In *IV Seminário Internacional de Pesquisa em Educação Matemática (SIPEM).* Brasília, DF: Sociedade Brasileira de Educação Matemática (SBEM).

Medeiros, C. F. (1987). Por uma Educação Matemática como Intersubjetividade. In M. A. V. Bicudo (Ed.), *Educação Matemática.* São Paulo: Ed. Cortez.

Mesquita, C. G. R. (2001). *A escrita matemática: Espaço para aprendizagens que fabricam conhecimentos.* Retirado de Agosto 6, 2010, partir de http://www.anped.org.br/reunioes/24/T1919506565133.doc

Moreira, M. A. (2001). *Aprendizagem significativa: A teoria de David Ausubel.* São Paulo: Centauro.

Moreira, P. C., & David, M. M. M. (2005). *S. A formação matemática do professor: Licenciatura e prática docente.* Belo Horizonte: Autêntica, 120p. (Tendências em Educação Matemática, 11)

Moretti, V. (2000). O conceito de função: os conhecimentos prévios e as interações sociais como desencadeadores da aprendizagem. In *I Seminário Internacional de Pesquisa em Educação Matemática (SIPEM).* Serra Negra, SP: Sociedade Brasileira de Educação Matemática (SBEM).

Nacarato, A. M., Bredariol, C. C., & Passos, M. P. F.. (2005). Trigonometria: uma análise de sua evolução histórica e da transposição didática desse conhecimento presente nos manuais didáticos e propostas curriculares. In: *V Congresso ibero-americano de educação matemática.* Porto. 1CD-ROM.

Nascimento, E. do., & Nascimento, E. C. da S. (2009). Matemática- Ferramenta interdisciplinar na construção da aprendizagem na educação Agrícola. In *IV Seminário Internacional de Pesquisa em Educação Matemática (SIPEM).* Brasília, DF: Sociedade Brasileira de Educação Matemática (SBEM).

Ninow, V., & Kaiber, C. T. (2015). Projetos de Trabalho: Possibilidades para o Ensino e Aprendizagem da Matemática. In *VI Seminário Internacional de Pesquisa em Educação Matemática (SIPEM).* Pirenópolis, GO: Sociedade Brasileira de Educação Matemática (SBEM).

Nishio, A. L., & Leal, W. da S. (2009). Matemática e o meio ambiente: o ensino de funções em análise de consumo de água e energia elétrica. In *IV Seminário Internacional de Pesquisa em Educação Matemática (SIPEM).* Brasília, DF: Sociedade Brasileira de Educação Matemática (SBEM).

Oliveira, A. L. de., Miranda, D. F. de., Laudares, J. B. (2012). Construção de objeto de aprendizagem para o reconhecimento de uma cônica. In *V Seminário Internacional de Pesquisa em Educação Matemática (SIPEM).* Petrópolis, RJ: Sociedade Brasileira de Educação Matemática (SBEM).

Oliveira, C. C. de. (2001). Ambientes Informatizados de Aprendizagem: Produção e avaliação de software educativo. Campinas: Papirus.

Oliveira, R. A. de., & Lopes, C. E. (2009). Construção do conhecimento matemático mediante um processo de intervenção com leitura e escrita no Ensino Médio. In *IV Seminário Internacional de Pesquisa em Educação Matemática (SIPEM).* Brasília, DF: Sociedade Brasileira de Educação Matemática (SBEM).

Otte, M. (2001). *Mathematical epistemology from a semiotic point of view.* Paper presented to the Discussion Group on Semiotics in Mathematics Education at the 25th PME International Conference, July 12–17 2001, University of Utrecht, Utrecht, The Netherlands.

Otte, M. (2003). Complementarity, sets and numbers. *Educational Studies in Mathematics, 53*(3), 203–228.

Pacheco, A. B., & Medeiros, C. F. de (2006). Uma investigação sobre as dificuldades no uso de estratégias para a resolução de problemas verbais no campo da Análise Combinatória. In *III Seminário Internacional de Pesquisa em Educação Matemática (SIPEM)*. Águas de Lindóia, SP: Sociedade Brasileira de Educação Matemática (SBEM).

Pereira, M. das G., Mattos, J. R. L., & da Nascimento, E. C. S. (2006). Desenvolvendo conceitos matemático no curso técnico em Zootecnia. In *III Seminário Internacional de Pesquisa em Educação Matemática (SIPEM)*. Águas de Lindóia, SP: Sociedade Brasileira de Educação Matemática (SBEM).

Pérez, D. G., Torregrosa, J. M., & Pérez, F. S. (1988). El fracaso en la resolución de problemas de física: Una investigacion orientada por nuevos supuestos. *Enseñanza de Las Ciencias, 6*(2), 131–146.

Perez, G. (1995). *A realidade sobre o ensino de Geometria no 1° e 2° graus no Estado de São Paulo*. Educação Matemática em Revista—Sociedade Brasileira de Educação Matemática, ano III, n. 4.

Pietropaolo, R. C. (2006). Demonstrações e Educação Matemática - uma análise de pesquisas existentes. In *III Seminário Internacional de Pesquisa em Educação Matemática (SIPEM)*. Águas de Lindóia, SP: Sociedade Brasileira de Educação Matemática (SBEM).

Pimm, D. (1999). *El lenguaje matemático en el aula* (2nd ed.). Madrid: Ediciones Morata.

Pinto, J. E., & Laudares, J. B. (2015). Objeto de Aprendizagem: Ensino dos Números complexos com Aplicações na Área Técnica em Eletroeletrônica. In *VI Seminário Internacional de Pesquisa em Educação Matemática (SIPEM)*. Pirenópolis, GO: Sociedade Brasileira de Educação Matemática (SBEM).

Pinto, M., & Tall, D. (1996). Student teachers' conceptions of the rational numbers. In L. Puig & A. Gutierrez (Eds.), *Proceedings of the 20th Conference of the Internal Group for the Psychology of Mathematics Education (PME)* (Vol. 4, pp. 139–146). Valencia, Spain: University of Valencia.

Pirola, N. A. (1995). *Um estudo sobre a formação dos conceitos de triângulo e paralelogramo em alunos de primeiro grau*. 1995. 108f. Dissertação (Mestrado em Psicologia Educacional)—Faculdade de Educação, Universidade Estadual de Campinas, Campinas.

Pirola, N. A., Brito, A. A. do S., Sparvoli, D. A. P., Alonso, P. E., Proença, M. C. de.; Costa, A. L. P. da., et al. (2006). Resolução de problemas com informações supérfluas: Uma análise do desempenho de alunos sob a ótica da teoria de Krutetskii. In *III Seminário Internacional de Pesquisa em Educação Matemática (SIPEM)*. Águas de Lindóia, SP: Sociedade Brasileira de Educação Matemática (SBEM).

Pirola, N. A., Quintiliano, L. de C., & Proença, M. C. de (2003). Um estudo sobre desempenho de alunos do ensino médio em tarefas envolvendo o conceito de polígono e poliedro. In *II Seminário Internacional de Pesquisa em Educação Matemática (SIPEM)*. Santos, SP: Sociedade Brasileira de Educação Matemática (SBEM).

Pompeu, C. C. (2012). A experiência escolar de alunos jovens e adultos e sua relação com a Matemática. In *V Seminário Internacional de Pesquisa em Educação Matemática (SIPEM)*. Petrópolis, RJ: Sociedade Brasileira de Educação Matemática (SBEM).

Pozo, J. I. (2000). A aprendizagem e o ensino de fatos e conceitos. In *Os Conteúdos na reforma: Ensino e aprendizagem de conceitos, procedimentos e atitudes*. Porto Alegre: Saraiva.

Proença, M. C. de., & Pirola, N. A. (2006). A formação conceitual em geometria: Uma análise sobre polígonos e poliedros. In *III Seminário Internacional de Pesquisa em Educação Matemática (SIPEM)*. Águas de Lindóia, SP: Sociedade Brasileira de Educação Matemática (SBEM).

Proença, M. C. de., & Pirola, N. A. (2009). Relações de inclusão entre quadriláteros: Conhecimento e desempenho de alunos do Ensino Médio. In *IV Seminário Internacional de Pesquisa em Educação Matemática (SIPEM)*. Brasília, DF: Sociedade Brasileira de Educação Matemática (SBEM).

Rezende, W. M. (2006). Um mapeamento das ideias fundamentais do Cálculo no ensino básico. In *III Seminário Internacional de Pesquisa em Educação Matemática (SIPEM)*. Águas de Lindóia, SP: Sociedade Brasileira de Educação Matemática (SBEM).

Ribeiro, M. R. R. C., & Júnior, A. T. (2012). Funções trigonométricas: Conhecimentos prévios dos estudantes do Ensino Médio. In *V Seminário Internacional de Pesquisa em Educação Matemática (SIPEM)*. Petrópolis, RJ: Sociedade Brasileira de Educação Matemática (SBEM).

Rosenbaum, L. S. (2012). Construtivismo no ensino de funções trigonométricas: Limites e possibilidades. In *V Seminário Internacional de Pesquisa em Educação Matemática (SIPEM)*. Petrópolis, RJ: Sociedade Brasileira de Educação Matemática (SBEM).

Sacristán, J. G., & Pérez-Gómez, A. I. (1998). *Compreender e transformar o ensino* (4th ed.). Porto Alegre: Artmed.

Santos, M. M. P., & Mattos, J R. L. de (2006). A construção dos conceitos de Função Polinomial do 1° Grau na realidade do mundo agrário: Numa visão educacional inclusiva através de métodos de projetos. In *III Seminário Internacional de Pesquisa em Educação Matemática (SIPEM)*. Águas de Lindóia, SP. Sociedade Brasileira de Educação Matemática (SBEM).

Santos, S. A. (2005). Explorações da linguagem escrita nas aulas de matemática. In A. M. Nacarato & C. E. Lopes (Eds.), *Escritas e leituras na educação matemática* (pp. 129–140). Belo Horizonte: Autêntica.

Saviani, D. (1980). *Educação: Do senso Comum à Consciência Filosófica*. São Paulo: Ed. Cortez.

Silva, C. V., & Perovano, A. P. (2012). Obstáculos na compreensão de frações por alunos da Educação Básica. In *V Seminário Internacional de Pesquisa em Educação Matemática (SIPEM)*. Petrópolis, RJ: Sociedade Brasileira de Educação Matemática (SBEM).

Silva, D. M. M. da., & Neto, M. O. T. (2006). Ensino e aprendizagem de razões trigonométricas no triângulo retângulo: Um estudo sobre os conhecimentos de estudantes do Ensino Médio. In *III Seminário Internacional de Pesquisa em Educação Matemática (SIPEM)*. Águas de Lindóia, SP: Sociedade Brasileira de Educação Matemática (SBEM).

Silva, M. A. da. (2006). Tradições e contradições no ensino da Relação de Euler. In *III Seminário Internacional de Pesquisa em Educação Matemática (SIPEM)*. Águas de Lindóia, SP: Sociedade Brasileira de Educação Matemática (SBEM).

Silva, M. A. da. (2012). Cálculo Diferencial e Integral no Ensino Médio: Um ensaio teórico. In *V Seminário Internacional de Pesquisa em Educação Matemática (SIPEM)*. Petrópolis, RJ: Sociedade Brasileira de Educação Matemática (SBEM).

Simon, M. A. (1995). Reconstructing mathematics pedagogy from a constructivist perspective. *Journal for Research in Mathematics Education, 26*(2), 114–145.

Smole, K. S., & Diniz, M. I. S. V. (2001). Ler e Aprender Matemática. In K. S. Smole & M. I. S. V. Diniz (Eds.), *Ler, escrever e resolver problemas* (pp. 69–86). Porto Alegre: Artmed.

Sobrinho, E. G. M., Lima, J. P., & Thomaz Neto, M. O. (2004). Desempenho de alunos do ensino médio na resolução de problemas verbais e nãoverbais em trigonometria. In: *VII Congresso norte/nordeste de educação em ciências e matemáticas* (pp. 225–233). Belém: UFPA/NPADC/CEJUP.

Tall, D. (1990). Inconsistencies in the Learning of Calculus and Analysis, 1,2, 3 & 4, 49–63. Focus.

Tall, D. (1993). Students' difficulties in calculus, plenary address. *Proceedings of Working Group 3 on Students' Difficulties in Calculus, ICME-7*, Québec, Canada, pp. 13–28.

Tall, D., & Rasslan, S. (2002). Definitions and images for the definite integral concept. In A. D. Cockburn & E. Nardi (Eds.), *Proceedings of the 26th Conference of the International Group for the Psychology of Mathematics Education*, Norwich, UK, Vol. 4, pp. 89–96.

Tall, D., & Vinner, S. (1981). Concept image and concept definition in mathematics with particular reference to limits and continuity. *Educational Studies in Mathematics, 12*(2), 151–169.

Teles, R. A. de M., & Bellemain, P. M. B. (2009). Uso de fórmulas para calcular a área e o perímetro de figuras geométricas planas. In *IV Seminário Internacional de Pesquisa em Educação Matemática (SIPEM)*. Brasília, DF: Sociedade Brasileira de Educação Matemática (SBEM).

Tinoco, L., & Nasser, L. (2000). A capacidade de argumentação e a familiarização com as demonstrações no ensino médio. In *I Seminário Internacional de Pesquisa em Educação Matemática (SIPEM)*. Serra Negra, SP: Sociedade Brasileira de Educação Matemática (SBEM).

Usiskin, Z. (1995). Concepções sobre a álgebra da escola média e utilizações das variáveis. In A. F. Coxford & A. P. Shulte (Eds.), *As ideias da álgebra*. São Paulo: Atual.

Van Hiele, P. (1984). Summary of Pierre Van Hiele's dissertation entitled: The problem of insight in connection with school children's insight into the subject – matter of geometry. In: Fuys, D. et al. (Eds.) *English translation of selected writing of Dina Van Hiele – Geodof and Pierre M. Van Hiele* (pp. 237–252).

Vergnaud, G. (1990). La théorie des champs conceptuels. *Recherches en Didactique des Mathématiques, Grenoble, 10*(23), 133–170.

Vigotsky, L. S. (1998). *Pensamento e da linguagem*. São Paulo: Martins Fontes.

Vigotsky, L. S. (2001). *A construção do pensamento e da linguagem*. São Paulo: Martins Fontes.

Vygotsky, L. S. (1989). *A Formação Social da Mente*. São Paulo-SP: Martins Fontes.

Wiley, D. A. (2000).Connecting learning objects to instructional design theory: A definition a metaphor, and a taxonomy. *The instructional use of learning objects*. Bloomington: Association for Educational Communications and Technology, 2000. Disponível em: Acesso em: 02 fev. 2014.

Xavier, C. M. dos S., & Lutaif, B. B. (2006). The Mathematics in the formation of a nursing technical course student: The subject relationship with the mathematical knowledge. In *III Seminário Internacional de Pesquisa em Educação Matemática (SIPEM)*. Águas de Lindóia, SP: Sociedade Brasileira de Educação Matemática (SBEM).

Chapter 5
Mathematics Education at University Level: Contributions from Brazil

Barbara L. Bianchini, Lilian Nasser, Lourdes Onuchic, and Sonia B. C. Igliori

Abstract This chapter presents an overview of the research in mathematics education in higher level from the production of Working Group 04 (WG04) of the Brazilian Society of Mathematics Education (SBEM), created in 2000. A historical summary of the WG is presented, abridging the set of themes and the main theoretical approaches addressed in papers by members of the WG in each edition of the International Seminar of Research (SIPEM) carried through, triennially, from 2000 to 2012. The research papers presented in the scope of the WG are finally described in the most recent edition of such seminar, occurred in 2015.

5.1 History of WG04

The processes of teaching and learning mathematics have been, for decades, subject of attention by part of the scientific community. According to Silva and Lima (2015), "in the beginning the focus was on research objects and phenomena of study fundamentally related to mathematics at school, being the elementary level present in research throughout the twentieth century and the secondary school, since the years 60". The same authors indicate that aspects relating to university education have been covered in surveys since the eighties. They point out that

> Mainly from the nineties, research about mathematics at superior education has shown a considerable increase, both in terms of quantity and quality. In 1997 the International Commission for Mathematics Instruction (ICMI) decided to organize a study about the teaching and learning of mathematics at the undergraduate level and, in 2001, Derek Holton

B. L. Bianchini (✉) · S. B. C. Igliori
Pontifícia Universidade Católica de São Paulo, São Paulo, SP, Brazil

L. Nasser
Universidade Federal do Rio de Janeiro, Rio de Janeiro, RJ, Brazil

L. Onuchic
Universidade Estadual Paulista Júlio de Mesquita Filho, São Paulo, SP, Brazil

A. J. Ribeiro et al. (eds.), *Mathematics Education in Brazil*,
https://doi.org/10.1007/978-3-319-93455-6_5

> published *The Teaching and Learning of Mathematics at University Level* (Holton, 2001), presenting personal reflections from such study. [….] Still in that decade the *International Conference on the Teaching of Mathematics* (ICMT) appeared, trying to supply the demand for more specific research at the higher level of teaching (Silva & Lima, 2015, p. 92).

In Brazil, in the 1980s, federal programmes incentivized support to research aiming at improving mathematics in basic school. For this reason, during this period, research related to mathematics education in higher education was not frequent. An important step for the consolidation of research work in this area was the creation, in 2000, of the International Seminar of Research in Mathematics Education (SIPEM). Amongst the twelve working groups, WG04 was dedicated, from the start, to discuss research focussed on higher education.

This theme of discussion stems from the international scope of mathematics education, reflecting the number of working groups in meetings such as the *Interamerican Conference on Mathematics Education*—CIAEM, with two discussion groups directly linked to the theme (Differential and Integral Calculus in Mathematics Education and Training of Teachers of Higher Education); and the *Congress of European Research in Mathematics Education*—CERME—, with one group, amongst the 24 working groups, addressing higher education (TWG 14 *University Mathematics Education*).

Frota, Bianchini and Carvalho state that "higher education has been established as a solid area of research in mathematics education, looking forward to explore the multiple aspects of the advanced mathematical thinking" (2013, p. 9). This situation is verified in Brazil, as research works carried in the scope of WG04 enabled the publication of two books and a thematic issue of a scientific magazine.

Another aspect worth mentioning is that

> [..]the paper in Mathematics Education at Higher Level has an interface with several interest areas, such as: New Technologies in teaching, Advanced Mathematical Thinking, Argumentation and Proofs, Functions and Graphs, Algebraic Thinking, Geometrical and Spatial Thinking, Visualization, Modelling, Training of Mathematics Teachers, Evaluation of Learning, beliefs and assumptions of students and teachers. (SIPEM I, 2000, p. 119).

We present below a synthesis of the research production by the group presented in SIPEM editions in the last 15 years, highlighting the trends addressed.

5.2 Mathematics Education at Higher Level in SIPEM

In SIPEM I, in 2000, 13 papers were presented with focus on: use of new technologies in the teaching of calculus and linear algebra and technologies of information and communication applied in distance teaching; functions of proof and argumentation in geometry; analysis of textbooks; understanding specific topics of calculus; analytical geometry and linear algebra; disciplines of the basic cycle of the exact sciences area; and the special treatment that must be given to the courses dedicated to the training of mathematics teachers.

Amongst the papers presented, debate around the teaching and learning of calculus and analysis (eight papers) was predominant. The theoretical frameworks were: theory of conceptual fields, registers of semiotic representations, tool-object dialectic, Van Hiele model, cognitivist theory, solidary assimilation, advanced mathematical thinking, conceptions (according to Balacheff), epistemological obstacles (according to Brousseau), net of knowledge and meanings, argumentation and proof (according to Balacheff), the place and the function of proof in mathematics (according to De Villiers), semantic fields, social representation, aspects related to technologies (according to Lévy) and anthropological theory of the didactic.

In SIPEM II, in 2003, 14 papers were selected for presentation. Most of the works addressed aspects related to the teaching and learning of calculus, innovative strategies for the teaching of linear algebra and geometry and probabilities. One of the papers focussed on the relation between calculus and analysis in the undergraduate course for the training of teachers, seeking to establish their specific differences. The references adopted were error analysis, textbook analysis, research on conceptual and epistemological difficulties; there was also a study discussing the creation of collaborative groups at university. The theoretical approaches adopted were: concept image and concept definition (according to Tall and Vinner), conceptual comprehension, conceptions (according to Sfard), advanced mathematical thinking, learning styles, registers of semiotic representations and mathematical literacy (according to Skvsmose).

In SIPEM III, in 2006, 16 papers were presented, grouped in three sessions: mathematics training, training in specific disciplines and teacher training. Almost half of these papers were related to issues inherent to differential and integral calculus. Mathematics in service courses is addressed in two articles, one related to administration and the other one to architecture. Another paper presented is about the estate of the art of research in mathematics education in two scientific journals. The use of games in mathematics teaching, mainly in teacher training, and error analysis are also referred to in the papers presented. Some of the theoretical frameworks quoted by the researchers are: hidden curriculum (according to Sacristan), theory of didactic situations and the three worlds of mathematics (according to Tall).

For SIPEM IV, in 2009, 23 papers were accepted. The most popular subjects were related to calculus, followed by teacher training and, finally, research on algebra. There were also several reflections on the teaching of mathematics in service courses, such as administration and engineering, as well as the question of the transition from basic to higher education, the social representations of the teacher, the students' styles of thinking, the use of technologies in teaching and learning and the relations between affectivity and mathematics. Some of the theoretical approaches used by the authors of the papers presented are: onto-semiotic approach, didactic contract, problem solving, conceptions on the algebraic thinking (Lins and Gimenez; Usinski, Kieran, Tall), model of Van Hiele, styles of thought, APOS theory, theory of significant learning (according to Ausubel), anthropological theory of the didactic, notion of frame (according to Duady), notion of point of view (according to Rogalski) and notion of levels of knowledge (according to Robert).

In SIPEM V, in 2012, 20 papers were accepted. The main subject discussed in the meetings of the group involved the process of teaching and learning of calculus, analytical geometry and algebra; the disciplines of analysis and geometry in the course of training of mathematics teachers; the ability of visualization in spatial geometry; mathematics education in the specialization of psychopedagogy; cognitive neuroscience and advanced mathematical thinking. The debates also included: the use of mathematical modelling in service courses and error analysis, concerning combinatory, pre-service and in-service training, and the training of mathematics teachers.

Moreover, the state of the art regarding inequations in higher education was presented. The theoretical approaches in this edition of the seminar were: registers of semiotic representation, mathematical research in the classroom, error analysis, taxonomy of educational objectives (according to Bloom), advanced mathematical thinking, the importance of visualization (according to Presmeg, Arcavi and Fischbein), epistemological theories of critics to scientific knowledge (Popper, Lakatos, Kuhn and Bachelard), anthropologic theory of didactics, research on the own practice (according to Bridge), sociocultural perspective of mediated action (according to Wertsch) and theoretical-conceptual theory relating psychology and knowledge (according to Piaget, Wallon, Vygotsky, Bruner, Moscovici, amongst others).

In the last SIPEM, in 2015, 16 papers were presented, and the subjects hardly changed. Certainly, the subject with the strongest presence in the research on higher education throughout the five SIPEM meetings was calculus teaching and learning, due, perhaps, to the high rate of school dropout and failure in this discipline. Many of those studies have explored the use of technology in education, mainly due to the convenience introduced by free software and to the increasing access of students to portable computers and tablets.

In the international scenario of research in mathematics education, calculus teaching and learning has also been predominant. Silva and Lima (2015) highlight also a working group on this subject, called *Students' difficulties in calculus*, established in the seventh International Congress of Mathematics Education (ICME), which took place in 1992. As Lima (2012, p. 234) observed, based on Marcolini and Perales's considerations (2005, p. 27), these concerns about the teaching and learning of calculus is perfectly justifiable because of the links that such content "keeps, both with the elementary and the advanced mathematics, as well as the role that it plays in other sciences (that) transform it into a set of knowledge with great theoretical and empirical value, indispensable for higher education".

The concern about the training of the future mathematics teachers has also been the subject of debates, addressing how the disciplines exclusively designed for the course of training of mathematics teachers must be approached, as against other undergraduate courses. Another important trend is the concern about students' beliefs and feelings, exploring a more human side of the hard disciplines in higher education.

5.3 Current Overview of Brazilian Research Works Focussed on Mathematics Education in Higher Education, Based on Papers Presented at WG04 in SIPEM VI

Research on mathematics education in higher education in Brazil started to gain strength especially in the first years of this century, as pointed out in the previous section. In SIPEM VI, interest in other subjects such as analytical geometry, geometry and probabilities arises progressively.

The use of technologies, the training of teachers, mathematics as a discipline of service and the transition from basic to higher education are subjects that are also gradually being explored, besides some research focussed on contents of discrete mathematics and analytical geometry. The new subjects appeared as preparation of material for students, in-service training of teachers; use of software in education; and transition from high school to university education.

It is important to point out the diversity of approaches that is not a phenomenon specific for WG04, but it is international, and has been paramount for several researchers (Artigue, 2016). It has been indicated that diversity is unavoidable, but it is necessary to be taken it into account for the progress of the field. For WG04, this dispersion of subjects makes it difficult, in a certain way, to organize this overview, that is, it does not favour the presentation of global and consensual results, as intended. Therefore, we have chosen to present the specificities of each research, representing the current stage of development of mathematics education in higher education based on papers from SIPEM VI.

Therefore, it is possible to separate the research works in two methodological aspects: the empirical and the theoretical approach. The empirical approach is less represented than the theoretical approach, but it is not possible to establish a reason for this. This may be related to the pragmatic hindrances that this type of research brings to the researchers.

With the empirical research, the specificity of the subjects investigated causes a relevant differentiation for this overview, because this differentiation interferes with a more global analysis of the research carried out, since the undergraduate course of the research can be one where mathematics is a discipline of support, or a discipline for the training of teachers, etc. And this is the way the research results will be produced.

In what follows, the articles are separated in empirical and theoretical categories.

5.3.1 *Empirical Research*

In the article by Trevisan, Borssoi and Elias, *Delineation of a sequence of tasks for an educational environment of calculus,* the subjects of the research are students of engineering, who have frequently been the focus in research on higher education, and the aspect under research is adapted to the subjects, as this audience does not have mathematics as their main interest, so the awakening of the interest towards this science is an essential point. For this matter, authors suggest the study of task sequences, based on the references in Ponte (2014), Stein and Smith (2009) and in Watson et al. (2013). We can highlight that this task dynamics allowed the authors to perceive that pupils had difficulties in dealing with the situation when it was presented broadly, as well as the need to, at first, consider a more directed task, where the student would have more information and little options to think.

In the article *Rate of change: How teachers in training understand the concept*, Bisognin, E. and Bisognin V. carry out the research on teachers under training, who are attending the discipline *Fundaments of Differential Calculus* of a master's degree course in mathematics teaching. The authors question how the subjects under research interpret and relate the information made explicit in different representations of the concept of rate of change. They follow Tall (1994), who deals with the different analytical and graphical representations of the concept of derivative in the classroom work, where technical aspects prevail. The author asks what teachers of a professional master course in mathematics education understand the concept of rate of change. How do those teachers interpret and relate the information as explicated in different representations of the concept of rate of change?

This research allows the authors to consider that even if the basic concepts of calculus have been explored in the undergraduate course of training of mathematics teachers and, in general, the concept of variation rate has been studied by the pupils during their graduation, the participants of the research had created very restricted images of the concept, little contributing to its understanding.

Sousa and Gomes, in the article *Impact of auxiliary programs in the discipline of differential and integral calculus I*, try to investigate if the insertion of auxiliary programmes could provide undergraduate students with a better performance in this discipline. This research is developed using data from the students' institution, which discloses that differential and integral calculus (CDI) is responsible for a great percentage of interruption, dropout and failure of pupils. Moreover, in some courses, this discipline is a prerequisite for others, which occasionally causes delays and unevenness amongst the pupils' learning at university.

The research was organized according to the research-action method, and it was developed in seven stages: data-collecting; interviews with professors; implementation of a questionnaire including open and closed questions to evaluate expectations and/or experiences of the students about ICT and the teaching of Calculus I; resolution of lists of exercises, contextualized in the area of the course where the discipline was taught and/or inside the discipline itself, in a way that the content was inserted in the context of Calculus I; after-class help previously scheduled and at a different time from regular lessons, in order to provide assistance with most common doubts about the contents of CDI I; creation and implementation of a sequence of activities with GeoGebra; dealing with potentialities and limitations of the sequence of activities. Data collection was based on the students' records during the development of the activities, their comments (collected in audio and transcripts), observations (in cards and daily notes of research), a questionnaire and semi-structuralized interviews, besides the observation of the marks of the students throughout each semester.

In *Visual exploration in the study of the behaviour of functions by means of their derivatives using learning objects in virtual environments*, Cunha and Laudares reflect, together with engineering students, on the challenges of introducing technologies in the international scene and they highlight, from Frota and Couy (2009), the importance of visualization and of visual thinking in the teaching and the learning of mathematics, as well as how much these aspects can be valued with the aid of the computer, considering that the relation between the didactic sequence and technology goes through learning objects (OA), developed on the basis of applets constructed by means of GeoGebra.

Collected data indicate, amongst others, that a virtual OA can allow the teaching and learning processes the possibility of visualizing some properties that are, in general, manipulated only algebraically.

In the article *A study regarding the three worlds of mathematics and the registers of semiotic representation: solutions of students to a task involving the concept of function*, Kirnev, Marins, Silva and Savioli intend to analyse, on the basis of the registers of semiotic representations (Duval, 2009, 2011), the difficulties presented for the research subjects in the solution of activities, and to identify indications of the path of the students along the theory of three worlds of mathematics (Tall, 2004).

From the conceptions of Tall (1991) and Tall and Vinner (1981) regarding aspects concerning the nature of advanced mathematical thinking (PMA) and of the model proposed by these researchers, called *Three worlds of the mathematics*, the authors consider to discuss the movement of the students through the three worlds (namely, conceptual embodied, proceptual symbolic and axiomatic formal).

The task was proposed to ten students who attended the discipline of real functions of one real variable, during the second semester of the course of training of mathematics teachers. The subjects had not yet studied the concept of function in that discipline, only in basic education; they had only worked with the concepts of sets and relations.

In the article *Mobilization of the styles of mathematical thinking by engineering students*, by Gomes, Lima and Bianchini, the authors analysed whether the strategy of cooperative learning can allow the student of this course to transit through various styles of mathematical thought. Such styles are not always explored by teachers in the classroom, nor by teaching materials, thus missing the chance to offer the students alternative insights in the understanding of a certain piece of specific mathematical content (Burton, 2013).

The authors had conducted a research on how beginner students of an engineering course mobilize the styles of mathematical thinking, when faced with routine questions in the discipline of analytical geometry, concerning the study of the straight line. Interviews with the students showed that they associate the algebraic formalism to the correct way to offer the solution to mathematical questions. Even those that predominantly mobilize the visual style, state that they know the solution expected by the teacher is the analytical one.

In the article *Circumferences or straight lines in the sphere? Visualizing them in the Cabri-3D*, Leivas presents a research on the use of the Cabri-3D, in the discipline of geometry, in a professional master degree course, in 2014, aiming to investigate how the pupils visualize curves and triangles in a spherical surface, called geodesic and spherical triangles. The author states that the use of media has significantly modified people's behaviour, especially in the school environment, where cellular phones and tablets are often present. Some ideas of some authors were used, such as Arcavi (1999) on visualization; Borba and Villarreal (2006) on technology, and ideas by Fayó (2012) regarding the participation of imagination in mathematical research. The research followed a qualitative approach in observational style and was based on the methodology of problem solving.

The author infers that the research proved innovation possible in the teaching of geometry and shows a methodological possibility, using the Cabri-3D, to develop visual abilities. The students ratify the impressions of Borba and Villarreal (2006) in experiments that promote discoveries. In relation to teachers, the will to innovate in the professional practice and to provide the students with such discoveries in the schools where they act, is a highlighted point of the research, which leads to the conclusion that the goal of the research has been reached.

The article by Nomura and Bianchini is inserted in the study of concepts of linear algebra and has the title: *The role of the concept image and concept definition in the constitution of the mathematical object eigenvalue and eigenvector: Narratives of an engineering student.* The guiding questions of the research were: Which conceptions (action-process-object-schema) are evidenced in pupils, after the study of the mathematical object eigenvalue and eigenvector in the initial and final phases of their academic formation in engineering courses? In these same phases, which concepts image and concepts definition are evidenced in the study of the mathematical object eigenvalue and eigenvector?

The article presents the analysis of the speech of a student in his last semester of engineering studies at a private institution of higher education with emphasis on his conception of eigenvalue and eigenvector mathematical object. The results displayed in this article evidence the concepts image and definition pertinent to the advanced mathematical thinking (Vinner, 1991) and the main ideas presented in the research by Domingos (2003).

5.3.2 Theoretical Research

The category of theoretical articles, which are summarized below, also includes a variety of theme, theoretical and methodological approaches. In the epistemological approach is the article by Lima, *In search for an identity for the discipline of calculus: First reflections*. The author states the research problem from a historical study developed by him, in which he analysed the introduction and development of calculus in the undergraduate course of mathematics at the University of São Paulo between 1934 and 1994.

The arguments handled are based on the theories of the instrumental and relational understanding and also on mathematics in the context of sciences (MCC) (Camarena, 2013) accepting two aspects that have been acknowledged as fundamental: (a) the aiding to pupils, in the sense defined by Skemp (1989), of the instrumental and relational understandings of mathematical objects; and (b) the promotion of an adequately contextualized approach for mathematical beings.

As a conclusion, the author highlights that, when trying to contextualize what is being worked in a way to make content become significant for the professional training of the pupil, there will be a greater possibility of the teacher to also value a relational and not only instrumental understanding of the calculus.

The article *Studies related to the fundamental concepts of calculus and analysis,* written by Jesus, intends to point out trends in research about difficulties in teaching and learning the basic concepts of differential and integral calculus and real analysis disciplines. It is included in the concern that arises from the high failure rate in the case of calculus and the high dropout rate in the case of the analysis and the inherent consequences of this situation, with special attention for the social role they fulfil, to the courses of training of teachers of mathematics.

It is a bibliographical research, with data collected in papers published in annals of events of the area; in topics of national and international literature, and brings the results of this research concerning the eighth edition of the CERME (Congress of the European Society for Research in Mathematical Education), thematic group University Mathematics Education (TWG 14).

Four general subjects are categorized for the organization of the data: Transitions; Affection; Teachers Practices and Mathematical Topics.

It was suggested that the paper by Monroy and Astudillo (2013) on reconstruction of definitions and its references could be studied in connection with the one by Lima and Silva (2012), regarding the rigour that must be adopted in the initial discipline of calculus. Also that the positioning of Baroni and Otero-Garcia (2012) about the real analysis would be enriched by the debate regarding the transition, approached by Winsøw (2013), of the students at university (until then students of the training of mathematics teachers course) for the schools of basic education, where they will act as teachers.

In the article *Development of material for the teaching of concepts of the differential calculus*, Igliori and Almeida, from PUC/SP, intend to contribute with the educational practice, in particular with the practice of teaching calculus. Taking into account the results of the research, they search in for difficulties and cognitive obstacles of the students; processes through which they learn a particular concept; effects of curricular and pedagogical innovations and, more recently, teacher's practices and beliefs. The research looks for answers about the duality between the theoretical maturity of mathematics education, and the situation of the teaching of mathematics, which seems to remain irregular and underdeveloped, regarding approaches that facilitate the learning. And still in the perspective of what Rasmussen, Marrongelle, and Borba (2014) consider, when they understand that, in sight of the depth of what is known about the learning of the pupils, it is necessary that researchers of mathematics education at higher level become engaged in the development of research projects that address questions about the teaching and learning of calculus, either of theoretical or of pragmatic nature. The proposal of the article in translating the theories into classroom practices is supported by the documentary genesis and by the theoretical constructs of Tall (1982) for the teaching of calculus.

The article by Queiroz and Barbosa, *Exercises of textbooks on financial mathematics and its borders with situations of the daily life and of work environments,* established characteristics for textbook exercises of financial mathematics that show possible borders with situations of daily life and of work environments. The analysed books had been selected from the bibliographies in the course programmes, collected in a Federal University and the State Universities of Bahia for business-oriented courses in the area. Other data had been collected by enquiring the professors who teach financial mathematics in these courses.

The authors have taken as a reference Barroso and Kistemann Jr (2013) and Skovsmose (2000). They have also quoted Rosetti Jr and Schimiguel (2011), who discuss teaching books on financial mathematics in Brazil, pointing out the distance between the contents studied in formal schooling environments and the knowledge concerning daily life and working environments. They state that the focus of teaching books for high school is in the solution of "semi-real" problems—those situations that are invented based on real situations, with direct application of presented formulas, without discussing the financial meanings related to them.

Onuchic and Allevato present a study entitled *Proportionality through problem solving in the higher course of training of mathematics teachers*, which has the goal of discussing the possibility to approach mathematical contents in the training of mathematics teachers course, starting from unifying ideas of the mathematics for the resolution of problems. The unifying ideas make possible that diverse concepts are approached in an integrated form, promoting more complete and deep understandings of mathematical contents.

Taking the proportionality concept as a unifying idea, they analyse the possibilities to use the methodology of teaching-learning-evaluating of mathematics by means of problem solving (Allevato & Onuchic, 2014). This methodology considers that the construction of the knowledge, in classroom, can be carried out based on problems that generate new concepts. The focus of the study was to present data from teachers, collected in initial and continuous training, aiming at the construction of knowledge concerning the proportionality concept, highlighting its relations to other contents, as well as the resolution of problems as a teaching methodology.

Defending that the reflections on learning must surpass its speculative considerations and become scientific, Onuchic and Allevato are supported in Bransford, Brown, and Cocking (2000), according to whom the emphasis in the understanding leads to one of the basic characteristics of the new science of learning, which is its focus in the knowledge processes. The approach presented and analysed in the article, based on the unifying ideas and on problem solving, according to the authors, favours this emphasis. Other authors were consulted, amongst them Donovan and Bransford (2005) and Van de Walle (2009).

The researchers Elias, Gereti and Savioli, in the article *"What a horror! Such a simple thing": a study on the production of meanings for mathematical questions,* intend to understand the production of meanings by teachers, from questions involving rational numbers.

The focus of this article is, mainly, on the notions of production of meaning and legitimacy of the model of the semantic fields, by Lins and Gimenez (1997), as a reference for the readings of the articulations of the teachers in a discussion arising from mathematical questions that involve rational numbers.

Students of one discipline of the programme of post-graduation in education of sciences and mathematics, of a State University of Londrina, had participated in this discussion. The speeches of the students had been explored, in the groups, which had been recorded, and the written productions registered in paper. It can be perceived that, when one speaks about meaning (understanding it as a knowledge that is what is said of an object and not what could be said, but what effectively is said in the context of an activity), to produce meaning is to speak regarding an object; being legitimacy related to the ways of producing meaning. Still, they point out that all knowledge being enunciated in the direction of an interlocutor is true, but the fact that something is stated does not mean it is true.

The central question of the research for the article by Biajone and Barolli entitled *The trajectory of production of curriculum of the discipline of discrete mathematics in a superior course of technology* is: Which trajectory is covered by disciplines of discrete mathematics (MD) in the production of its curriculum in the contexts of the influence, the production of texts and the practice in a superior course of technology in analysis and development of systems (ADS)?

The authors seek to understand the purposes and the reason for these disciplines to be part of the curriculum for the superior course of technology.

According to them, the trajectory of the production of a curriculum of MD in ADS starts by establishing MD as a higher education discipline; it transits for its stabilization as lapsing in the matrix of that course and reaches its appropriation as clipping or curricular perspective produced by the professor at the moments of creation and experiencing of this lapsing. From there comes the theoretical choice of Ball, Bowe, and Gold (1992) about cycle of curriculum policies.

The data collected were organized around two axes of analysis: (1) the production of a curriculum for a discipline of MD in its constitution and stabilization in the course of ADS and (2) the production of a curriculum for a discipline of MD in its appropriation by the ADS course. Some of the preliminary analyses evidence an increasing positioning that the MD seems to have been constituted and legitimated in a conflicting land of curricular interests, contested by groups that search the priority of what is expected from this disciplines for the university formation of courses in the area of computation, which affected the context of the influence of the training of the technologist in ADS.

Nasser, Vaz and Torraca present an article as a part of a larger research, entitled *Transition from high school to higher education: Investigating difficulties in analytical geometry*. Specifically, they intend to investigate the knowledge brought by students of a group in the first period of a course of training of mathematics teachers, concerning vectors and the representation of straight lines in the plane. They used the theoretical framework of the theory in the registers of semiotic representations (Duval, 2003). Research activities were implemented, categorized in accordance with the theory of error analysis (Cury, 2007).

The authors take as a base the research, amongst others, of Tall (1991) and Pallis (2010) on the difficulties with the infinitesimal calculation, the concept of limit and derivative and its applications. They elaborated a research activity, aiming to diagnose what the pupils know about vectors and straight lines when they get to university. Such activity was implemented in a group of 30 pupils who attended a discipline of analytical geometry, during the first week of lessons, in the course of training of mathematics teachers. Amongst the results, the authors highlight that the important contents of analytical geometry are not studied in high school. It was not clear for the authors if pupils can make, correctly, a distinction between a straight line and a segment of a straight line.

We synthesize below the main subjects addressed in the papers presented in this last edition of the SIPEM. They reflect, although partially, the current concerns of the members of WG04 and bring in their core the main current questions concerning the teaching and the learning of mathematics in university courses. We also present a synthesis of the main theoretical and methodological frameworks that were at the base of the research concerning each one of these subjects.

5.3.3 Theoretical and Methodological Aspects

The production of curriculum for mathematical disciplines of service courses, that is, that do not aim at the training of mathematicians, is one of the current concerns of the researchers of the WG. The research in this line presented in the last SIPEM had moments of theoretical discussions, regarding curriculum, taking into special consideration the characteristics of a course in computer science, of the structure of the pedagogical speech and the sociology of education. The methodology employed was the case study.

The question of the importance that university students mobilize different styles of mathematical thinking was investigated from theoretical frameworks of cognitive psychology and studies concerning the structure of mathematical thinking. The research was carried out through an intervention developed according to a strategy of cooperative learning, in which the relation of the student with a mathematical content was analysed.

Aspects of visualization have been studied, based on theoretical ideas about this *construct*, especially with regard to mathematical visualization, the use of technology in education and in mathematics learning and imagination. Observational methodology has been used, based on problem solving.

The creation of educational material based on research results was discussed through inputs referring to the work of the professor in the creation of materials for the classroom and in cognitive aspects. As a methodological approach, the documentary genesis was used.

Another object of research was *The delineation of a sequence of tasks for a computational environment for calculus*. Such research was based on concepts of realistic mathematics education and on reflections regarding the design of tasks. The developed sequence was implemented in an undergraduate course and the productions of the students were analysed.

The impact of auxiliary programmes in disciplines of calculus was investigated taking as a base the recommendations to use information and communication technologies (ICT) in education. The research-action has been used as a methodological approach.

Reflections on the search of an identity for the discipline of calculus was the theme chosen by one of the members of the WG in the most recent meeting of the group. The research was carried out based on theoretical issues related to two ways of understanding the mathematical concepts: relational and instrumental.

Moreover, ideas regarding the need to contextualize the teaching of calculation in the different undergraduate courses in which these disciplines are present were also mobilized. From the methodological point of view, such research consists of a bibliographical study.

The subject of transition from basic to higher education was approached based on semiotic theories, error analysis, analysis of activities and of cognitive aspects. The methodology adopted was the application of tests.

The learning of concepts of linear algebra was also object of a research that was guided by cognitive theories. The methodology was the analysis of the speech.

The research concerning *Visual exploration in the study of the behaviour of a mathematical object* was based on reflections regarding the challenges of the introduction of technologies in the teaching and learning processes and on the importance of visualization and visual thinking. The methodology adopted was the application of a teaching sequence that would have to be developed by the students with the aid of computational tools, particularly with a learning object based on applets.

The use of theories in the analysis of task resolutions was object of a research developed on the basis of the concept of PMA and on the theory of the three worlds of mathematics. The protocols with the student resolutions were evaluated based on the analysis of the speech.

The pre-service and in-service training of teachers is also an aim of the research by the members of the WG. In this last edition of the SIPEM, the paper concerning in-service training had as a reference theoretical cognitive constructs related to the analytical and graphical representations of a mathematical concept in conflict with the technical aspects that prevail in the classroom. The methodology adopted in this work was the application of tests. The research regarding the initial training of teachers was based on the idea of, by means of the methodology of problem solving, constructing knowledge by creating problems on new contents. The collection of data was carried out through participant's observation and documentary analysis.

In relation to the *Production of meanings for mathematical concepts,* the research presented was based on the emphasis of understanding as one of the basic characteristics of the new science of learning. The analysis method was based on subjects' speeches that had been recorded, and in their written productions recorded on paper.

One of the pieces of research presented had as its object of research the trends in the studies related to the fundamental concepts of calculus and analysis, published in annals of two important events in the field. It is handled, therefore, as a bibliographical study.

The analysis of textbook exercises in comparison with the situations faced by students in everyday life and in working environments was also the aim of one of the research papers present in the scope of WG04 in the last SIPEM. Such study was supported by the ideas of several authors who discuss textbooks, pointing out the distance between the school and everyday life.

5.3.4 Trends of the Research in Mathematics Education at Higher Education Level

The investigation on mathematics education in higher education follows the same trend of the inquiries in mathematics education in the other levels of education, which is the search for approaches that can be fertile to indicate ways that take account of the analysis of the phenomena that involve the teaching and the learning of mathematics.

This search implies in variations of thematic, theoretical frameworks and methodological approaches. The relation of higher education with the other levels in the formation of students fosters a symbiosis with the research in other levels. As an example, the research that involves the function concept can be indicated. However, there are things to be distinguished in the particularities of the research concerning higher education. And the trends of this branch will be highlighted in this item of the overview. We can say that what took the mathematics educators to dedicate themselves to the research of learning in higher education was students' failure in certain disciplines, especially calculus and linear algebra. The research, of cognitive basis, concerning the learning of calculus and linear algebra, those objectifying the elaboration of material for the teaching of these disciplines, the use of technology, modelling, and didactic engineering, are trends of this area. The specific studies for the training of teachers are also trends, in which special theoretical frameworks are used. The error analysis, which goes through all the education levels, is also a trend in higher education. Today the philosophical perspectives, including semiotics, have presented a fertile way of analysis of phenomena in the learning of mathematics and research works that do not distinguish levels of education have also gained space, but they focus on the agreement and on the formation of mathematical thinking. This is a source that is going through a significant amount of research in mathematics education.

The history of the mathematics must continue to be explored and the neuroscience must be reinforced in the next years. Another trend that can be highlighted is the one of the studies that strengthens the interdisciplinarity in the construction of the knowledge and how to include the school mathematics in this perspective. The role of mathematics in the formation of the citizen, in the conception of realistic mathematics, brings light to approaches, throwing its attention to the disciplines of this area, which will be attended by the youngsters that enter university in the area of exact sciences.

5.4 Last Considerations

The main consideration to be highlighted is that WG04 has grown in quantity of research, indicating both, that the area is getting a good theoretical base for this, and that the difficulties of the students of this level of education have been calling the attention of the researchers. It also indicates that it is important to treat them scientifically, so that mathematics is not in detriment of the studies in the area of exact sciences. There is a significant amount of research by the students of service courses in which mathematics is present.

Another important consideration is the increasing amount of research in the scope of the training of teachers, and in the transition from basic to higher education. Reflections about these two subjects are vital for the development of the Brazilian education. We highlight as an advance the quality of the research based on cognitive scientific theories, as the ones by Tall and Vinner on the scope of semiotics; Duval's, on the attribution of meanings, and Lins's, amongst many others. We also consider as an advance the increasing number of disciplines in higher education of interest for the researchers.

What has not yet been identified in WG04 research is research addressing classroom management, the use of more advanced technologies, interdisciplinary studies, the relationship between international and national results, curricular analyses for courses of higher education, objectives of mathematical disciplines in the different courses, etc.

The diversity of approaches, since each researcher chooses his/her subject, target audience and theoretical framework, makes it difficult to look at the production of the WG as a whole. What is analysed is the research and not its results. This is indicative that it is necessary to continue in this direction, with the accomplishment of overviews that point out more general findings of this WG that, in a way, represent the national production of research in mathematics education in higher education. The approach of the Brazilian research must be a concern of the WG, so that the national studies can impact on the international research.

References

Allevato, N. S. G., & Onuchic, L. R. (2014). Ensino-Aprendizagem-Avaliação de Matemática: Por Que Atravės da Resolução de Problemas? In L.R. Onuchic et al. (Org.), Resolução de Problemas: Teoria e Prática (pp. 35–52). Jundiaí: Paco Editorial.

Arcavi, A. (1999). The role of visual representation in the learning of mathematics. In *Proceedings of the XXI Conference on the Psychology of Mathematics Education,* North American Chapter. Mexico. Retrieved from http://www.clab.edc.uoc.gr/aestit/4th/PDF/26.pdf.

Artigue, M. (2016). The challenge diversity of theories in Mathematics. In *Proceedings of the13th International Congress on Mathematical Education* (pp. 24–31). Hamburg, Germany.

Ball, S., Bowe, R., & Gold, A. (1992). *Reforming education & changing school: Case studies in policy sociology*. London: Routledge.

Baroni, R. L. S., & Otero-Garcia, S. C. (2012). Dois Vieses para a Disciplina de Análise em Cursos de Licenciatura em Matemática. *Anais do V Seminário Internacional de Pesquisa em Educação Matemática*. Rio de Janeiro: SBEM.

Barroso, D. F., & Kistemann, M. A., Jr. (2013). Uma Proposta de Curso de Serviço para a Disciplina Matemática Financeira. *Educação Matemática Pesquisa, 15*(2), 465–485.

Borba, M. C., & Vilarreal, M. (Orgs.) (2006). Tendências Internacionais em Formação de Professores de Matemática. Belo Horizonte: Autêntica.

Bransford, J. D., Brown, A. L., & Cocking, R. R. (Eds.). (2000). *How people learn: Brain, mind, experience, and school: Expanded edition*. Washington: National Academies Press. https://doi.org/10.17226/6160.

Burton, L. (2013). Mathematics and their epistemologies—And the learning of mathematics. In I. Schwank (Ed.), *European research in mathematics education I* (Vol. 1). Retrieved from http://www.fmd.uni-osnabrueck.de/ebooks/erme/cerme1-proceedings/cerme1-proceedings-1-v1-0-2.pdf.

Camarena, P. A. (2013). Treinta Años de la Teoría Educativa "Matemática en el Contexto de las Ciencias". *Innovación Educativa, 13*(62), 18–44.

Cury, H. N. (2007). *Análise de Erros: O Que Podemos Aprender com os Erros dos Alunos*. Belo Horizonte, MG: Autêntica Editora.

Domingos, A. D. M. (2003). *Compreensão dos Conceitos Matemáticos Avançados – a Matemática no Início do Superior.* Doctoral thesis in sciences of education, Universidade de Nova Lisboa, Lisboa.

Donovan, M. S., & Bransford, J. D. (2005). *Introduction. National Research Council. How students learn mathematics in the classroom*. Washington, DC: The National Academy Press.

Duval, R. (2003). Registros de Representações Semióticas e Funcionamento Cognitivo da Compreensão em Matemática. In S. D. A. Machado (Org.), Aprendizagem em Matemática: Registros de Representação Semiótica (pp. 11–33). Campinas: Papirus.

Duval, R. (2009). *Semiósis e Pensamento Humano: Registros Semióticos e Aprendizagens Intelectuais*. São Paulo: Livraria da Física.

Duval, R. (2011). Registros de Representações Semióticas e Funcionamento Cognitivo da Compreensão em Matemática. In S. D. A. Machado (Org.), Aprendizagem em Matemática: Registros de Representação Semiótica (8th ed.). São Paulo: Papirus.

Fayó, A. N. (2012). Imaginação, Geometria Dinâmica e Cabri. In F. U. Guerra & H. Z. A. Caracciolo (Eds.), *VI Congresso Iberoamericano de Cabri. ACTAS* (pp. 15–22). Lima: Pontificia Universidad Católica del Perú.

Frota, M. C. R., Bianchini, B. L., & Carvalho, A. M. T. (Orgs.) (2013). Marcas da Educação Matemática no Ensino Superior. Campinas, SP: Papirus.

Frota, M. C. R., & Couy, L. (2009). Estratégias para o Ensino-Aprendizagem de Funções com um Foco no Pensamento Visual. In *Anais do IV Seminário Internacional de Pesquisa em Educação Matemática* (pp. 1–20). Brasília: SBEM.

Holton, D. (2001). *The teaching and learning of mathematics at university level. New ICMI study series* (Vol. 7). Dordrecht: Kluwer Academic Publishers.

Lima, G. L. (2012). *A Disciplina de Cálculo I do Curso de Matemática da Universidade de São Paulo: Um Estudo de seu Desenvolvimento de 1934 a 1994*. Doctoral thesis in mathematics education, Pontifícia Universidade Católica, São Paulo.

Lima, G. L., & Silva, B. A. (2012). O Ensino de Cálculo na Graduação em Matemática: Considerações Baseadas no Caso da USP. In *Anais do V Seminário Internacional de Pesquisa em Educação Matemática*. Rio de Janeiro: SBEM.

Lins, R. C., & Gimenez, J. (1997). *Perspectivas em Aritmética e Álgebra para o Século XXI*. Campinas: Papirus.

Marcolini, M., & Perales, J. (2005). La Noción de Predicción: Análisis y Propuesta Didáctica para la Educación Universitaria. *Revista Latinoamericana de Investigación en Matemática Educativa (RELIME), 8*(1), 25–68 Retrieved from http://www.clame.org.mx/relime/200502a.pdf.

Monroy, A. A. & Astudillo, M. T. G. (2013). Interactive construction of a definition. In *Proceedings of the 8th Conference of European Research in Mathematics Education (CERME8)*, Ankara, Turkey.
Pallis, G. A. (2010). Transição do Ensino Médio para o Ensino Superior [C.D.]. In *Anais do X Encontro Nacional de Educação Matemática*, Salvador, BA.
Ponte, J. P. (2014). Tarefas no Ensino e na Aprendizagem da Matemática. In J. P. da Ponte (Org.), Práticas Profissionais dos Professores de Matemática. Lisboa: Instituto de Educação da Universidade de Lisboa.
Rasmussen, C., Marrongelle, K., & Borba, M. C. (2014). Research on calculus: What do we know and where do we need to go? *ZDM, 46*(4), 507–515.
Rosetti, H., Jr., & Schimiguel, J. (2011). Estudo de modelos de Matemática Financeira em bibliografia básica. In *Anais da 13ª Conferência Interamericana de Educação Matemática (CIAEM)*, Recife, Brazil.
Silva, B. A., & Lima, G. L. (2015, December). Os Cursos de Cálculo Difundidos pela USP e as Preocupações Didáticas Presentes em Livros Adotados e em Práticas Docentes. *UNIÓ: Revista Iberoamericana de Educación Matemática, 43*, 88–111.
Skemp, R. R. (1989). *Mathematics in the primary school*. Londres: Routledge.
Skovsmose, O. (2000). Cenários para investigação. *Bolema, 14*, 66–91.
Stein, M. H., & Smith, M. S. (2009). Tarefas Matemáticas como Quadro para Reflexão. *Educação e Matemática, 105*, 22–28.
Tall, D. (1982). The Blancmange function continuous everywhere but differentiable nowhere. *The Mathematical Gazette, 66*(435), 11–22.
Tall, D. (1991). The psychology of advanced mathematical thinking. In D. Tall (Org.), Advanced mathematical thinking. Dordrecht: Kluwer.
Tall, D., & Vinner, S. (1981). Concept image and concept definition in mathematics, with special reference to limits and continuity. *Educational Studies in Mathematics, 3*(12), 151–169.
Tall, D. O. (1994). Understanding the processes of advanced mathematical thinking. *Invited talk given at the International Congress of Mathematicians*. Zurich. Edited version published in *L'Enseignement Mathématiques*, 1996, 42, 395–415.
Tall, D. O. (2004). Thinking through three worlds of mathematics. In M. J. Hoines & A. Fuglestad (Eds.), *Proceedings of the 28th Conference of the International Group for the Psychology of Mathematics Education* (Vol. 4, pp. 281–288). Bergen: Bergen University College.
Van De Walle, J. A. (2009). *Matemática no Ensino Fundamental: Formação de Professores e Aplicação em Sala de Aula*. Trad. Paulo H. Colonese (6th ed.). Porto Alegre: Artmed.
Vinner, S. (1991). The role of definitions in teaching and learning. In D. Tall (Org.), Advanced mathematical thinking (pp. 65–81). Dordrecht: Kluwer.
Watson, A., Minoru, O. K., Ainley, J., Frant, J. B., Doorman. M. & Kieran, C. et al. (2013). Task design in mathematics education. In C. Margolinas et al. (Eds.), *Proceedings of the ICMI Study 22* (pp. 9–16). Oxford: ICMI.
Winsøw, C. (2013). The transition from university to high school and the case of exponential functions. In B. Ubuz, C. Haser, & M. A. Mariotti (Eds.), *Proceedings of the 8th Conference of European Research in Mathematics Education* (pp. 2476–2485). Ankara: Middle East Technical University.

Chapter 6
History of Mathematics and Culture: Moments and Movements in Brazilian Mathematics Education

Cristiane Coppe de Oliveira, Cláudia Regina Flores, Daniel Clark Orey, and Maria Cristina Araújo de Oliveira

Abstract This chapter presents moments and movements discussed by the Working Group WG05-History of Mathematics and Culture of the Brazilian Society of Mathematics Education. Considering the proceedings and the books that have resulted from the six editions of the International Seminar of Research, SIPEM, we conducted a survey to map the current research and explore the discussions, drawing conclusions concerning the theoretical and methodological perspectives of the group. In the study, three axes have been characterized: ethnomathematics, history of mathematics, and history of mathematics education. We observed a variety of approaches that can be summarized around the area of multicultural education, the role of history in teaching and teacher education, and the importance of developing historical knowledge about the creation of knowledge, methodologies, and pedagogical practices. Here, the importance of research that incorporates diverse sociocultural methods to produce mathematical and historical knowledge in school mathematics and its pedagogical practices cannot be overstated.

6.1 Introduction

The recognition that many diverse cultures have developed a variety of techniques to work with calculations, measurements, comparisons, and volumes, among others throughout history has led researchers, nationally and internationally, to wonder

C. C. de Oliveira (✉)
Universidade Federal de Uberlândia, Uberlândia, MG, Brazil

C. R. Flores
Universidade Federal de Santa Catarina, Florianópolis, SC, Brazil

D. C. Orey
Universidade Federal de Ouro Preto, Ouro Preto, MG, Brazil

M. C. A. de Oliveira
Universidade Federal de Juiz de Fora, Juiz de Fora, MG, Brazil

A. J. Ribeiro et al. (eds.), *Mathematics Education in Brazil*,
https://doi.org/10.1007/978-3-319-93455-6_6

about a variety of ways in which mathematical knowledge is produced. According to Bishop (1994), this form of research is based on an anthropological-ethnographic perspective and seeks to understand mathematical knowledge from three major focuses.

The first, with an emphasis on traditional cultures, shows the existence of knowledge and practices experienced by members of different sociocultural groups. The second, centres on ancient documents of non-western societies, and is based on historical values. Finally, the third focuses on different groups in a society to enable investigations related to the production of mathematical knowledge in a socio-psychological way.

According to D'Ambrosio (2005), in the early 1970s, many Brazilian researchers, along with researchers from the United States, took an interest in those issues, bringing their perspectives and findings to the field of mathematics education. D'Ambrosio (1985) initiated a research line he named the "ethnomathematics program", which discusses the mathematics produced by a diversity of sociocultural groups, including the importance, appreciation, and awareness of those practices for mathematics education.

From that line of research, a working group called Mathematics Education, History and Culture was formed under the coordination of Prof. Ubiratan D'Ambrosio. In 2000, the group held its first meeting and developed several works, which were synthesized into a *Book of Abstracts* from SIPEM I held in Serra Negra, São Paulo, Brazil.

Since then, members of this group have continued their work around two key words or themes: history and culture. Thus, by including an ethnomathematics perspective, researchers started a dialogue between culture and the production, generation, institutionalization, and dissemination of knowledge related to diverse ways of counting, sorting, ordering, locating, modelling, explaining, and inferring the mathematics both found and used in cultural contexts that often break with classical education paradigms.

D'Ambrosio (1999b) stated that history shows that the evolution of academic mathematics reflects changes of cultural, linguistic, social, political, economic, ideological, and religious factors. Regarding this, many investigations have been conducted concerning the history of mathematics and mathematics education, where the researchers investigated interfaces between the history of mathematics and its contributions to mathematics education. This included work that sought to look at how the teaching and learning processes occur in the context of the historical understanding of school mathematics.

This chapter presents the ideas shared and movements discussed by the members of this group, as well as the analysis of proceedings and books from the former six editions of SIPEM. In the study, three axes have been characterized: ethnomathematics, history of mathematics, and history of mathematics education. We are pleased to see a variety of approaches articulated with multicultural education, with the role of history in teaching and teacher education, and the importance of historical knowledge on the constitution of knowledge and practices, methodologies, and pedagogical practices.

A remarkable trend has been growing steadily concerning research in the history of mathematics education that continues to engage the group in finding new boundaries between history and culture in mathematics education. Despite the many different approaches and problems, this WG focuses on history and culture specifically in the context of mathematics education.

6.2 History and Culture: Theory and Methodology

Since the research work of this group is concerned with both the history and culture of mathematics, it is necessary to understand how the researchers have used those terms in theoretical and methodological ways. According to the *Dictionary of Etymology*,[1] history comes from the Latin *historia* "narrative of past events, account, tale, story", from the Greek *historia* "a learning or knowing by inquiry; an account of one's inquiries, history, record, narrative", from *historein* "inquire", from *histor* "wise man, judge". From this, it follows that history as a scientific activity is the search for, and the study of traces of, past events, achievements, tragedies, and ideas that ensure the memory of a society and place humanity in a certain time. History, in this sense, is the methodical narration of facts that are important to a society.

During the 1960s and 1970s, a new way of conceiving history and historical research emerged, giving emphasis to studies through new approaches and objects that address new questions considered relevant to society. Within this new conception of history, De Certeau (2005) inspires researchers involved in this WG through the notions of deviations, ownership and the very question of the subject, i.e. understanding the speaker and to whom he speaks. So does the work of Chartier (1990), by a new theoretical horizon, which argues that cultural objects are produced "between practices and representations", and that the subjects that produce and receive culture correspond to "ways of doing" and "ways of seeing".

The notion of culture, in its broadest sense, goes along with the research in this WG. However, just like history, culture encompasses complex concepts that guide the way we do research. According to the *Dictionary of Etymology*, the word originates from the Latin *cultura* "cultivating, agriculture". However, since 1805, the word appears to mean "the intellectual side of civilization," and since 1867, the "collective customs and achievements of a people".

According to Laraia (1999, p. 69), "men of different cultures use various lenses and, therefore, have divergent views of things". The author also considers that the:

> way of seeing the world, the assessments of moral and value order, different social behaviours and even body postures are thus products of a cultural heritage, i.e., the result of the operation of a particular culture (1991, p. 70).

In this WG, culture has been, identified, on the one hand, by its:

---

[1] Retrieved from: http://www.etymonline.com. Access: October 21, 2016.

> systems of explanations, philosophies, theories, and actions and the daily behaviour. All of this is based on processes of communication, representations, classification, comparison, quantification, counting, measurement, inferences. These processes take place in differently in different cultures and transform over time. They always reveal the influences of the environment and are organized with an internal logic, they codify and formalize themselves. Thus, knowledge is born (D'Ambrosio, 2005, p. 101–102).

Certeau's discussions (2005), on the other hand, proposes that culture is a set of practices that establishes a field in which meanings are produced and shared through the processes of objectification, subjectivity, and identification. These processes generate representations that articulate a domain, a clash, a resistance, a speech, a value, and a rule. Hence, mathematical knowledge can be problematic in its practices, representations, and schooling. It is noticeable that researchers both connect favourably to, and allow for a certain focus on ethnomathematics, as well as seek to provide a connection between history and context as they are articulated by this notion of culture.

6.3 Ethnomathematics: Motivations

From the perspective of multicultural education and ethnomathematics existing in the proposal, and the work of WG5 in SIPEM I, Knijnik's research (2000) stands out in particular. The author discussed an ethnomathematics approach[2] from her seminal ethnographic research in relation to the landless movement in southern Brazil. The relations of ethnomathematics work with pedagogical implications arose from SIPEM II, held in the city of Santos, SP, with research that presents possibilities in different cultural contexts. Among the works, Monteiro (2003), for example, discussed components of mathematical modelling from the perspective of ethnomathematics, listing the possibilities and limits in school processes. Damasceno and Gomes (2003) focused on pedagogical possibilities that consider work with the cassava flour culture and found similarities between traditional knowledge and academic knowledge that helped them to identify mathematical thinking in this process.

Youth and Adult Education (EJA) is contemplated in Knijnik's work (2003) and continues to consider the relationship between culture, curriculum, and oral mathematics, in order to understand this connection as a sociocultural practice of the field. Indigenous knowledge, in turn, begins as a research strand by Mendes's work (2003), where the author analysed the number of meanings for the Kaiabi indigenous context and its consequences for the development of a mathematics textbook in their mother tongue.

[2] In 2000, the author conceptualized, as an ethnomathematics approach, the "research of mathematical traditions, practices and conceptions of a social group and the educational work that is developed, aiming for the group to interpret and decode its knowledge, acquire knowledge produced by academic mathematics, make comparisons between its knowledge and academic knowledge, analysing the power relations involved in the use of these two knowledges" (Knijnik, 2000, p. 178).

The third edition of SIPEM, held in the city of Curitiba, Paraná, Brazil, brought new kinds of research strands from an ethnomathematics perspective. Knijnik (2006) studied the dialogue between present-day ideas and Michel Foucault's discourse on peasant culture in EJA. Passos (2006) developed articulations between ethnomathematics and critical mathematics education, and Costa (2006) discussed the importance of considering the logical, mathematical, and mythological thinking in the cultural context of the African-Brazilians and indigenous peoples. Focus on research making a connection between teacher education and ethnomathematics was also emphasized. The work by Domingues (2006) discussed the issue of training indigenous teachers in the state of São Paulo, against the desire/need/complexity of acquisition of school mathematics. Domite (2006), in turn, moved from traditional teacher training paradigms as social and intellectual beings to the perspective of teachers' proximity to the culture that each student carries within him/herself. The legitimization of different forms of mathematical knowledge in teaching practices and in the mathematics curriculum was also highlighted, notably by the works presented by Mafra and Fossa (2006), Fantinato and Santos (2006), Gonçalvez and Monteiro (2006), and Santos (2006).

Pedagogical practices were discussed in the works of SIPEM IV, in the city of Brasília, Brazil. For example, Costa (2009) discussed the implementation of Federal Law 11645/08, including as mandatory African, African-Brazilian, and indigenous culture in the school curriculum. Mendonça and Pinto (2009) emphasized the practices of indigenous Xacriabá teachers in mathematics classes. The work conducted by Oliveira (2009a, 2009b) delineated new paths towards a dialogue between the Gilbert Durant's imaginary theory and ethnomathematics. The results of these studies reveal that the inclusion of cultural aspects in the mathematics curriculum has long-term benefits for students' mathematical achievements because these aspects contribute to their perception that mathematics is part of their daily lives. They also deepen the understanding of its nature by enhancing their ability to make meaningful connections. Thus, the pedagogical action of an ethnomathematics programme added value when compared to other editions of SIPEM because, generally speaking, indigenous training work gained greater prominence.

The main theme of SIPEM V, held in Petrópolis, Rio de Janeiro, Brazil, was the indigenous context. It brought the mathematical knowledge developed by members of the communities of Serra da Moça (Voltolini & Kaiber, 2012) and Paresí (Nascimento & Silva, 2012), while the research conducted by Bernardi, Caldeira, and Duarte (2012) presented contributions to the understanding of the continuous education process of indigenous Kaingang teachers in the city of Chapecó, in the state of Santa Catarina. Ethnomodelling also arose as an aspect of the ethnomathematics programme (Rosa & Orey, 2012), and Miarka (2012) brought important contributions to our thinking about the ethnographic possibilities in our work with ethnomathematics.

At SIPEM VI, the articles submitted and approved that had connection with ethnomathematics were classified in three categories: indigenous contexts and knowledge, countryside/rural contexts and knowledge, and theoretical research and reflections.

In the category of indigenous contexts and knowledge, Nascimento and Silva (2015) proposed strategies for the process of teaching and learning in geometry based on geometric patterns observed in the handicrafts developed by the Tikuna indigenous people of the Umariaçu community, in the state of Amazonas. The results of this study showed that the handicrafts, which play an integral part in the sociocultural context of the students, can help them understand academic curricular geometric concepts.

Similarly, within the school context, the results of the study conducted by Tomaz (2015) has exposed the tensions that emerge from activities related to notions of probability learned by indigenous students enrolled in an intercultural education training course for teachers at the Federal University of Minas Gerais. The analysis of the data collected in this research showed that the tensions involved in these activities occurred when these students did not recognize the application of probability in their cosmological beliefs. Tensions during the contact of two different cultural perspectives were softened when the power relationship between the teacher and the indigenous students became horizontal, allowing both indigenous and non-indigenous worldviews to be shared and incorporated.

Regarding the use of local knowledge developed by indigenous Brazilians in pedagogical actions of the ethnomathematics programme, the results of the study conducted by Costa and Mattos (2015) with members of the Karipuna ethnicity in the village of Manga, in Oiapoque, in the state of Amapá, showed that members of this cultural group developed specific community forms related to the productive process of cassava fields. From the cultural knowledge that these members collectively developed to organize the plantations, the researchers focused on the contributions of ethnomathematics regarding the teaching of mathematics. Thus, students were able to realize that this human and contextual knowledge helped them understand reality better.

Since the first Portuguese settlers arrived in Brazil, culture and customs, including mathematical knowledge and practices developed by Brazilian indigenous peoples, such as the Guaranis, were considered inferior and worthless. For example, considering the diverse multicultural characteristics of these indigenous peoples, Silva and Caldeira (2015) discussed the importance of the counting system and the graphic symbols developed by members of two Guarani groups: the *Itaty*, of Morro dos Cavalos, and the *M'Biguaçu*, located between the cities of Palhoça and Biguaçu, in the state of Santa Catarina. The results of this study show how the numbering and the graphic symbols systems the Guarani developed are intrinsically associated with the characteristics of their culture.

In the category of field/rural contexts and knowledge, the importance of establishing a connection between the school and practical knowledge that students acquire in educational institutions as well as in their own communities was verified. In this sense, Barbosa (2015) investigated how six rural education students confirmed the importance of bringing their reality to the mathematics classrooms. The results of this study showed that it was not possible to use the knowledge coming from a students' reality to establish a bridge between two different epistemologies, but it was possible to verify the influence of this knowledge and the languages of the students to incorporate their reality to the school curriculum.

Regarding the use of local mathematical knowledge in a real context, the research conducted by Brito and Mattos (2015) analysed the mathematical knowledge produced by six farmers at Colônia Agrícola do Matapí, in Porto Grande, state of Amapá. The ethnomathematics perspective concerning the production, storage and commercialization of agricultural products and was used to analyse and interpret the results, and to establish their relationship with school knowledge.

With this in focus, Machado (2015) showed the contributions of rural producers to the processes of agricultural modernization and investigated the ways in which small producers deal with marketing their production. The results of this study showed that the mathematical knowledge used by these farmers enables them to play the role of traders in the agricultural activity. From this, a new agricultural training was proposed to adjust the model established by modernization from the very need of farmers to remain in rural areas. Other results showed that these producers were often not concerned with business accounting but used creative measures to mitigate the challenges posed by agricultural modernization.

The countryside and rural regions of Brazil present diverse and unique cultures with an increasing potential for the documentation of insights of the connections between mathematics, mathematics education, and culture. In this sense, Silva (2015) discussed the cultural approach of mathematization and measurement units that emerged from social practices developed in a rural settlement in the state of Maranhão. For that, they needed to verify possible historical convergences and cultural influences between Portugal and France in relation to the system of measures to identify the units used by rural workers in order to seek possible dialogues with school mathematics.

To discuss the principles of the ethnomathematics programme, it is necessary to propose a theoretical basis to discuss how the social, economic, and political contexts are important for the development of mathematical ideas. It is also important to know how and why different individuals reveal diverse interests, preferences, talents, and abilities, as well as specificities and strategies that we use to generate, organize, disseminate, and share mathematical knowledge to solve the problems we face in our daily lives.

In the area of research and theoretical reflections, Monteiro and Mendes (2015) investigated the ethnomathematics from studies that show of the variety of ways in which this programme in mathematical education is understood. Their goal was to analyse this movement from Foucault's perspective. The results of their work demonstrated the link between ethnomathematics and mathematics education based on the concept of a counter-conduct proposed by Foucault. Continuing the research on an ethnomathematics programme, Oliveira (2015) investigated emerging perspectives that arise in the discourse of a mathematics teacher in a school community, in a Costa da Caparica neighbourhood, in Portugal. This research, which is related to a project called Fronteiras Urbanas (Urban Frontiers—2012–2014), was used to develop an ethnomathematics perspective that involved the process of students' mathematics teaching and learning in this school community. This study used the communicative, analytical, and technological instruments of D'Ambrosio's trivium curriculum for mathematics (1999a).

Regarding the theoretical reflections for this programme, Rosa and Orey (2015) argued that ethnomathematics shares several characteristics with Lakatos's methodological scientific research programmes, since its main components are a firm nucleus, positive and negative heuristics, and the protective belt of auxiliary hypotheses, which facilitate the analysis of empirical phenomena. Therefore, in the Lakatosian sense, the main objective of the ethnomathematics programme is to develop and strengthen the theories that make up its protective belt, expanding it and making it more precise in relation to the empirical predictions that aim to strengthen its nucleus. Consequently, ethnomathematics is a Lakatosian research programme composed of irrefutable theories that make possible the theoretical and methodological decisions that enable its progressiveness.

In another theoretical reflection, Marchon and Fantinato (2015) proposed the construction and development of a sociocultural basis for ethnomathematics as proposed by D'Ambrosio (1985) from his writings in mathematics education through a textual production that shows possible conceptual changes and transformations when adapting the theoretical base of this programme. These authors also presented some readings that contributed to the reflection on the sociocultural foundations of mathematics education and made possible to rethink fundamental theoretical, social, and philosophical aspects of ethnomathematics.

Continuing with the research that began in SIPEM V, in 2012, Rosa and Orey (2015) deepened the theoretical reflection on the application of particular techniques of ethnomathematics in conjunction with the tools used in modelling. Ethnomodelling, which is a tool that provides a holistic view of the nature of mathematical knowledge, connects the cultural and academic aspects of mathematics through a dialogical approach. In the ethnomodelling process, the emic and etic approaches to mathematical knowledge facilitate the translation of problem situations present in the systems drawn from the reality of members from distinct cultural groups. Emic knowledge is essential to understand ideas, procedures, and mathematical practices of the members of these groups intuitively, while etic knowledge is essential for a comparison. The dialogical perspective between the views uses the emic and etic approaches to obtain a more comprehensive understanding of the mathematical knowledge developed by members of diverse cultural groups.

The discussions presented above demonstrate that research in ethnomathematics appears in all editions of SIPEM as a kind of "firm nucleus", according to the Lakatosian conception of the programme. It is the motivating axis of the working group that deepens, but also renews itself, opening gaps for the exploration of other problems related to culture, and to history itself—of the characters, of the contents. This is the case, for example, of the research in history of mathematics that we now present.

6.4 History of Mathematics, History in Teaching: Configurations

According to D'Ambrósio (2000), the history of mathematics "aims to reflect on basic questions, leading to the construction of a historiographic project of the history of mathematics in Brazil" (p. 172). This would imply the investigation of ideas and achievements of renowned mathematicians, which were brought and disseminated in Brazil. Following this theoretical-methodological orientation, Sad and Dynnikov (2003) presented work that deals with the first scientific missions encouraged by agreements between Brazil and Europe, bringing Italian mathematicians to Brazil in the 1930s to teach at the University of São Paulo, in the city of São Paulo, in the state of São Paulo, Brazil. There was a concern about what and how to research in history of mathematics. Thus the work of Sad and Dynnikov (2003) presents questions pertaining to the delimitation of research, consideration of facts and sources and types of research.

However, the research in the area of the history of mathematics is shy in editions of SIPEM, despite the strong recognition of the need for a historical understanding of mathematics in Brazilian mathematics education. Two interest groups support this: one focuses on the historical study of influential characters in Brazilian education; and the other tries to understand mathematics contents historically.

In the first case, the work of Trentin (2009), which analyses Manoel Ferreira de Araújo Guimarães's translation of the work *Éléments de Géométrie*, by Adrien Marie Legendre, appears in SIPEM IV. Then, in SIPEM V, two works can be characterized in this axis, namely: Santos (2012) studies the historical and traditional story of Dedekind, who suggests the construction of real numbers from cuts, and Araújo (2012) presents an investigation about the life and works of Joaquim Gomes de Souza, a person known in the history of Brazilian mathematics as Souzinha (1829–1864).

The investigation of concepts and disciplines, from other perspectives, already appears in SIPEM I, as in Moura and Sousa's work (2000). The authors discuss historical elements that constitute the conceptual nexus of algebraic thought. Meneghetti (2003) investigated the intuitive and logical aspects in the development of calculation, and which were considered educational trends. Abdounur (2003) studied the mathematical concept of incommensurability through analogies with music and astronomy. However, considering new approaches in history that see the production of knowledge as the result of social practices, Lannes (2003), for example, dwells on the empirical construction of mathematical knowledge, discussing it as a network of meanings within a given collectivity. And Flores (2006a, 2006b) analyses social practices on how to draw and represent the space of the eighteenth century fortifications on Santa Catarina Island, in the city of Florianópolis, state of Santa Catarina, Brazil, using it to understand a form of knowledge that is delineated by geometry and perspective. From SIPEM IV, Silvai (2009) sought a historical (re) construction in the conceptual development of differential and integral calculus, looking at it as a construction of models.

Through two major centres of research, renowned mathematicians and mathematical concepts, researchers in mathematics history reflect on the development of a Brazilian mathematics education that started with foreign scholars and theories, but that was articulated within the specificities of Brazilian culture. Furthermore, the history of mathematics in mathematics teaching has become a subject of research in Brazil, mainly because it recognizes that it "provides a good opportunity to develop our view of 'what mathematics is' or that the history of mathematics allows us to have a better understanding of concepts and theories" (Baroni, Teixeira, & Nobre, 2004, p. 165).

In fact, we can say that there is a delimitation of internationally articulated trends. Heeffer (2006), for example, has raised three good reasons to argue that mathematics education benefits from the history of mathematics:

> The first is epistemological and addresses a contextual view on mathematical knowledge. The second concerns the phylogenic aspects of the development of mathematics. Conceptual difficulties with teaching children mathematics often correspond with historical periods of conceptual crisis in mathematics. A third, historiçal argument, draws on the vast repository of experience in mathematics education. We provide examples for each of these arguments from the history of algebra (Heeffer, 2006, p. 1).

Thus, from the third edition of SIPEM, the research focused on this relational aspect of the history of mathematics and mathematics education as presented. Batista (2003) discussed the inclusion of the history of mathematics in teaching, evidencing a study on the systems of equations from mathematicians Seki Kowa and Leibnz. Bayer (2006) deals with the use of the history of mathematics as a motivational and contributing factor to teaching and learning. Scheide (2006) also discussed the contribution of the history of mathematics to improve the teaching and learning process of the students in order to make them more critical and active citizens.

Maciel, Cardoso, and da Fonseca (2012) promoted meaningful learning of the concept and function through history. Roque (2012) identified pedagogical potentialities of the history of mathematics in activities related to integers. Dias and Saito (2009) justified and proposed an approach that favours the construction of an interface between the history and teaching of mathematics, based on new historiographical and methodological trends. Finally, Sá (2015), presented research on the use of history of mathematics in the classroom from an experience with teaching graph theory.

On the other hand, teacher education regarding the knowledge of mathematics history with the possibility of knowing methods and techniques that can aid in his/her practice (Baroni et al., 2004) has also been a research problem since the first edition of SIPEM. This is the case for Nobre (2000), who discusses the implications of the history of mathematics for the training of mathematics professionals. In practical situations, Brolezzi (2000), for example, proposed workshops with teachers to discuss the conceptual tension of the discrete/continuous pair, and to rethink the construction of the idea of number, and the birth of differential and integral calculus. Then, during SIPEM III, Motta and Brolezzi (2006) discussed the role of teachers in regard to the integration of history of mathematics in the process of teaching and learning in practical situations. Similarly, Abdounur (2006) organized an exhibition to address historical-didactic aspects of the relationship between mathematics and music, which characterized a space for teachers to experience cultural and extension activities in their curricular tasks.

It is remarkable that a small research sample focusing on history allows for the circulation and delimitation of diversified research objects. This aspect led Miguel (2003), in SIPEM II, to propose a distinction between the investigative research fields of the history of mathematics, history of mathematics education, and history in mathematics education. Therefore, it is noteworthy that there is the configuration of three mixing strands that help to define their own ways to deal with history and culture, which allowed the dismemberment of these important investigation fields such as the case of the Brazilian history of mathematics education.

6.5 History of Mathematics Education: Dismemberment

During SIPEM II, held in 2003, the first works on what would be characterized later as the history of mathematics education started to appear. Here we comprise the works that are concerned with the historicity of the processes of teaching, learning, and teacher training in the context of mathematics. Therefore, historical studies on textbooks, teaching renewal movements, teacher training courses, legislation, pedagogical journals, notebooks, that is, all documents geared towards teaching or learning mathematics are included in this framework.

In 2003, four studies were approved, two of which used comparative perspectives in history. One of the authors, Valente (2003), sought to understand the appropriations of Brazilian and Argentinean proposals for the renewal of mathematics teaching from the first international movement that had the creation of ICMI, in 1908, as a milestone. Soares (2003) compared two national programmes for the evaluation of textbooks: The National Textbook Commission, created in 1938, during the management of Gustavo Capanema as a Minister of Education and Health during Getúlio Vargas's government, and the National Textbook Programme, created by the Ministry of Education, in 1985. Duarte (2003) investigated the mathematics proposed in the *Journal of Mathematics and Physics Notes*, published between 1953 and 1954, by students of the Mathematics and Physics Section of the Sciences and Letters Philosophy School of University of São Paulo that aimed at disseminating mathematical knowledge for students in college, teachers, and high school students. In a similar context, Ferreira (2003) analysed the mathematical content included in the school curricula of the teacher formation school, Escola Normal do Espírito Santo, from 1892 to 1971.

The third edition, held in 2006, contained works that explain the history of mathematical education as a keyword or major theme and other papers that, even without being explicit, could be considered in this framework, when identified by similarity of contexts.

Five studies were developed from textbooks as main sources. Carvalho (2006) compares editions of the work *Moyens d'apprendre à compter sûrementet avec facilité* by Condorcet in Brazil and Portugal. School mathematics in German-Brazilian schools founded in the late nineteenth and early twentieth centuries, in Rio Grande do Sul, is Mauro's object of research (2006). The author focuses on the process of producing textbooks and newspapers to serve that community. Another study analysed the teaching of mathematics in *General Geography*, by Varenius, published in 1650, highlighting the relations in knowledge in those times (Brito, 2006).

Two studies analysed specific topics in textbooks: Thales's theorem, in the collection of mathematics course *Matemática Ginasial—1ª série* written by Euclides Roxo, Cecil Thiré, Júlio César de Mello, and Souza,[3] published between 1940 and 1942 (Pereira, 2006); and the insertion of the decimal metric system in Brazil from 1856 (Zuin, 2006). Although both consider textbooks as sources for research, the way they are analysed is different. Whereas for the study on the Thales theorem the emphasis is on the mathematical treatment given in the work in which the segments are incommensurable, for the decimal metric system, the book is analysed as a vehicle for insertion of new contents that parameterize new practices in society.

The historical study of the education of teachers who teach mathematics was the subject of three studies dealing with different levels—elementary, middle and high school, and higher education. Regarding elementary education, the study by Soares (2006) analysed rules for recruiting and selecting teachers for the first schools in Rio de Janeiro, from the turn of the eighteenth century to the nineteenth century. The work by Baraldi (2006) investigated the Campaign for Improvement and Diffusion of Middle and High School Education (CADES) as a training course for middle and high school mathematics teachers based on oral history as a methodology. Oliveira (2006c) investigated the learning of mathematical analysis that Ubiratan D'Ambrosio had during his undergraduate time in mathematics in the 1950s, by associating his training to the activity of research in mathematics. The sources mobilized in the study ranged from sheets made by Ubiratan himself as a student, interviews with him and his teacher, Elza Gomide, and the books included in his course bibliography. The last two studies were both conducted in the 1950s and represent the contrasts in Brazilian educational reality. CADES course, the training offered aimed at preparing teachers in a short period, many times during vacations, to work in middle and high school. The mathematics course attended by Professor Ubiratan, on the other hand, aimed at formation for research in mathematics.

Two studies investigated knowledge and practices for mathematics teaching in Brazil's northern and northeastern states. Neto and Braga (2006) investigated mathematical practices and knowledge that guided the official curriculum for public schools in the state of Pará, between 1900 and 1920. The research, which uses

[3] The three authors taught at the secondary school Colégio Pedro II, an institution founded in 1837 in the colonial period, which, until the 1930s, served as an official reference for other Brazilian educational institutions, until the creation of the Ministry of Education, in the same decade.

written documents and image records, identifies the influence of Comte's positivist ideas in curricular organizations. Gutierre (2006) discussed mathematics education in middle and high school that was disseminated by *The Athenaeum* in the state of Rio Grande do Norte, Brazil, between the 1930s and 1960s, a period in which proposals for educational reform by Francisco Campos and Gustavo Capanema stabilized. The two works cover the first half of the twentieth century and are emblematic of the process of strengthening national references, particularly teachers and authors of mathematics textbooks. Two studies investigated specific knowledge areas from an ahistorical perspective. Miguel and Souza (2006) investigated the obsolescence process of the litmus test, an arithmetic method to confirm the result of the addition operation, in the Brazilian school context. Finally, Oliveira (2006a) investigated infinitesimal calculus taught at the polytechnic school, Escola Politécnica de São Paulo, in 1904. The studies have in common the way they analyse the specific contents inserted in the school/institutional, cultural, and social context that allow the maintenance or the rejection of certain knowledge in the scope of education.

The modern mathematics movement is also the subject of two studies. Duarte (2006) discussed the view of the mathematician, teacher, and author of textbooks Benedito Castrucci, on the movement in Brazil. Fischer and Carpes (2006) investigated the experience with—pilot and experimental—classes organized by the Mathematics Education Study Group of Porto Alegre (GEEMPA), starting in 1972. Using written documents and statements from teachers involved, they discuss how this experience is considered positive and marked by modern mathematics ideas as carried out at a time of criticism and decline of this movement.

Three studies analyse trajectories and productions of characters directly or indirectly related to Brazilian mathematics education and to the field of mathematics. A study by Oliveira (2006b) analysed proposals for the teaching and learning of mathematics published in the *Al-Karismi Magazine*, organized by MalbaTahan,[4] in the 1940s. Forner and Lopes (2006) investigate, through oral history, the interpretations and uses mathematics educators made of Paulo Freire's work, an important reference for Brazilian education. Using documents in archives in England and Berlin as sources, the process of historical recognition by English mathematician, Arthur Cayley, is the objective of the research by Mattos (2006a, 2006b).

In SIPEM IV, in 2009, the number of studies identified as history of mathematics education reduces to eight. The modern mathematics movement is a representative theme in this universe because it dealt with three of them. Oliveira (2009) discussed the growth of a modern mathematics programme for middle and high school education disseminated by G.E.E.M[5] (Mathematics Education Study Group) and outlined

[4] Pseudonym of the professor of mathematics, author of didactic textbooks, Júlio César de Mello e Souza.

[5] This group of mathematics teachers from the state of São Paulo, with different levels of education, played a prominent role in the dissemination of proposals of modern mathematics throughout Brazil. The group had the professor and consecrated author of mathematical textbooks, Osvaldo Sangiorgi, as a president, among other distinguished names of teachers and authors of books of

in pedagogical journals that circulated between the 1950s and 1960s in Brazil. Geometry in the perspective of this movement is analysed by Leme da Silva (2009), through didactic texts produced by the School Mathematics Study Group (SMSG), of the discussion in the Royaumont and Dubrovnik Seminars, promoted by the European Organization for Economic Cooperation (OECE) and the First Inter-American Conference on Mathematics Education (CIAEM). Batista et al. (2009) investigated the influences of this movement in the curricular organization of the Brazilian Federal District, inaugurated in the 1960s, at the same time as the ideas of modern mathematics were disseminated in the country.

The influence of the book *International Conferences on Public Instruction*, from 1934 to 1963, published by the Ministry of Education and Culture, National Institute of Pedagogical Studies, in 1965, on the elaboration, organization, curriculum constitution, teaching methods, and textbooks of mathematics in Brazilian schools for the middle and high school levels are the objective Ribeiro's study (2009).

The forms of mathematics taught at a middle and high school levels is the objective of the research carried out by Valente (2009), from the analysis of textbooks, indicating a differentiated process compared to that in middle school. The text conjectures the existence of two sequential school subjects—one for middle school and another for high school—with a certain autonomy. The process of professionalization of mathematics teachers is dealt with by Dias and Bertani (2009) in the context of the mathematics course in the state of Bahia, Brazil. The authors prioritized primary sources and notebooks between 1943 and 1965, and registers of actual practices.

Brito (2009) analysed the teaching of mathematics in the seventeenth century at the Akademisches Gymnasium in Hamburg and in the universities of Königsberg, through the analysis of the correspondence between Bernhard Varenius and his teacher, Joachim Jungius. Miorim, Brito, and Faria (2009) presented a survey of articles that studied mathematics history and education work published in journals, categorizing them according to the approach used.

In SIPEM V, held in 2012, five studies related to the history of mathematics education strand were presented. Three of them used a very similar theoretical-methodological contribution based on the history of the school subjects. Two were related to the teaching of mathematics in elementary school, one by Villela (2012), from the analysis of reports issued by the direction of a school in Vassouras, in the state of Rio de Janeiro, Brazil, in the late nineteenth century; and another by Borges and Duarte (2012), who investigated the magazine *A Escola*, published between 1923 and 1924. The third study, carried out by Ferreira (2012), investigated the historical trajectory of the mathematics teaching methodology disciplines in the teacher training courses in three universities in the state of São Paulo, Brazil.

The appropriation of innovation proposals for the teaching of mathematics disseminated from two international movements—the first, with the creation of ICMI in 1908, and the second, the movement of modern mathematics in the 1960s and 1970s—by two textbook authors representative of each of the periods is the theme

mathematics in Brazil, who integrated the group.

of the work done by Silva and Silva (2012). Mathematics education in colonial Brazil was addressed by Magalhães and Silva (2012).

In SIPEM VI, referenced researchers in the area of mathematics education began to appear with publications in the history of mathematics education, such as the work done by Ubiratan D'Ambrosio, Wagner Valente, Gert Schubring, and Maria Ângela Miorim.

During the SIPEM VI meeting, held in Pirenópolis, Goiás, in November 2015, many presentations dealt with studies in the history of mathematics education. Therefore, the coordination of the WG subdivided the works into axes, one of them on that topic. Six out of the 22 studies presented belonged to the history of mathematics education axis. Half of these studies drew on references, but discussed this knowledge from different perspectives. Two studies analysed the presence of the discipline and its purposes in different educational and historical contexts. The first context is the training of elementary school teachers during the first decades of the First Republic, between 1889 and 1930 (Oliveira, 2015), and the second is the context of elementary education at the end of the Empire (1883). It was institutionalized in the country at that time, in the framework of proposals by Rui Barbosa, an influential intellectual of the period (Guimarães, 2015). The third text discussed theoretical-methodological perspectives for the writing of a history of drawing, not getting restricted to the history of sciences or education, but crossing different fields of knowledge (Machado & Flores, 2015).

The other three works dealt with a wide scattering of themes comprising the historical study of mathematics education research in Brazil, from studies resulting from graduate courses in science and mathematics teaching between 1975 and 1984, which are currently considered milestones for the constitution of the mathematics education as a professional field in the country (Miranda, 2015); from the analysis of mathematics teachers' representations about the modern mathematics movement in a countryside city in the state of São Paulo (Rodrigues, 2015); and from the investigation of mathematics conceptions underlying professional and technological education between the 1940s and 1980s, in the federal technical institute, Instituto Técnico Federal do Espírito Santo (Pinto, 2015).

During SIPEM VI, the works used documentary and oral sources and, for the most part, used explicitly contributions of cultural history as a theoretical-methodological perspective, referenced in authors such as Peter Burke and Antoine Prost. Studies that include historical perspectives found in school subjects are also mobilized in most texts. In a specific way, one of the studies used the contributions of oral history, and another, Foucault's philosophical perspective.

From this perspective of research on the history of mathematics education, represented since SIPEM II, it was manifest that the research was concerned with a variety of ways of learning and teaching mathematics that occurred in the midst of changes and permanencies, at different levels of teaching and in different times and spaces in Brazilian education.

6.6 In Conclusion: Emerging Aspects, New Designs

At the beginning of the WG in 2000, D'Ambrosio's proposed delimitation brought in its essence the richness of many research perspectives, which, due to their very diversity, led to diverse and divergent theoretical and methodological frameworks. Throughout its trajectory, the group added more research concerned with themes involving history and culture in its wide and very diverse relations to mathematics, most notably: ethnomathematics, history of mathematics, history of mathematics education, and the history in mathematics education. The space has continued to prove to be quite fertile in terms of ongoing discussion, production, and dissemination of the topics, which can be observed through national and international scientific meetings and publications.

Particularly, the role of research in ethnomathematics can be affirmed by recent trends in education because it envisions hope for the transformation of the role of the school, specifically, for example, the school in rural and countryside contexts. In Brazil, there has been modest growth in relation to the discussion and the conduction of research related to these areas of study, due to the recognition of research contribution in the development of rural and countryside education.

In this sense, approximately two decades ago, Bishop (1994) argued that rural education can be considered as a potential construct to be applied in ethnomathematics research. In this case, the study of ethnomathematics helps educators to connect school mathematics with students and their communities. Extensive global work with ethnomathematics has been conducted in rural and countryside contexts; however, concepts related to rurality itself are rarely considered to have a relevant influence on mathematics education since there is a gap in the conduct of the investigations related to the problem of rural and countryside education.

In addition, an important aspect of the ethnomathematics research programme is the connection between indigenous knowledge and the curriculum of mathematics. In this case, it is considered necessary to explore the possibilities to use indigenous knowledge as a pedagogical action for mathematics education. Also required is the identification of issues and problems that occur in complex contexts of indigenous groups (such as natural, cultural, political, and social), so that different approaches to diverse educational situations can be discussed (D'Ambrosio & Rosa, 2008). In this sense, it is important to emphasize that Brazilian indigenous peoples have their rights guaranteed by the Federal Constitution, as well as by federal and state laws that govern indigenous school rights and education.

In this sense, it is important to promote discussion in relation to the dialogue between local (indigenous) and scientific (academic) knowledge, as well as the influence of mathematics on indigenous schools and teacher training. In initial indigenous teacher training courses, specifically in intercultural degrees, future teachers reflected on their own pedagogical practices and their actions, which enable them to understand various aspects of the teaching and learning process in the context of an indigenous school.

In order to promote intercultural and bilingual education in these schools, one of the main objectives of the discussions was to analyse some of the many challenges different training courses currently meet for indigenous teachers in the field of mathematics education. According to Rosa and Gavarrete (2016), these challenges involve the relationship between traditional and scientific knowledge during the development of pedagogical actions used in classrooms by indigenous teachers.

In order to debate the principles of an ethnomathematics programme, it is necessary to propose a theoretical basis to discuss how the social, economic, and political contexts are important for the development of mathematical ideas, a fact observed in the diversity of the work presented through all SIPEM's meetings. It is also important to know how and why different individuals have come to reveal their diverse interests, preferences, talents, and abilities, as well as specificities and strategies to generate, organize, disseminate, and share mathematical knowledge to solve the problems they face in their daily lives.

Therefore, when the focus of a study is on the nature of mathematics, attention should be centred on both the legitimization of the students' knowledge that develops from the experiences built up by their own experiences, and the study of pedagogical possibilities about how it is possible to work with the teaching and learning process that takes place inside and outside the school environment. For Shirley and Palhares (2016), discussions related to the educational aspects of ethnomathematics assist educators in the establishment of cultural models of beliefs, thoughts, and behaviours, to contemplate the pedagogical potential of the work that considers the prior knowledge of the students, as well as meaningful and empowering mathematical learning.

The pedagogical action of ethnomathematics can be considered as a contextualization of the ideas, procedures, and mathematical practices developed by the members of distinct cultural groups. However, how this pedagogical work can be done in classrooms is problematic (D'Ambrosio & Rosa, 2008). In this context, it is necessary to understand how ethnomathematics can contribute to the development of activities contextualized in multicultural classrooms.

In this regard, D'Ambrosio (1999b) stated that the growing trend towards multiculturalism recognizes ethnomathematics as a valid school practice that enhances creativity, reinforces cultural self-respect, and offers a broad view of humanity. In everyday life, an ethnomathematics perspective increasingly recognizes systems of knowledge, which offers the possibility of a more favourable and harmonious relation in human behaviour and between humans and nature.

The result of our discussions in these presentations continues to show that ethnomathematics is alive and embraces the ideas, thoughts, and mathematical practices developed by members of all cultures. In this perspective, an anthropological body of research focuses on diverse forms of intuitive mathematical thinking, as well as the development of equally diverse cognitive processes, which are widely developed by members of distinct cultural groups. Thus, ethnomathematics is a programme that seeks to study how students understand, articulate, process and use ideas, concepts, and mathematical practices that can solve problems related to their daily activities.

Seen in this context, the focus of ethnomathematics consists essentially of a critical analysis of the generation and production of the mathematical knowledge and intellectual processes, the social mechanisms in the institutionalization of knowledge; and the diffusion of that knowledge (Rosa & Orey, 2007). In the *holistic* context of mathematics that uses an anthropological perspective to include diverse perspectives, patterns of thought, and histories, the study of the *systems* taken from reality help students to reflect, and understand extant relations among all of the components of the system.

From this reflection, it is our opinion that these systems address the issues regarding mathematics education in non-western cultures by bringing the cultural background of students into the mathematics curriculum in order to connect it to the local-cultural aspects of the school community into the teaching and learning of mathematics. This alternative approach helps promote intellectually innovative ideas in mathematics education by both deepening and widening the western-academic understanding of mathematics (Rosa, 2015).

The unique cultural background of the students in Brazil represents a set of values and diverse worldviews that are diffused across generations. The studies presented in WG5 show that the ideas, procedures, and mathematical practices developed by members of different cultural groups favour our understanding of the internal logic and beliefs of members of these groups.

Regarding the history of mathematics, school knowledge, methodologies, and mathematical practices, it is considered that, in the same way as with ethnomathematics, researchers continue to contribute to mathematics education in its various social, political, and cultural dimensions, and is reflected by graduation courses and in the training and development of future mathematics teachers.

In particular, the development of mathematics education does not arise in isolation: it is part of an educational system, which, in turn, interacts with a certain socio-political system (Schubring, 2006). Understanding the reasons why certain mathematical contents become objects of teaching over others, or certain methodologies are more adequate than others, according to the research presented in this WG, seems to be fundamental for teacher education. In addition, this facet of the research was developed especially in the group requiring new outlines for the group itself from the researchers.

Although it is recognized, as Karp (2014) says, that “the history of mathematics education is a branch of research that is still only taking shape, and consequently its methodology, too, is still only in its formative stage” (p. 9), Brazilian researchers have been finding different ways to investigate historically mathematics education in Brazil, whether through documentary sources, textbooks, pedagogical journals, and even through what has been named as oral history.

Moreover, this research modality has attracted national and international attention of researchers in mathematics education, contributing to the rapid growth of the research. In addition, the creation of this WG brings together the research working specifically with aspects of the history of mathematics education in Brazil. Just at the international level, the 29 chapters in the *Handbook on the History of Mathematics*

Education (Karp & Schubring, 2014) demonstrate the amplification and need for this specific WG.

As far as the history of mathematics is concerned, from the discussion described above, research focuses on contents or characters that, in one way or another, are linked to either issues of ethnomathematics or the history of mathematics education. This does not mean that the history of mathematics in Brazil is no longer interesting, but that, because of the work involved by the group, it has grown enormously and has begun to answer questions emerging from education. It is true that research in education history has sparked investigations since the 2000s, but its production discussed in the group is still growing and should be done under the auspices of a separate strand.

In this way, by the very movement of research involved in it, WG5 has been characterized from emerging themes that deal with both history and culture of mathematics to address issues in Brazilian mathematics education, in a way that a new configuration was required. Just as history and culture are always in motion, the Group changes and transforms itself with every edition of SIPEM.

Acknowledgment The authors wish to thank Prof. Dr. Milton Rosa at the Federal University of Ouro Preto for his time taken and his ideas for the betterment of this chapter. Milton, we deeply appreciate your time and feedback!

References

Abdounur, O. J. (2003). Música, astronomia e Matemática: uma analogia envolvendo escalas, calendários e incomensurabilidade. *Livro de Resumos, II SIPEM*. Retrieved from http://www.sbembrasil.org.br/files/sipemII.pdf.

Abdounur, O. J. (2006). Aspectos histórico-didáticos da relação matemática/música sob forma de uma exposição. *Anais do III SIPEM*. Retrieved from http://www.sbembrasil.org.br/files/sipemIII.pdf.

Araújo, I. C. (2012). Os diferentes discursos na formação de uma imagem de Joaquim Gomes de Souza (1829-1864). *Anais do V SIPEM*. Retrieved from http://www.sbembrasil.org.br/files/v_sipem/?page=publications&language=br.

Baraldi, I. M. (2006). Um estudo em Educação Matemática sobre a CADES: uma pesquisa em constituição. *Anais do III SIPEM*. Retrieved from http://www.sbembrasil.org.br/files/sipemIII.pdf.

Barbosa, L. N. S. C. (2015). Sobre trazer a "realidade" para as aulas de matemática na educação do campo: O que algumas pessoas têm a dizer. VI Anais do VI Seminário Internacional de Pesquisa em Educação Matemática - VI SIPEM (pp. 1–12). Pirenópolis, GO, Brazil: SBEM.

Baroni, R. L. S, Teixeira, M. V., & Nobre, S. (2004). A investigação científica em História da Matemática e suas relações com o Programa de Pós-Graduação em Educação Matemática. In M. A. V. Bicudo & M. de C. Borba (Orgs.), Educação Matemática: pesquisa em movimento (pp. 164–185). São Paulo: Cortez.

Batista, C. O., Souza, D., Costa, E. S., Silva, E. B., Bertoni, N. E., Carvalho, R. P. F. et al. (2009). Instâncias que atuaram na aprendizagem-ensino de matemática e na educação matemática escolar no distrito federal – etapas institucionais e influências externas. *Anais do IV SIPEM*. Retrieved from http://www.sbembrasil.org.br/files/sipemIV.pdf.

Batista, I. L. (2003). Abordagem histórico-filosófica e a Educação Matemática: estudo de uma investigação. *Livro de Resumos, II SIPEM*. Retrieved from http://www.sbembrasil.org.br/files/sipemII.pdf.

Bayer, A. (2006). A contribuição da história da matemática para o ensino da matemática na 2ª CRE. *Anais do III SIPEM*. Retrieved from http://www.sbembrasil.org.br/files/sipemIII.pdf.

Bernardi, L. T. M., Caldeira, A. D., & Duarte, C. G. (2012). Posição de fronteira e produção de significados na educação matemática indígena. *Anais do V SIPEM*. Retrieved from http://www.sbembrasil.org.br/files/v_sipem/?page=publications&language=br.

Bishop, A. J. (1994). Cultural conflicts in mathematics education: Developing a research agenda. *For the Learning of Mathematics, 14*(2), 15–18.

Borges, R. S. S., & Duarte, A. R. S. (2012). A Revista Escola e a Matemática no ensino primário (1923–1924). *Anais do V SIPEM*. Retrieved from http://www.sbembrasil.org.br/files/v_sipem/?page=publications&language=br.

Brito, A. J. (2006). O estudo de um manual de ensino de Matemática: o livro de Geografia Geral. *Anais do III SIPEM*. Retrieved from http://www.sbembrasil.org.br/files/sipemIII.pdf.

Brito, A. J. (2009). As cartas de Varenius e o ensino de matemática. *Anais do IV SIPEM*. Retrieved from http://www.sbembrasil.org.br/files/sipemIV.pdf.

Brito, D. R., & Mattos, J. R. L. (2015). Saberes matemáticos produzidos e praticados por agricultores de uma comunidade agrícola. VI Anais do VI Seminário Internacional de Pesquisa em Educação Matemática - VI SIPEM (pp. 1-10). Pirenópolis, GO, Brazil: SBEM.

Brolezzi, A. C. (2000). Discreto e Contínuo: explorando uma tensão na história e no ensino da matemática. *Livro de Resumos, I SIPEM*. Retrieved from http://www.sbembrasil.org.br/files/sipemI.pdf.

Carvalho, J. P. (2006). Edições de Condorcet no Brasil e em Portugal. *Anais do III SIPEM*. Retrieved from http://www.sbembrasil.org.br/files/sipemIII.pdf.

Chartier, R. (1990). *A História Cultural: entre práticas e representações*. São Paulo: Difel.

Costa, W. N. G. (2006). A tessitura de uma perspectiva teórica para estudos que tomem como foco o pensamento lógico/matemático e o mítico. *Anais do III SIPEM*. Retrieved from http://www.sbembrasil.org.br/files/sipemIII.pdf.

Costa, W. N. G. (2009). Uma abordagem das culturas negras e das indígenas nas aulas de matemática. *Anais do IV SIPEM*. Retrieved from http://www.sbembrasil.org.br/files/sipemIV.pdf.

Costa, R. S., & Mattos, J. L. (2015). A Etnia Karipuna do Oiapoque: cultura indígena e Etnomatemática. *Anais do VI SIPEM*. Retrieved from http://www.sbembrasil.org.br/visipem/anais/story_html5.html.

D'Ambrosio, U. (1985). Ethnomathematics and its place in the history and pedagogy of mathematics. *For the Learning of Mathematics, 5*(1), 44–48.

D'Ambrosio, U. (1999a). Literacy, matheracy, and technoracy: A trivium for today. *Mathematical Thinking and Learning, 1*(2), 131–153.

D'Ambrosio, U. (1999b). In Focus... mathematics, history, ethnomathematics, and education: A comprehensive program. *The Mathematics Educator, 9*(2), 34–36.

D'Ambrósio, U. (2000). Cultura e História da Matemática. *Livro de Resumos, I SIPEM*. Retrieved from http://www.sbembrasil.org.br/files/sipemI.pdf.

D'Ambrosio, U. (2005). Sociedade, cultura, matemática e seu ensino. *Educação e Pesquisa, 31*(1), 99–120.

D'Ambrosio, U., & Rosa, M. (2008). A dialogue with Ubiratan D'Ambrosio: A Brazilian conversation about ethnomathematics. *Revista Latinoamericana de Etnomatemática, 1*(2), 88–110.

Damasceno, A., & Gomes, R. (2003). A cultura de farinha: uma proposta de ensino de matemática através dos saberes dessa tradição. *Livro de Resumos, II SIPEM*. Retrieved from http://www.sbembrasil.org.br/files/sipemII.pdf.

De Certeau, M. (2005). *A cultura no plural*. (E. A. Dobránszky, Trad). (4a ed.). Campinas, SP: Papirus.

Dias, A. L. M., & Bertani, J. A. (2009). O curso de Matemática e o curso de Didática na Bahia: a história da constituição de um corpo profissional. *Anais do IV SIPEM*. Retrieved from http://www.sbembrasil.org.br/files/sipemIV.pdf.

Dias, M. da S., & Saito, F. (2009). Interface entre história da matemática e ensino: uma aproximação entre historiografia e perspectiva lógico-histórica. *Anais do IV SIPEM*. Retrieved from http://www.sbembrasil.org.br/files/sipemIV.pdf.

Domingues, K. C. M. (2006). Educação e cultura: a escola tradicional e a escola indígena. In Anais do III Seminário Internacional de Pesquisa em Educação Matemática - III SIPEM (pp. 80). Águas de Lindóia, SP, Brazil. SBEM.

Domite, M. C. S. (2006). Formação de professores de etnomatemática: compreendendo para pedir mudanças. In Anais do III Seminário Internacional de Pesquisa em Educação Matemática - III SIPEM (pp. 77). Águas de Lindóia, SP, Brazil. SBEM.

Duarte, A. R. S. (2003). Notas de Matemática e Física: um ele de integração entre pesquisa e ensino. *Livro de Resumos, II SIPEM*. Retrieved from http://www.sbembrasil.org.br/files/sipemII.pdf.

Duarte, A. R. S. (2006). O Movimento da Matemática Moderna na perspectiva de Benedito Castrucci. *Anais do III SIPEM*. Retrieved from http://www.sbembrasil.org.br/files/sipemIII.pdf.

Fantinato, M. C. C. B., & Santos, R. K. (2006). Etnomatemática e prática docente: legitimando saberes. *Anais do III SIPEM*. Retrieved from http://www.sbembrasil.org.br/files/sipemIII.pdf.

Ferreira, V. L. (2003). Uma Análise da Matemática nos Currículos da Escola Normal. *Livro de Resumos, II SIPEM*. Retrieved from http://www.sbembrasil.org.br/files/sipemII.pdf.

Ferreira, V. L. (2012). A metodologia do ensino de Matemática: história e formação de professores. *Anais do V SIPEM*. Retrieved from http://www.sbembrasil.org.br/files/v_sipem/?page=publications&language=br.

Fischer, M. C. B., & Carpes, F. (2006). Das classes-piloto às classes experimentais: investigando a experiência organizada pelo GEEMPA, ao tempo da Matemática Moderna. *Anais do III SIPEM*. Retrieved from http://www.sbembrasil.org.br/files/sipemIII.pdf.

Flores, C. R. (2006a). Representação do espaço das fortificações da Ilha de Santa Catarina do século XVIII: saber militar e imagem perspectiva. *Anais do III SIPEM*.

Flores, C. R. (2006b). Representação do espaço das fortificações da ilha de Santa Catarina no século XVIII: saber militar e imagem perspectiva. *Anais do III SIPEM*. Retrieved from http://www.sbembrasil.org.br/files/sipemIII.pdf.

Forner, R., & Lopes, J. A. (2006). O olhar de educadores matemáticos sobre a obra de Paulo Freire. *Anais do III SIPEM*. Retrieved from http://www.sbembrasil.org.br/files/sipemIII.pdf.

Gonçalvez, E. C. S., & Monteiro, A. (2006). Medidas e práticas sociais. *Livro de Resumos, II SIPEM*. Retrieved from http://www.sbembrasil.org.br/files/sipemII.pdf.

Guimarães, M. D. (2015) Rui Barbosa e a matemática presente no parecer da Reforma do Ensino Primário (1883): o estudo particular do desenho. *Anais do VI SIPEM*. Retrieved from http://www.sbembrasil.org.br/visipem/anais/story_html5.html.

Gutierre, L. S. (2006). Colégio Estadual do Ateneu norte rio-grandense: algumas reflexões. *Anais do III SIPEM*. Retrieved from http://www.sbembrasil.org.br/files/sipemIII.pdf.

Heeffer, A. (2006). The methodological relevance of the history of mathematics for mathematics education. Retrieved from http://logica.ugent.be/albrecht/thesis/Thailand2006.pdf.

Karp, A. (2014). The history of mathematics education: Developing a research-methodology. In A. Karp & G. Schubring (Eds.), *Handbook on the history of mathematics education*. New York: Springer.

Karp, A., & Schubring, G. (2014). *Handbook on the history of mathematics education*. New York: Springer.

Knijnik, G. (2000). O político, o social e o cultural no ato de educar matematicamente as novas gerações. In: Matos, J. F., FERNANDES, E. (Eds.). Actas do POROFMAT 2000 (pp. 48-60). Lisboa, Portugal: Associação de professores de Matemática de Portugal.

Knijnik, G. (2003). Cultura, currículo e matemática oral na Educação de jovens e adultos do campo. *Livro de Resumos, II SIPEM*. Retrieved from http://www.sbembrasil.org.br/files/sipemII.pdf.

Knijnik, G. (2006). Regimes da verdade sobre a Educação Matemática de Jovens e Adultos do campo: um estudo introdutório. *Anais do III SIPEM*. Retrieved from http://www.sbembrasil.org.br/files/sipemIII.pdf.

Lannes, W. (2003). A matemática como produto cultural. *Livro de Resumos, II SIPEM*. Retrieved from http://www.sbembrasil.org.br/files/sipemII.pdf.

Laraia, R. B. (1999). *Cultura: Um Conceito Antropológico* (12a ed.) Ed. Rio de Janeiro: Jorge Zahar.

Leme da Silva, M. C. (2009). O ensino de geometria no cenário internacional do Movimento da Matemática Moderna. *Anais do IV SIPEM*. Retrieved from http://www.sbembrasil.org.br/files/sipemIV.pdf.

Machado, R. B., & Flores, C. R. (2015) Entre práticas (historiográficas e de desenho): um movimento metodológico de pesquisa. *Anais do VI SIPEM*. Retrieved from http://www.sbembrasil.org.br/visipem/anais/story_html5.html.

Machado, V. L. (2015). Saberes e fazeres matemáticos integrados ao cotidiano do produtor rural. VI Anais do VI Seminário Internacional de Pesquisa em Educação Matemática - VI SIPEM (pp. 1-10). Pirenópolis, GO, Brazil: SBEM.

Maciel, P. R. C., Cardoso, T. F. L., & Fonseca, C. S. da (2012). O conceito de função através da história da matemática. *Anais do V SIPEM*. Retrieved from http://www.sbembrasil.org.br/files/v_sipem/?page=publications&language=br.

Mafra, J. R. S., & Fossa, J. A. (2006). Espaços de transversalidade no ensino de matemática. In Anais do III Seminário Internacional de Pesquisa em Educação Matemática - III SIPEM (pp. 79). Águas de Lindóia, SP, Brazil. SBEM.

Magalhães, C. M. R., & Silva, V. C. (2012). História do desenvolvimento da Matemática e cultura no Brasil colonial. *Anais do V SIPEM*. Retrieved from http://www.sbembrasil.org.br/files/v_sipem/?page=publications&language=br.

Marchon, F. L., & Fantinato, M. C. (2015). As Bases Socioculturais para Educação Matemática a Partir da Trama Textual da Etnomatemática. VI Anais do VI Seminário Internacional de Pesquisa em Educação Matemática - VI SIPEM (pp. 1-11). Pirenópolis, GO, Brazil: SBEM.

Mattos, A. C. (2006a). Processo de reconhecimento histórico de um matemático. In Anais do III Seminário Internacional de Pesquisa em Educação Matemática - III SIPEM (pp. 70). Águas de Lindóia, SP, Brazil. SBEM.

Mattos, A. C. de (2006b). Processo de reconhecimento histórico de um matemático. *Anais do III SIPEM*. Retrieved from http://www.sbembrasil.org.br/files/sipemIII.pdf.

Mauro, S. (2006). Uma história da matemática escolar desenvolvida por comunidades de origem alemã no Rio Grande do Sul entre os séculos XIX e XX. *Anais do III SIPEM*. Retrieved from http://www.sbembrasil.org.br/files/sipemIII.pdf.

Mendes, J. R. (2003). Os significados do "número" no contexto indígena Kaiabi: aspectos simbólicos e políticos na elaboração do livro de Matemática em língua indígena. *Livro de Resumos, II SIPEM*. Retrieved from http://www.sbembrasil.org.br/files/sipemII.pdf.

Mendonça, A. A. N., & Pinto, M. M. F. (2009). As formas de participação nas aulas de matemática nas escolas indígenas Xacriabá. *Anais do IV SIPEM*. Retrieved from http://www.sbembrasil.org.br/files/sipemIV.pdf.

Meneghetti, R. C. G. (2003). O papel do empirismo e racionalismo na história do desenvolvimento do cálculo. *Livro de Resumos, II SIPEM*. Retrieved from http://www.sbembrasil.org.br/files/sipemII.pdf.

Miarka, R. (2012). Escrevendo uma prática matematicamente: uma possibilidade etnográfica para a Etnomatemática. *Anais do V SIPEM*. Retrieved from http://www.sbembrasil.org.br/files/v_sipem/?page=publications&language=br.

Miguel, A. (2003). Perspectivas teóricas no interior do campo de investigação "História na Educação Matemática". *Livro de Resumos, II SIPEM*. Retrieved from http://www.sbembrasil.org.br/files/sipemII.pdf.

Miguel, A., & Souza, E. S. (2006). Um estudo sobre o processo de obsolescência de uma prática cultural: a prova dos nove. *Anais do III SIPEM*. Retrieved from http://www.sbembrasil.org.br/files/sipemIII.pdf.

Miorim, M. A., Brito, A. J., & Faria, J. C. (2009) Relações entre história e educação matemática: um olhar sobre periódicos produzidos por associações científicas brasileiras. *Anais do IV SIPEM*. Retrieved from http://www.sbembrasil.org.br/files/sipemIV.pdf.

Miranda, G. A. (2015) Passos embrionários da pesquisa em Educação Matemática no Brasil: personagens e produções do primeiro mestrado em ensino de ciências e matemática da Unicamp. *Anais do VI SIPEM*. Retrieved from http://www.sbembrasil.org.br/visipem/anais/story_html5.html.

Monteiro, A. (2003). A modelagem matemática na perspectiva da Etnomatemática: possibilidades e obstáculos no processo de escolarização. *Livro de Resumos*, II SIPEM. Retrieved from http://www.sbembrasil.org.br/files/sipemII.pdf.

Monteiro, A., & Mendes, J. R. (2015). Etnomatemática como movimento de contraconduta na mobilização de saberes em práticas culturais. VI Anais do VI Seminário Internacional de Pesquisa em Educação Matemática - VI SIPEM (pp. 1–11). Pirenópolis, GO, Brazil: SBEM.

Motta, C. D. van B., & Brolezzi, A. C. (2006). História da Matemática na Ed. MTM: espelho ou pintura? *Anais do III SIPEM*. Retrieved from http://www.sbembrasil.org.br/files/sipemIII.pdf.

Moura, A. R. de L., & Sousa, M. do C. (2000). O lógico-histórico da álgebra não simbólica e da álgebra simbólica: dois modos de ver. *Livro de Resumos, I SIPEM*. Retrieved from http://www.sbembrasil.org.br/files/sipemI.pdf.

Nascimento, E. C. S., & Silva, M. P. S. (2012). Da aldeia para a cidade: a matemática da etnia Paresí e a inserção escolar indígena. *Anais do V SIPEM*. Retrieved from http://www.sbembrasil.org.br/files/v_sipem/?page=publications&language=br.

Nascimento, E. C. S., & Silva, R. C. (2015). A arte indígena como instrumento para o ensino da geometria. VI Anais do VI Seminário Internacional de Pesquisa em Educação Matemática - VI SIPEM (pp. 1–11). Pirenópolis, GO, Brazil: SBEM.

Neto, M. O. T., & Braga, O. F. (2006). Saberes e práticas em matemática nas escolas públicas paraenses no início do século XX: entre discursos e imagens. *Anais do III SIPEM*. Retrieved from http://www.sbembrasil.org.br/files/sipemIII.pdf.

Nobre, S. (2000). História da Matemática e a formação do profissional em matemática. *Livro de Resumos, I SIPEM*. Retrieved from http://www.sbembrasil.org.br/files/sipemI.pdf.

Oliveira, A. S. V. (2006a). O cálculo infinitesimal na Escola Politécnica de São Paulo, no ano 1904: uma análise documental. *Anais do III SIPEM*. Retrieved from http://www.sbembrasil.org.br/files/sipemIII.pdf.

Oliveira, C. C. (2006b). O discurso pedagógico de Malba Tahan na revista Al-Karismi: reflexões para a Educação Matemática brasileira. *Anais do III SIPEM*. Retrieved from http://www.sbembrasil.org.br/files/sipemIII.pdf.

Oliveira, M. C. A. (2006c). "Assistindo" o curso de Análise Matemática com Ubiratan D'Ambrosio. *Anais do III SIPEM*. Retrieved from http://www.sbembrasil.org.br/files/sipemIII.pdf.

Oliveira, M. C. A. (2009). Revistas pedagógicas e a produção de história da educação matemática no Brasil: o caso da gestação de um programa de matemática moderna para o secundário. Anais do IV Seminário Internacional de Educação Matemática. - IV SIPEM (pp. 182–183). Taguatinga, DF, Brazil: SBEM.

Oliveira, C. C. (2009a). Uma perspectiva transdisciplinar para as pesquisas que envolvem discursos em Educação Matemática. *Anais do IV SIPEM*. Retrieved from http://www.sbembrasil.org.br/files/sipemIV.pdf.

Oliveira, M. C. A. (2009b). Revistas pedagógicas e a produção de história da educação matemática no Brasil: o caso da gestação de um programa de Matemática Moderna para o secundário. *Anais do IV SIPEM*. Retrieved from http://www.sbembrasil.org.br/files/sipemIV.pdf.

Oliveira, M. C. A. (2015). O desenho nas prescrições oficiais para a formação de normalistas nas primeiras décadas do Brasil republicano. *Anais do VI SIPEM*.

Passo, C. M. (2006). Articulação entre etnomatemática e educação matemática crítica: algumas contribuições para a sala de aula de matemática. *Anais do III SIPEM*. Retrieved from http://www.sbembrasil.org.br/files/sipemIII.pdf.

Pereira, A. C. C. (2006). Uma análise da abordagem do teorema de Thales na coleção Curso de Matemática de Euclides Roxo, Cecil Thiré, Júlio César de Mello e Souza. *Anais do III SIPEM*. Retrieved from http://www.sbembrasil.org.br/files/sipemIII.pdf.

Pinto, A. H. (2015) Experiência Matemática e Educação Profissional: História e Memória. *Anais do VI SIPEM*.

Ribeiro, D. F. C. (2009). Conferências internacionais de instrução pública: um estudo inicial da trajetória histórica da constituição do currículo de Matemática. *Anais do IV SIPEM*. Retrieved from http://www.sbembrasil.org.br/files/sipemIV.pdf.

Rodrigues, Z. G. M. (2015). O Movimento da Matemática Moderna: algumas compreensões a partir das falas dos professores de matemática da região de Ribeirão Preto. *Anais do VI SIPEM*. Retrieved from http://www.sbembrasil.org.br/visipem/anais/story_html5.html.

Roque, A. C. C. (2012). Potencialidades pedagógicas da História da Matemática em atividades relacionadas aos números inteiros. *Anais do V SIPEM*. Retrieved from http://www.sbembrasil.org.br/files/v_sipem/?page=publications&language=br.

Rosa, M., & Orey, D. C. (2007). Cultural assertions and challenges towards pedagogical action of an ethnomathematics program. *For the Learning of Mathematics, 27*(1), 10–16.

Rosa, M., & Orey, D. (2012). Etnomodelagem: as perspectivas êmica e ética na pesquisa em Etnomatemática e modelagem. *Anais do V SIPEM*. Retrieved from http://www.sbembrasil.org.br/files/v_sipem/?page=publications&language=br.

Rosa, M. (2015). Aspectos de insubordinação criativa nas pesquisas em etnomatemática. In B. S. D'Ambrosio, & C. S. Lopes. (Orgs.), *Vertentes da subversão na produção científica em educação matemática* (pp. 325–346). Campinas, SP: Mercado de Letras.

Rosa, M., & Orey, D. C. (2015). Etnomatemática: um programa de pesquisa lakatosiano. VI Anais do VI Seminário Internacional de Pesquisa em Educação Matemática - VI SIPEM (pp. 1-9). Pirenópolis, GO, Brazil: SBEM.

Rosa, M., & Gavarrete, M. E. (2016). Polysemic interactions between ethnomathematics and culturally relevant pedagogy. In M. Rosa, U. D'Ambrosio, D. C. Orey, L. Shirley, P. Palhares, & M. E. Gavarrete (Orgs.), *Current and future perspectives of ethnomathematics as a program* (pp. 23–30). Hamburg: Springer Open.

Sá, L. C. e (2015). História da Matemática no Ensino de Teoria de Grafos: Uma Experiência à Luz do Jogo de Vozes e Ecos. *Anais do VI SIPEM*. Retrieved from http://www.sbembrasil.org.br/visipem/anais/story_html5.html.

Sad, L. A., & Dynnikov, C. M. S. da S. (2003). A pesquisa em História da Matemática: reflexões teórico-metodológicas. *Livro de Resumos, II SIPEM*. Retrieved from http://www.sbembrasil.org.br/files/sipemII.pdf.

Santos, L. F. da C. (2012). Cortes de Dedekind? uma discussão sobre as abordagens de Dedekind e Tannery. *Anais do V SIPEM*. Retrieved from http://www.sbembrasil.org.br/files/v_sipem/?page=publications&language=br.

Santos, M. (2006). Unidades de medida cotidianas em assentamentos sergipanos: varas, tarefas e celamins. *Anais do III SIPEM*. Retrieved from http://www.sbembrasil.org.br/files/sipemIII.pdf.

Scheide, T. de J. F. (2006). O papel da história da matemática no processo de ensino e aprendizagem. *Anais do III SIPEM*. Retrieved from http://www.sbembrasil.org.br/files/sipemIII.pdf.

Schubring, G. (2006). Researching into the history of teaching and learning mathematics: The state of the art. *Paedagogica Historica, XLII*(IV–V), 665–677.

Shirley, L., & Palhares, P. (2016). Ethnomathemtics and its diverse pedagogical approaches. In M. Rosa, U. D'Ambrosio, D. C. Orey, L. Shirley, P. Palhares, & M. E. Gavarrete. (Orgs.), *Current and future perspectives of ethnomathematics as a program* (pp. 13–17). Hamburg: Springer Open.

Silva, M. A., & Silva, J. N. D. (2012). Movimento modernizador da Matemática secundária nos livros didáticos de Stávale e Sangiorgi. *Anais do V SIPEM*. Retrieved from http://www.sbembrasil.org.br/files/v_sipem/?page=publications&language=br.

Silva, F. J. F. (2015). Matematizações em um assentamento rural: um olhar histórico sobre unidades universais de contar e medir. VI Anais do VI Seminário Internacional de Pesquisa em Educação Matemática - VI SIPEM (pp. 1–12). Pirenópolis, GO, Brazil: SBEM.

Silva, S. F., & Caldeira, A.D. (2015). Etnomatemática do sistema de contagem Guarani: símbolos gráficos. VI Anais do VI Seminário Internacional de Pesquisa em Educação Matemática - VI SIPEM (pp. 1-13). Pirenópolis, GO, Brazil: SBEM.

Silvai, M. D. F. da (2009). A História do Cálculo sob a ótica da construção de modelos: contribuições dos gregos a Newton. *Anais do IV SIPEM*. Retrieved from http://www.sbembrasil.org.br/files/sipemIV.pdf.

Soares, F. (2003). Passado e Presente: O PNLD e a Comissão Nacional do Livro Didático – uma análise comparativa. *Livro de Resumos, II SIPEM*. Retrieved from http://www.sbembrasil.org.br/files/sipemII.pdf.

Soares, F. (2006). Virada do século XVIII para o XIX. *Anais do III SIPEM*. Retrieved from http://www.sbembrasil.org.br/files/sipemIII.pdf.

Tomaz, V. S. (2015). Uma abordagem intercultural para o ensino de probabilidades na formação de educadores indígenas. VI Anais do VI Seminário Internacional de Pesquisa em Educação Matemática - VI SIPEM (pp. 1-13). Pirenópolis, GO, Brazil: SBEM.

Trentin, P. H. (2009). A tradução realizada por Manoel Ferreira de Araújo Guimarães da obra Éléments de Géométrie de Adrien Marie Legendre, no século XIX: apontamentos preliminares. *Anais do IV SIPEM*. Retrieved from http://www.sbembrasil.org.br/files/sipemIV.pdf.

Valente, W. R. (2003). Rumos da Educação Matemática na Argentina e no Brasil: elemento para uma história comparativa. *Livro de Resumos, II SIPEM*. Retrieved from http://www.sbembrasil.org.br/files/sipemII.pdf.

Valente, W. R. A. (2009). Matemática do colégio através dos livros didáticos: subsídios para uma história disciplinar. *Anais do IV SIPEM*. Retrieved from http://www.sbembrasil.org.br/files/sipemIV.pdf.

Villela, L. M. A. (2012). Vestígios sobre a aritmética em escolas primárias da região de Vassouras (RJ), de 1880 a 1900. *Anais do V SIPEM*. Retrieved from http://www.sbembrasil.org.br/files/v_sipem/?page=publications&language=br.

Voltolini, L., & Kaiber, C. T. (2012). Conhecimentos Matemáticos no contexto da Comunidade Indígena Serra da Moça. *Anais do V SIPEM*. Retrieved from http://www.sbembrasil.org.br/files/v_sipem/?page=publications&language=br.

Zuin, E. S. L. (2006). O sistema métrico decimal nos Elementos de Arithmetica de José Joaquim D'Avila pela defesa de um novo sistema de pesos e medidas no Brasil dos oitocentos. *Anais do III SIPEM*. Retrieved from http://www.sbembrasil.org.br/files/sipemIII.pdf.

Chapter 7
Digital Technologies and Mathematics Education: Interlocutions and Contributions Based on Research Developed in Brazil

Maurício Rosa, Marcelo Bairral, Verônica Gitirana, and Marcelo Borba

Abstract This chapter articulates research developed within the working group Mathematics Education: Digital Technologies and Distance Education of the Brazilian Mathematics Education Society (SBEM). To address the research foci of this group, we highlight theoretical and methodological aspects that form the framework of the studies that currently inform the work in this area in Brazil. More specifically, the paper is organized around the perspectives which are shaping the research activities of the group and include: the humans-with-media construct, which highlights the centrality of media in the productions of mathematical knowledge; computer-supported collaborative learning and the exploration of collaboration as a principle for cognitive mathematical development; factors involved in the design of digital resources for mathematics education; the use of touchscreen devices in the mathematical activity and the embodied nature of mathematical cognition; and the initial and continuing development for teachers who work with these technologies, both in the classroom and in distance education. In considering the results of these ongoing projects, collectively they contribute to a conception of education that treats digital technologies as transforming/empowering the production of mathematical knowledge, and not as auxiliaries of this process.

M. Rosa (✉)
Universidade Federal do Rio Grande do Sul, Porto Alegre, Brazil

M. Bairral
Universidade Federal Rural do Rio de Janeiro, Seropédica, Brazil

V. Gitirana
Universidade Federal de Pernambuco, Recife, Brazil

M. Borba
Universidade Estadual Paulista, Rio Claro, Brazil

A. J. Ribeiro et al. (eds.), *Mathematics Education in Brazil*,
https://doi.org/10.1007/978-3-319-93455-6_7

7.1 Introduction

This chapter discusses the research that has been developed in Mathematics Education: Digital Technologies and Distance Education in the framework of Working Group 6 (WG6) of the Brazilian Mathematics Education Society (SBEM), based on the papers presented at the International Symposium on Research in Mathematics Education (SIPEM), in particular, in SIPEM VI, held in 2015. From this, we seek to discuss this production by means of the study of the theoretical bases, in order to show to those who are outlining, generally, the understanding of the use of technologies in our country. We highlight the results and main issues that have arisen in the scope of this production and, mainly, theoretical-methodological bases that have been developed by our researchers and their relations with international research. We consider that the main themes linked to our group, and that portray the dialogue and the advances of our research in terms of contribution to mathematics education in the world are the role of the media in terms of the production of mathematical knowledge; the collaboration between the participants of mathematical educational spaces with the technological resources; the development of digital resources focused on mathematics education; the consideration of the body together with the resources used; and the education of mathematics teachers under the focus of the use of digital technologies (DTs).

7.2 First Ideas

This chapter articulates research that has been developed within the framework of WG6 of the Brazilian Mathematics Education Society (SBEM). However, in order to address the focus of this group's research, which is called "Mathematics Education: Digital Technologies and Distance Education", we begin by telling how the group was formed, indicating the institutions and distribution in Brazil of the community that has been composing WG6 since its foundation during SIPEM I.

There was a strong movement in research on digital technologies from mathematics education graduate programmes. In this way, a research panel was organized on the topic at the WG for all conference participants. At that time, the Internet was still quite precarious in the country, and research on online courses would be conducted only 3 years later. A central theme was the use of software programs, and the way students learned from them. Concomitantly, the way mathematics transformed itself by having as its ally the digital technologies was assessed.

With 15 years of existence, in SIPEM VI (2015) ten scientific articles were approved and published from several Brazilian research groups. These articles gained prominence in this text to highlight the group's history during those 15 years of existence, which is going to be presented by other research. In this way, we consider five large strands that relate digital technologies and mathematical education through the theoretical bases that have been used and constructed in the group, as well as the emergence of different digital resources and concerns in terms of their use.

We highlight some theoretical aspects that form, in general lines, that line of research in Brazil. The humans-with-media construct (HWM), for example, articulates the idea that there is no knowledge produced by the human only on the object, or even only between humans, but humans with media, in a non-hierarchical way. That is, the media participate in the process. Currently, this construct delineates new articulations with the activity theory, seeking new convergences with this theory. Similarly, computer-supported collaborative learning (CSCL) stands out in Brazilian research, focusing on collaboration as a principle of cognitive mathematical development. However, this theoretical approach in our group supports the collaboration from the integration of computational media, not exactly being part of the cognitive process as it presents HWM, but as an effective support to the learning that occurs in the collaboration between subjects and/or cooperation between them. In this perspective, different understandings of the role of digital technologies in mathematics education in terms of learning are assumed, and end up reflecting other actions as the way of thinking the development of digital resources for the teaching of mathematics. Thus, the didactic engineering-computer of educational software gains space and discusses how the development of educational software focused on mathematics happens and should happen. The action of developing assumes collaboration beforehand as an educational principle and aims at the products to manage this process in the student's action.

In another theoretical aspect, also focused on mathematical cognition, the study of the use of touchscreen devices becomes the core of the process of producing mathematical knowledge. The touch and the interference of this touch in learning make the difference in the research that comes back to thinking geometrically with tablets, for example. In this way, the embodied cognition reveals itself sustaining that, which until the moment was not discussed specifically. Although, on the one hand, the media participates in the production of knowledge and, on the other hand, the collaboration between the peers is highlighted, possibly, the action of "touching" the screen is taken naturally in both. But, the research of this specific process, which assumes the body as conditioning of thinking geometrically, opens new perspectives to Brazilian mathematics education.

In addition, among the perspectives of the production of mathematical knowledge with digital technologies, the focus of WG6 research is the mathematics education of teachers who work with these technologies, both in the classroom and in distance education. Therefore, mathematics teacher education, which happens in different modalities and aims that teachers come to act in these educational modalities, has been perceived within the group, since the technologies are taken as means that participate effectively or even support this education. In this way, the research at the centre of this group goes to a conception of education that treats the digital technologies as transforming/potentiating the production of mathematical knowledge, and not as simple auxiliaries of this process, in terms that also change the way to conceive, to plan, and to articulate the very education of teachers in face of the use of digital technologies. In this sense, the conception of cybereducation with mathematics teachers emerges and gives indications of how the work of teachers with digital resources, be it mobile, touch, image, construction, dynamic geometry, etc. will be considered in the near future.

With this, we understand that this articulation within WG6 studies with national and international research contribute to the field of mathematics education with technologies, which dialogues with different temporal moments lived in our country in order to project a future not far from conceptions of use in terms of teaching, learning, and education of mathematics teachers.

7.3 Humans-with-Media and Its Transformations in Mathematics Education

A historically constituted concern on the construction of theories was observed among the works presented in WG6, such as the development of the theoretical construct called "Humans-with-Media" proposed by Borba (1999), which had as one of its first discussions the article presented in SIPEM I (Borba, 2000), consolidated in Borba & Villarreal's book (2005). This construct comes from the theoretical view that discusses how computers affect human cognition (Tikhomirov, 1981), together with the idea of collective intelligence (Lévy, 2000), in which the dichotomy between human and technique is considered "outdated". Thus, this view identifies the production of knowledge in a space shared by humans and non-human actors. In this perspective, the media also are in the knowledge production process, as well as the human beings, without a separation or hierarchy between them. Borba and Villarreal (2005, p. 3) reveal:

> We do, however, propose ideas, expressed in the form of theoretical constructs, about how we can overcome the dichotomy between humans and technology that underlines many of the difficulties that we, as a community, have experienced in implementing the use of technology in schools in ways that are not domesticated.

Borba and Villarreal (2005) understand the terms (media and technologies) practically as synonyms, because they believe that technology is always used for communication, and so the technology can be seen as media. Thus, they use the term media to emphasize the aspects of communication of intelligence technologies, which is an expression used by Lévy (1994) to characterize three major techniques linked to memory and knowledge: orality, writing, and computing.

This construct, which was already based on activity theory authors in the first generation of discussions about humans-with-media and in technical philosophy authors, has turned over for 16 years of SIPEM. WG6 has witnessed, in particular, those of the construct's transformations. For example, the article that Souto and Borba (2015) presented in the fourth edition of this event interweaves the ideas of humans-with-media with the principles of Activity theory strand, advocated by Engeström (1987), aiming to propose a model analysis of the mathematics learning process in online environments, "minicyclones of expansive transformations".

The conception of learning that appears in Souto & Borba's discussion (2015) can be considered to complement Engeström's work (1999) because these authors consider it as a process with emotional, social, historical, and cultural dimensions, including new ways of thinking, beyond which is the result of a collective production of human and non-human actors. The construction course of this analytical model relied initially on empirical data analysis, through which the authors examined whether some changes in learning have been driven by tensions "*generated based on feedback from the media that influenced mathematical reasoning and actions of learners*" (Souto & Borba, 2015, p. 2).

One of the first changes in those ideas is present in the theoretical discussion that the authors set to analyse the humans-with-media construct as an activity system (Souto & Borba, 2015). In this work, the "leitmotiv"[1] was based on a bias of the historical evolution of the activity theory, in which the construct was analysed initially from the perspective of Leontiev (1978) and, in the sequence, from the point of view of Engeström (1987). In the results, some questions were raised: "What is the role of media in humans-with-media theoretical construct analysed as an activity system? Can media as artefacts (according to the activity theory) transform themselves in this system?" (Souto & Borba, 2015, p. 2).

This analytical model is part of a theoretical and methodological perspective, in which the authors also proposed the establishment of "humans-with-media systems". This type of system resulted from the interaction of the third generation of the activity theory with the humans-with-media theoretical construct, so that the media are seen as things that can play different roles, going beyond the condition of artefact as presupposes the activity theory. Summing up, a humans-with-media system has triangular representations that resemble features of Engeström and Vygotsky. The purpose of the authors is to emphasize that in the process of learning there is a collective education between human and non-human actors who "mix" and relate dialectically to each other without any hierarchy and makes the idea of mediation more "fuzzi" (Souto & Borba, 2016). Thus, the authors overcome the rigidity attributed to the triangular representation and that has been the target of criticism by the very activity theory theorists. We understand, therefore, that humans-with-media construct is transformed by the systemic way in which the activity theory sees learning and, at the same time, the activity theory is transformed by those who do research using digital technologies. Therefore, the activity of humans-with-media collectives also brings contributions to the activity theory.

Thus, it was in order to analyse the "minicyclone expansive transformation" system that this was being constructed. For SIPEM VI, the design of the analytical model was presented in a more refined form. According to Souto (2015), it has reinterpretation of features of what Engeström (1999) understands as zone of proximal development, as it favours the understanding of movements that indicate the

[1] Leitmotif also leitmotiv (līt′mō-tēf′) n. (1) A melodic passage or phrase, especially in Wagnerian opera, associated with a specific character, situation, or element. (2) A dominant and recurring theme, as in a novel. Retrieved from http://www.thefreedictionary.com/leitmotif [2016].

passage of internalization (or reproduction of culture) for externalization (search for new).

The metaphor "minicyclone" was not chosen at random. In different opportunities, the authors point out that learning does not occur in a linear form and, therefore, should not be considered as such. Thus, the idea is to link the proposal to this concept in the natural phenomenon called cyclone.

According to Souto & Borba (2015, p. 7), the minicyclones expansive transformations

> usually begin when the apprentices, even without awareness, mobilize themselves in search of the breakup of a dominant pattern of mathematical production. In other words, we can say that they were in a stable situation, which can be described by a linear behaviour conducive to the reproduction of knowledge of how learned (reproduction of culture) and, at a given moment arises a desire or need to drives the search for 'new'. This can be seen in doubts, questions, critical, self-critical that originates stresses in the system, which can be considered as expanding possibilities. The insertion of a new element, such as a media, can trigger this beginning.

The initial development of these minicyclones, as described in the previous citation, can be caused by any media. Therefore, at this moment, it has already possibly seen some interrelationships between human and non-human actors. Souto and Borba (2015) point out that the media involved in the process can be considered as protagonists in the development of their own minicyclone. This is because, according to these authors, media can act as mobilizing agents and, therefore, play distinct roles in the system.

With the advancement in the development of a minicyclone, the mobilizations may intensify, and assume the emergence of tensions and the need to overcome them. Souto and Borba (2015) point out that there are several possibilities in which learners must overcome a particular limitation imposed by these tensions and, therefore, they can generate different segments within their own minicyclone.

Because of these complexities, the authors believe it is not possible to predict accurately the beginning of a minicyclone nor the direction that it will take or the tensions and transformations that will occur. "What we can say is that it does not return to the place where it has already been" (Souto & Borba, 2015, p. 7).

Souto and Borba (2015) also analyse data on what teachers developed, in a virtual learning environment, on mathematical character studies. The conclusion of these authors is that the understanding of the movements, stagnations, tensions, and transformations in the online learning process is favoured by the "lens" of minicyclones' expansive transformations. However, by using the notion of humans-with-media that the "division between humans and technology may be inappropriate for understanding the role of computers in society, and in education, in particular" (Borba & Villarreal, 2005 p. 18), linked to the activity theory, this work of the WG contributes originally to the development of the third generation of this theory. Then, for us, it brings an important sense to what the working group has redefined regarding the use of digital technologies in mathematics education.

7.4 CSCL, Collaboration, and Integration of Media

Research on integrating digital resources in teachers' practices follow some theoretical perspectives that are interconnected in the context of online teaching. On the one hand, some studies theoretically founded in the instrumental approaches of Verillon and Rabardel (1995) are aware on how teachers transform artefact into instruments by using and creating schemes of instrumented actions. Thus, teachers undertake a process called instrumental genesis in their practice, and students pass by an instrumental genesis, while dealing with the artefacts to solve the tasks. In the online education, in which mediation is made through computational systems and the Internet, studying the instrumental genesis is revealing to be even more important, mainly considering the need for group work and collaborative learning among students and teacher. On the other hand, collaborative learning and computational systems to support it become central to research on online education, particularly, to mathematics education, in which teaching and learning involve using symbolic, graphic, and algebraic representations, and interacting with colleagues through them.

During SIPEM IV, Bittar (2009) brings into WG6 the discussion on teacher's instrumental genesis, pointing out contributions of action-research to teacher training, mainly to integrate media into their practice. She addresses the importance of the notion of integration of media in teacher practice and the transformations teacher undergoes throughout a process of reflection about their actions.

From this viewpoint, in SIPEM VI, Lucena and Gitirana (2015) uses the instrumental orchestration (OI) proposed by Trouche (2005), who discusses how teachers integrate digital resources in classes, as a theoretical background to analyse online teacher practice in courses of Brazilian Open University (UAB—Brazilian Open University). Thus, they expanded the field in which OI is usually used. In doing so, they pointed out that the didactic configurations of the online mathematics courses analysed were overemphasizing mathematics information, and pointed out a lack of resources and orientations for online mediation that considers communications within mathematical languages. They also mapped the resources and their scheme of utilization for online education into four spaces, considering an analytic geometry course developed within *moodle* platform: *an individual space*, subdivided into external and internal spaces; *a sharing space*, in which resources are shared with class, but one has no access to colleague's actions with resources; *a collaborative space* where interactions between students and with colleague's actions with resources are possible. The collaborative space, by its turn, is divided into synchronous and asynchronous. This helps us understand the model of distance education studied, and the composition of the environment in use, which goes further to think CSCL. It reveals how important it is to realize the composition of a virtual environment according to the support for sharing resources, interactions with resources, interactions with colleagues' productions through the mathematic representations using resources, while planning each module of the course, in what Trouche (2005) calls "didactic configuration".

Studies from WG6 have approached the perspective of computer supported collaborative work (CSCL). Besides aiming at invigorating the teaching and learning processes, that perspective presents a conceptualization for the collaborative learning that we agree, for instance, as a process of sharing meaning. The individuals in the group need to work, learn, and seek for answers together. The main factor is the interaction, the exchange, and the experience together. Although collaboration involves individual learning, it is not limited to this. It integrates individuals as members of the group, but also articulates phenomena such as negotiation and sharing of understandings—including the construction and the maintenance of shared strategies for solving tasks—that are fulfilled interactively through group processes (Stahl, Koschmann, & Suthers, 2006).

Based on CSCL, Araújo Filho and Gitirana (2015) presented a study about the professional knowledge mobilized by prospective mathematics teacher when they are planning lessons about function according to the CSCL theory. The authors also analysed the relations of that knowledge with the levels of collaboration. Their analysis was based on Baker (2002), who considers three dimensions: symmetry, agreement, and alignment. Symmetry refers to the role of the subjects involved in the session (propositor or reactor). In the agreement dimension, the positioning of the reactor in relation to the proposer's message is considered as in agreement or in disagreement. Finally, the alignment considers the group as a whole. It is observed whether the group is working in a uniform way for the solution of a determined problem, even when they have divergent ideas. Collaboration requires agreement or disagreement, but with symmetry between the subjects and group alignment.

They also discussed the creation of CSCL environments through media integration. Nowadays, new informatics, potentialities, and artefacts are emerging in the digital technology scene. The researchers ask two questions: should we consider the requirements necessary to compose a CSCL environment by integrating the media, or should the requirements of an environment (or software) be considered to be integrated with other media? The creation of the environment through the integration of different media and platforms was a differential of the research from Araújo Filho and Gitirana (2015), and a CSCL environment was created, integrating different media of available sharing, collaboration and interaction, for example, Gdrive environment tools, Teamviewer, Modellus, and e-mail.

The creation of collaborative sessions requires a structure that seeks to engage subjects in the proposal. Dillenbourg (2002) defends the concept of script, which is composed of a session structuring guide. Araújo Filho and Gitirana (2015) adopted the script of instruction, which offers participants instructions that can be followed or not, with the presence of a mediator. The analysis identifies collaboration in only one of the groups, the one with the lowest number of participants. However, the participants took more time in researching material and choosing the most appropriate content. Applying more time searching resources, the group does not execute the requested class plan in the script, which reveals something already discussed by Cress, Stahl, Ludvigsen, and Law (2015) about the time needed to establish the collaboration. Even in the group where collaboration was observed, collaborative work did not occur during the entire session. However, the results point out benefits in collaboration and cooperation. Even groups in which only cooperation was observed, presented relevant knowledge in the discussion about teaching and learning concerning function.

7.5 Development of Digital Resource

Human beings (re)create tools that, by its turn, modify human (activities). Motivated by activities demands, they modify the tools, both regarding the artefact itself and their schemes of utilization. In the dialectic movement, the importance of reflecting on the development of digital resources arises, as well as the development methodologies, as research focus on digital technology and distance education in the context of mathematics education. Some of the papers discussed in the group deal with the development of educational software (Bellemain, Ramos, & dos Santos, 2015; Carvalho, 2015), models of educational software engineering (Bellemain et al., 2015), and requirements to the development of these resources (Araripe & Bellemain, 2015; Bairral, Assis, & Silva, 2015b).

In this perspective, an interesting point raised from WG6 research are models of educational software engineering specific to mathematics teaching. This discussion has been presented in WG6 discussion since SIPEM I, with Bellemain (2000), which is aligned with the notion of informatic transposition of mathematics concepts (Balacheff 1994a). Informatic transposition focuses on changes each mathematics knowledge passes in the development of an educational software exploring it. Decisions taken while programming sometimes leads mathematics knowledge in the software in a different way than mathematics theory. In continuity to research on the topic, Bellemain et al. (2015) bring to the discussion an engineering model for the development of educational software specific to digital technologies to teach mathematics that incorporates principles of didactic engineering (Artigue, 1990) and of educational software engineering (Galvis, 1992). They draw their attention to "elaborate theoretic-methodologic principals to conceive and develop educational software" (Bellemain et al., 2015, p. 2) for mathematics education. Thus, the study is aligned with the engineering of the *Environmement Informatique d'Apprentissage Humain*—EIAH (informatic environment of human learning), the objective of which is construction of knowledge related to EIAH7 conception (Tchounikine, 2009, p. 14), of the notion of computational didactic, of informatic transposition, or in CKc (conception, knowing, concept) from Balacheff (1994a, 1994b, 1995), and in instrumental orchestration from Trouche (2005), theoretical constructs of an transdisciplinary approach to conceive computational resources to mathematics teaching.

Thus, Educational Software Engineering (ESE) is pointed out to be a transdisciplinary process. "ESE is not only a methodological problem, but also a theoretical one, which leads to the elaboration of original and specific notions and knowledges" (Bellemain et al., 2015, p. 3). The main point is the development of capacity to find answers for theoretical, technological, and technical questions that are raised by the design of educational software within the principles and circumscription systems, of each domain of knowledge involved, particularly those related to teaching and learning mathematics, from epistemological, cognitive, and didactic perspectives. The processes and techniques of validation of educational software in the area are also expanded, with the inclusion of theoretical validation, as a process carried out by a theoretical and a semi-theoretical expert, based on laboratory experiments (with teachers and students selected).

Starting by analysing and adaptating a game to mathematics teaching (Melo, Montenegro, Santos, Moraes, & Bellemain, 2013) developed within a project to develop games using recycling materials (Gitirana et al., 2013), they investigated the didactic study that led to the development of a game to study rational numbers—called "Bingo of Rational Numbers". Principles of didactic engineering are integrated to computational problematic, allowing to rethinking the didactic analyses built within their principles. The step of previous analysis, in its three (epistemological, didactical, and cognitive) dimensions, is integrated, and the computational dimension was thought to connect itself with these three dimensions. Thus, they (re) create a game with functionalities based on both technological contributions and effective didactical contributions from them.

Moreover, the analyses of requirements and development of environments are also addressed in other researchers of WG6. Understanding gesture as a materialization of students' thinking allows teachers to improve their practice, as defended by Hostetter and Alibali (2008). Bairral, Assis, and Silva's work (2015b) also leads us to think about the requirements for geometric software and environments that uses touchscreen technology. Their findings are also important to the field of software development. Students' actions and their mathematics narratives are important to implement tools with better interpretations of human manipulations within touchscreen technology. Also, Lima's study (2015) reveals the importance of including dynamic representations, articulated to semiotic representation systems, in order that some aspects built within a narrative thinking can be linked to the paradigmatic thinking, and to be connected to concepts to be developed in the study of functions. Thus, she points out some important characteristics that are viewed as requirements to mathematics educational software development, such as: the inclusion of knowledge representations more close to the reality, given "wings" to narrative productions and to mathematics semiotic representation, forging connections.

The software development is also discussed in Carvalho's study (2015). She brings into discussion some results of a software developed called *Consecutivo*. Its development aimed to explore dynamic representations to engage learners within situations of systematic observations, perceptions of patterns, and elaboration of formal justification. As requirements to facilitate students' development of proofs, the analyses undertaken are linked to dynamic connection of figural and algebraic language.

7.6 Touchscreen Devices and Mathematics Learning

The mobile digital technologies, particularly touchscreen devices, promise to bring great changes in mathematics teaching and learning processes. Brazilian research is still focused on the use of tablets by teachers, specifically, on UCA[2] project (Scherer & Silva, 2014) or in undergraduate courses (Carvalho, 2015). Bairral, Assis, and Silva (2015a) raise reflections and possibilities of cognitive or epistemological changes that may occur through different manipulations on the screen from a mobile device. Considering touch on screen different from mouse manipulation (Arzarello, Bairral, & Danè, 2014), the authors emphasize that the interaction on these interfaces is a new field of embodied and multimodal knowledge production.

Bairral, Assis, and Silva analysed the use of touchscreen technology in geometry tasks with high school students (2015a) and prospective mathematics teachers (2016) using the sketchometry device. The researchers seek to identify singularities and effects of touchscreen manipulation as a new form of gestural expression on (or with) the screen. According to them, in the same way that the dynamic geometry environments modified forms of learning with paper and pencil, touchscreen devices are transforming our cognition and our ways of interaction and interpersonal communication.

Theoretically the study relates the embodied cognition perspective with the communicative modality in mathematic classrooms. Sensory cognition (Radford, 2014) and embodied cognition (Bolite Frant, 2011) support the idea that thinking is intrinsically developed with our gestural expressions (including touch on screen) and allow a new look for the knowledge production. With Botzer and Yerushalmy (2008), those authors seek to understand the relationship between gestures, manipulations on screen, and mathematical narratives in the construction of geometric concepts.

Bairral et al. (2015b) observed that prospective mathematics teachers using sketchometry also applied two domains (constructive and relational) of manipulation on screen (Arzarello et al., 2014) to solve the proposed tasks. The users' reflection and solution moved continuously from constructive to relational and vice versa. While in the construction domain the manipulation is often for the construction of the mathematical object itself, in the relational domain the user manipulation deepens its line of reasoning and refines conjectures and emerging geometric properties. In the relational realm, the organization of ideas, clarification of concepts or properties is more frequent. Since touch device provides more flexibility and simultaneously object movement (angles, sides, shapes, etc.) on screen, mathematical thinking on relational domains seems to be powerful for improving reasoning and proving process in dynamic geometric environment (Sinclair & Robutti, 2013) with touchscreen device.

Knowing the importance of this kind of manipulation (tap, double tap, hold, drag, flick, rotate, etc.) on the screen is essential for the teacher work out on

[2] One Computer per Child.

proposing tasks in which the student can use them, and this use empowers some kind of learning process (conceptualization, representation, geometric construction, justification, proving, etc.) related with some aspects from teaching planning.

Bairral et al. (2015a) based on Baccaglini-Frank and Mariotti (2010) recognize that a solution for a problem by open conjectures in dynamic geometry environments involves two phases: a phase of conjecture, in which students engage in the exploration of a figure and its argument leads to written formulation of a statement; and a test phase, in which the students try to prove the conjecture. The proof is a product of this second phase. The analyses carried out by both Bairral et al. (2015b), using sketchometry, and Arzarello et al. (2014), using geometric constructer, observed that the approach *handling by drag* (touch-approach) has proved to be fruitful when learners are expanding their abilities to explore, argue, and justify certain geometric property (Leung, 2011).

The touch-approach way of manipulation on screen, highlighted by Bairral et al. (2015a), is located in a border region of the two (conjecture and proof) phases proposed by Baccaglini-Frank and Mariotti (2010). Within the conjecture phase, the touch-approach functions as a manipulation that assists the exploration and emergency properties, observing various geometric shapes. In relation to proof, this manipulation contributes to a deeper and more detailed analysis of what has emerged in the exploration of conjecture. This manipulative moment is as if the user stopped to select and organize their ideas in order to structure its line of argument. Those domains are dynamic, intertwined and its movements of exploration, discovery, and mathematic learning have continuous feedback (Assis et al., 2016).

Besides cognitive singularities, another contribution from mobile technologies with touchscreen and app is the possibility of working without Internet connection. For instance, MyAppShared device allows users to share their device. That possibility does not demand physical informatics spaces (laboratories), licensed software or hardware. For poor countries or schools settings, it can contribute for changing the quality providing the same resources and new ways for both mathematics teaching and learning. Therefore, transformations using such mobile devices with touchscreen also need new pedagogical proposals and, for that, it is always necessary to invest in innovative program for the professional development of teacher with digital technologies.

7.7 Mathematics Teachers Education and Digital Technologies: Towards Cybereducation

Since SIPEM I, 14 articles on digital technologies, distance education, and mathematics teachers education have been analysed, discussed, and presented in Working Group 06 (WG6), among the 54 articles. All papers are derived from completed or partially completed studies in different parts of Brazil, on these themes, and depict the theoretical and methodological perspectives of a period.

At the first moment of the discussion in this Working Group, for example, the research on mathematics teachers' education portrayed the "*Continuous Teachers Education of Mathematics Communicators: from the classroom to the Internet*" (Itacarambi, 2000, p. 01). The discussion presented a new complex question, the teacher's learning and managing the teaching of mathematics generated by the acquisition of computers, peripherals, and software in schools. The research, then, fell behind study developed in other parts of the world when compared to Brazil's actual situation. Arnold, Shiu, and Ellerton (1996), researchers in Australia and the UK, had already discussed distant teaching of mathematics and mathematics education, for example, years before. They said:

> The possibilities offered by hypertext facilities are challenging, essentially redefining existing approaches to instructional design. It is easy to get "lost in hyperspace", and good page design and links must serve to minimize this problem. In addition to hypertext and cross-platform facilities, a web browser such as Netscape offers powerful multimedia capabilities (especially sound and video) which can serve to enhance the learning experience. In addition to the learning of mathematical concepts, of course, videos options greatly enhance access to the teaching process for preservice teachers (Arnold et al., 1996, p. 733).

In Brazil, however, the absence of use of information and communication technologies (ICT) in initial mathematics teachers education caused the investigation about the reasons for a future inclusion of these technologies by teachers in the classroom. Among the reasons, the use of computers at school was legitimized by social and cultural demand. Some mathematics teachers even said they feared losing their jobs if they did not decide for this insertion. In addition, they believed in the potential that these software programs have in mathematics teaching. In both cases, the continued education of these teachers became the key to their performance with technologies. However, the practice of reproduction in terms of following models, solving problems or mimicking close professionals, had repercussions in the form of insertion of technology in the continuing education of mathematics teachers and, consequently, in the classroom. The movement passes the investigation of "why" and "how" we can educate teachers with the use of ICT to the research of effective uses of these technologies, in distance and blended ways. This movement constituted itself in the framework of WG6, and the discussion in terms of teachers' education went towards distance education, reflecting, for example, about how the rescue through a virtual community, of different cultures in the school environment, could influence the continuing education process of mathematics teachers (Miskulin, Silva, & Rosa, 2006). In this sense, evidences pointed to aspects from many cultures that make up the teaching culture and contributed in reframing of teaching practice. Aspects from the multitude of cultures that intersect in the constitution of the teaching culture interfered directly in the constitution of the virtual community. Thus, there was a dialectical movement in the constitution of teacher's knowledge and in this relation is that was constructed one of the identities of this teacher, who becomes the protagonist of their own story. The explanation of this relation between cultural diversity, present in teaching culture, and the implicit aspects in the creation of a virtual community made possible the understanding and possible reframing of teaching practice at the time. This makes us say that the research conducted in Brazil during this period had similar interests to those that were in the world. If we consider the research of McGraw, Lynch, Koc, Budak, and Brown (2007, p. 95), we can say that they

> [...] consider the potential of multimedia cases as tools for teacher professional development. Specifically, we examined online and face-to-face discussions that occurred within groups composed of pre-service mathematics teachers, in-service mathematics teachers, mathematicians, and mathematics teacher educators. Discussions within these heterogeneous groups tended to focus on issues of classroom implementation of the tasks shown in the multimedia case. Secondary foci of discussion included task characteristics and appropriateness of tasks for engaging students in thinking about mathematical concepts and processes. Analysis of contributions to discussions across group member type revealed differences that suggest that the variety of backgrounds and experiences of group members can blend in ways that support rich and critical discussions of mathematics, teaching, and learning.

Therefore, the variety of origins and experiences from the members of the research group can mix themselves in order to support the rich and critical discussions about mathematics, teaching, and learning, which suggests the influence of aspects from the multitude of cultures that intersect in the constitution of the teaching culture.

In this way, we understand that the subject "Mathematics Teachers Education" has advanced, within the group that discusses technology and distance education, on results, because it has taken characteristics that concern online environments. Among them, for example, there are environments that take ownership of videoconference use in distance education. In this sense, the study of Llinares and Valls (2009, p. 247) "explores how preservice primary teachers became engaged in meaning-making mathematics teaching when participating in online discussions within learning environments integrating video-clips of mathematics teaching". Even with the constant pursuit of knowledge about technological resources and processes like this, in our view, interaction and dialogue will always be fundamental to the production of knowledge (Borba, Malheiros, & Zulatto, 2007) and this means that research in Brazil has sought increasingly advanced technological resources to keep these aspects in the mathematics teacher education.

Araújo Filho and Gitirana (2015), for example, analyse the knowledge mobilized during the planning of a mathematical functions lesson in an online collaborative environment for mathematics teachers education. They highlight that the knowledge organized by collaborative groups brought contributions to the interaction between subjects. Collaboration, in the present case, justifies a range of educational knowledge, including the detailed planning of the class. Thus, we advanced in research on mathematics teacher education in distance education. Theoretically, the roles outlined in this mode of education, and the importance given to online collaboration stood out. Besides that, we surpassed the reproduction model that was seen before, in terms of following models, solving problems or acting, even often mimicking close professionals, when Rosa and Pazuch (2015) discuss, for example, the formation of a collaborative space as a constituent action of relation with knowing, among teachers who teach mathematics, in a process called blended cybereducation.

Through a totality, cybereducation with mathematics teachers (Rosa, 2010, 2015) includes the education seen in the (mathematics) specific, pedagogical, and technological dimensions assuming the use of digital technologies (DT), in particular, cyberspace in distance education environment from the perspective of being-with, think-with, and know-how-to-do-with-DT. Thus, the being-with-DT includes the creation of a new world or microworlds, where users are necessarily "plugged" with the technological means; thinking-with-DT can allow the construction of mathematical knowledge in relations with the world and with others, including the (trans)formation of mathematical ideas with technological means (computer, software, video, etc.); and the know-how-to-do-with-DT shown by intentional actions taken with the world, with myself and others, actions that are performed in the activity in the construction of a product in practice. Thus, we understand the DT as a possible means of cognitive changes, rather than a "forced insertion" in the school context. In the study by Rosa and Pazuch (2015), participants reflected, planned, and developed activities-of-geometry-with-digital-technologies for use with their own students of elementary school. Thus, the constitution of a blended collaborative space contributed to the continuation of the initial mathematics education and the reorganization of teacher planning, showing other modes of relation to knowing in mathematical, pedagogical, and technological terms.

With this, we realize that WG6 transitioned by different theoretical references in terms of initial and continuing teacher education for and with the use of technology. Besides that, constructionism (Papert, 1980) served as an important theoretical research support on teacher education with the use of DT. In this bias, studies used the research by Valente (2005), dealing with the cycle and the spiral of learning, for example. These studies made it possible to understand the computer's role in the construction of knowledge in the education that was carried out, and understand the teacher education with DT in the constructionist perspective. It was in addition to understanding what the teacher thought about digital resources, but how the teacher thought-with-DT (Rosa, 2010, 2015). For this, it was considered a philosophical assumption that identifies the use of DT in the Heideggerian phenomenological perspective of being-there, and/or being in the world with (Heidegger, 1996), which, in Rosa (2008), is presented in terms of connection with cyberspace as being-with, think-with, know-how-to-do-with-technologies. It is important to highlight that among the different theoretical bases on teacher education and digital technologies, the changes that the media suffer constantly generate new questions about how to educate teachers to address these changes while they transform their own education and teaching practice. Thus, we have evolved a lot in terms of thinking about the mathematics teachers education, in line with the use of resources and technological environments in order to no longer characterize and invest in a reproductive practice, but in an expansion/transformation of existing mathematical thinking and that to be encouraged.

7.8 Some Remarks

The discussion shows a diversity within the different theoretical ideas that support the research presented in WG6. It seems true that in most research this working group we summarized above, however, there is one common point: all of them seem to look for, or state at front, that mathematics may be changing as different technology become present in the classroom. This was not always the case, in the early days of the working groups, there seems to be much stronger a research line that would claim that the use of software, for instance, was only an instance, so that afterwards mathematical content—as it was learned in schools without digital technology—should be learned in an intact way. It seems that most, or all of the research presented, would claim that mathematics is in change as touch technology, or different ways of using Internet are present in the production of knowledge by students or teachers, or by collectives that involve both of them.

This does not mean that new theorems, or new mathematical results, are coming out of the mathematical investigations developed with software, touchscreen, videos, collaboration, etc., but it does mean, as stated by many participants of the group, that the process has changed. Mathematics in this perspective should not be considered as a result, but as process, mathematics in change. The above thesis has more than 10 years, but it can be used to describe this working group in their last face-to-face meeting. As D'Ambrosio and Borba (2010) would say, trends in Brazil are intertwined. So, we may see in this way of conceptualizing mathematics in change, roots of the research in ethnomathematics, which strongly emphasized the notion that mathematics change as cultural groups differ. We may see in this group, with their different theories, mathematics changing as technology differs.

This way of seeing the interface of mathematics and digital technology may be a contribution of Brazilian research for the international mathematics education. If this research, as a group go in the direction of contributing to established theories in mathematics education, as some papers point out, we may be giving different steps in originality in the next one or two meetings of this working group. On the other hand, it is easy to see that studying the relationship between innovations developed in small groups and a larger network of schools is still a challenge in Brazil, not only because of the strong epistemological research focus of most researchers, but also because of material conditions regarding the use of digital technology in different schools.

References

Araripe, J., & Bellemain, F. (2015). Interconnecting digital interfaces and semiotic representation registers: Analysis of interactions for collaborative learning supported by linear algebra objects computer. *Anais do VI SIPEM*. Retrieved from http://www.sbembrasil.org.br/visipem/anais/story_html5.html

Araújo Filho, R. M., & Gitirana, V. (2015) Teaching skills mobilised and collaboration in online training environments. *Anais do VI SIPEM*. Retrieved from http://www.sbembrasil.org.br/visipem/anais/story_html5.html

Arnold, S., Shiu, C., & Ellerton, N. (1996). Critical issues in the distance teaching of mathematics and mathematics education. In A. J. Bishop, K. Clements, C. Keitel, J. Kilpatrick, & C. Laborde (Eds.), *International handbook of mathematic education*. Netherlands: Kluwer Academic.

Artigue, M. (1990). Ingénierie didactique. *Recherches en Didactique des Mathématiques, 9*(3), 281–308.

Arzarello, F., Bairral, M., & Dané, C. (2014). Moving from dragging to touchscreen: Geometrical learning with geometric dynamic software. *Teaching Mathematics and Its Applications, 33*(1), 39–51. https://doi.org/10.1093/teamat/hru002.

Assis, A. R., da Silva, B. C. C. C., & Bairral, M. A. (2016). Toques em tela de tablets e domínios de aprendizagem em geometria. *Educação Matemática em Revista, 51*, 6–14 Retrieved from http://www.sbem.com.br/revista/index.php/emr/article/view/677/pdf.

Baccaglini-Frank, A., & Mariotti, M. A. (2010). Generating conjectures in dynamic geometry: The maintaining dragging model. *International Journal of Computers for Mathematical Learning, 15*(3), 225–253.

Bairral, M. A., Assis, A. R., & Silva, B. C. C. (2015a). *Mãos em ação em dispositivos touchscreen na educação matemática*. Rio de Janeiro: Edur.

Bairral, M. A., Assis, A. R., & Silva, B. C. C. (2015b). Toques para ampliar interações e manipulações touchscreen na aprendizagem em geometria. *Anais do VI SIPEM*. Retrieved from http://www.sbembrasil.org.br/visipem/anais/story_html5.html

Baker, M. (2002). Forms of cooperation in dyadic problem-solving. *Revue d'intellingence Arttificielle, 16*, 587–620.

Balacheff, N. (1994a). La transposition informatique. Note sur un nouveau problème pour la didactique. In Artigue et al. (Eds.), *Vingt ans de didactique des mathématiques en France. Hommage à Guy Brousseau et Gérard Vergnaud* (pp. 364–370). Grenoble: La Pensée Sauvage.

Balacheff, N. (1994b). *Didactique computationnelle, évocation d'un projet de recherche. Prépublication 94-07*. Rennes, France: IRMAR.

Balacheff, N. (1995). Conception, connaissance et concept. In D. Grenier (Ed.), *Séminaire Didactique et Technologies Cognitives en Mathématiques* (pp. 219–244). IMAG, Grenoble: IMAG.

Bellemain, F. (2000). A transposição didática na engenharia de softwares educativos. *Livro de Resumos, I SIPEM* (pp. 198–204). Retrieved from http://www.sbembrasil.org.br/files/sipemI.pdf

Bellemain, F. G. R., Ramos, C. S., & dos Santos, R. T. (2015). Engenharia de Software Educativos, o caso do Bingo dos Racionais. *Anais do VI SIPEM*. Retrieved from http://www.sbembrasil.org.br/visipem/anais/story_html5.html

Bittar, M. (2009). Integração da tecnologia nas aulas de matemática: Contribuições de um grupo de pesquisa-ação na formação continuada de professores. *Anais IV SIPEM*. Retrieved from http://www.sbembrasil.org.br/files/sipemIV.pdf

Bolite Frant, J. B. (2011). Linguagem, Tecnologia e Corporeidade: Produção de significados para o tempo em gráficos cartesianos. Retrieved from http://www.redalyc.org/articulo.oa?id=155019936014

Borba, M. C. (1999). Tecnologias Informáticas na Educação Matemática e Reorganização do Pensamento. In M. A. V. Bicudo (Ed.), *Pesquisa em Educação Matemática: Concepções e Perspectivas* (pp. 285–295). São Paulo, Sao Paulo: UNESP.

Borba, M. C. (2000). GPIMEM e UNESP: Pesquisa, extensão e ensino em Informática e Educação Matemática. In M. G. Penteado & M. C. Borba (Eds.), *A Informática em Ação: formação de professores, pesquisa e extensão* (pp. 47–66). Rio Claro, Sao Paulo: Olho D'água.

Borba, M. C., Malheiros, A. P. S., & Zulatto, R. B. A. (2007). *Educação a distância online*. Belo Horizonte, Minas Gerais: Autêntica.

Borba, M. C., & Villarreal, M. V. (2005). *Humans-with-media and the reorganization of mathematical thinking: Information and communication technologies, modeling, experimentation and visualization* (Vol. 39). New York: Springer.

Botzer, G., & Yerushalmy, M. (2008). Embodied semiotic activities and their role in the construction of mathematical meaning of motion graphs. *International Journal of Computer in Mathematics Learning, 13*, 111–134. https://doi.org/10.1007/s10758-008-9133-7.

Carvalho, M. (2015). Formação inicial do professor de matemática: utilização das TIC, dispositivos touchscreen dos tablets, no Estágio Supervisionado. *Boletim Gepem, 67*, 71–83.

Cress, U., Stahl, G., Ludvigsen, S., & Law, N. (2015). The core features of CSCL: Social situation, collaborative knowledge processes and their design. *International Journal of Computer-Supported Collaborative Learning, 10*(2), 109–132.

D'Ambrosio, U., & Borba, M. C. (2010). Dynamics of change of mathematics education in Brazil and a scenario of current research. *ZDM Mathematics Education, 42*, 271–279.

Dillenbourg, P. (2002). Over-scripting CSCL: The risks of blending collaborative learning with instructional design. In P. A. Kirschner (Ed.), *Three worlds of CSCL: Can we support CSCL?* (pp. 61–69). Heerlen, Nederland: Open Universiteit.

Engeström, Y. (1987). *Learning by Expanding: An activity-theoretical approach to developmental research*. Orienta-Konsultit: Helsinki.

Engeström, Y. (1999). *Learning by expanding: Ten years after*. Retrieved from http://lchc.ucsd.edu/MCA/Paper/Engestrom/expanding/toc.htm

Galvis, A. H. (1992). *Ingeniería de Software Educativo*. Santafé de Bogotá: Ediciones Uniandes.

Gitirana, V., Teles, R. A. M., Bellemain, P. M. B., Castro, A. T., Almeida, I. A. C., Lima, P. F., & Bellemain, F. (2013). *Jogos com Sucata na Educação Matemática*. Recife, Pernambuco: Editora Universitária da UFPE.

Heidegger, M. (1996). *Being and time*. Albany: State University of New York Press.

Hostetter, A. B., & Alibali, M. W. (2008). Visible embodiment: Gestures as simulated action. *Psychonomic Bulletin & Review, 15*(3), 495–514.

Itacarambi, R. R. (2000). Formação contínua de professores comunicadores de matemática: da sala de aula à internet. *Livro de Resumos, I SIPEM*. Retrieved from http://www.sbembrasil.org.br/files/sipemI.pdf

Leontiev, A. N. (1978). O desenvolvimento do psiquismo. Livros Horizonte, Lisboa.

Leung, A. (2011). An epistemic model of task design in dynamic geometry environment. *ZDM Mathematics Education, 43*(3), 325–336. https://doi.org/10.1007/s11858-011-0329-2.

Lévy, P. (1994). *As Tecnologias da Inteligência: o futuro do pensamento na era da informática*. (Carlos Irineu da Costa, Trans.). São Paulo, Sao Paulo: Editora 34. (Original work published 1990)

Lévy, P. (2000). *A Inteligência Coletiva*. (Luiz Paulo Rouanet, Trans.). São Paulo, Sao Paulo: Edições Loyola. (Original work published 1997)

Lima, R. (2015). Window of SimCalc World: A possibility for the teaching of function concept. *Anais do VI SIPEM*. Retrieved from http://www.sbembrasil.org.br/visipem/anais/story_html5.html

Llinares, S., & Valls, J. (2009). The building of pre-service primary teachers' knowledge of mathematics teaching: Interaction and online video case studies. *Instructional Science, 37*(3), 247–271.

Lucena, R., & Gitirana, V. (2015). The online tutoring didactic settings: The choice of resources for didactic tutor mediation. *Anais do VI SIPEM*. Retrieved from http://www.sbembrasil.org.br/visipem/anais/story_html5.html

McGraw, R., Lynch, K., Koc, Y., Budak, A., & Brown, C. A. (2007). The multimedia case as a tool for professional development: An analysis of online and face-to-face interaction among mathematics pre-service teachers, in-service teachers, mathematicians, and mathematics teacher educators. *Journal of Mathematics Teacher Education, 10*(2), 95–121.

Melo, M. S. L., Montenegro, G. M. M., Santos, L. S., Moraes, M. D., & Bellemain, P. M. B. (2013). *Bingo dos Números Racionais – Indicações Didáticas. Projeto Rede na Educação Matemática*. Recife, Pernambuco: NEMAT/UFPE.

Miskulin, R. G. S., Silva, M. R. C., & Rosa, M. (2006). Comunidade Virtual como lócus do Resgate da Cultura Docente: contribuições para a formação continuada do professor de matemática. *Anais do III SIPEM*. Retrieved from http://www.sbembrasil.org.br/files/sipemIII.pdf

Papert, S. (1980). *Mindstorms: Children, computers and powerful ideas*. New York: Basic Books.

Radford, L. (2014). Towards an embodied, cultural, and material conception of mathematics cognition. *ZDM Mathematics Education, 46*(3), 349–361.

Rosa, M. (2008). *A Construção de Identidades Online por meio do Role Playing Game: relações com o ensino e aprendizagem de matemática em um curso à distância*. Unpublished doctoral dissertation in Mathematics Education – Instituto de Geociências e Ciências Exatas, Universidade Estadual Paulista, Rio Claro, Brazil.

Rosa, M. (2010). Cybereducation: Focusing the education of teachers towards Online Mathematics Education. In M. Pinto & T. Kawasaki (Eds.), *Proceedings of the 34th Conference of the International Group for the Psychology of Mathematics Education*. PME Group: Belo Horizonte, Minas Gerais.

Rosa, M. (2015). Cyberformação com professores de matemática: interconexões com experiências estéticas na cultura digital. In M. Rosa, M. A. Bairral, & R. B. Amaral (Eds.), *Educação Matemática, Tecnologias Digitais e Educação a Distância: pesquisas contemporâneas* (pp. 57–96). São Paulo, Sao Paulo: Livraria da Física.

Rosa, M., & Pazuch, V. (2015). Cybereducation with mathematics teachers in a b-learning course: Constituting a collaboration space. *Anais do VI SIPEM*. Retrieved from http://www.sbembrasil.org.br/visipem/anais/story_html5.html

Scherer, S., & Silva, L. Q. D. (2014). Formação de professores para o uso de laptops educacionais: reflexões sobre o ensino de geometria (Abstract). *RIE Digital, 66*(2). Retrieved from http://rieoei.org/rie_contenedor.php?numero=6354&titulo=Forma%C3%A7%C3%A3o%20de%20professores%20para%20o%20uso%20de%20laptops%20educacionais:%20reflex%C3%B5es%20sobre%20o%20ensino%20de%20geometria.

Sinclair, N., & Robutti, O. (2013). Technology and the role of proof: The case of dynamic geometry. In M. A. K. Clements, A. J. Bishop, C. Keitel, J. Kilpatrick, & F. K. S. Leung (Eds.), *Third international handbook of mathematics education* (pp. 571–596). New York: Springer.

Souto, D. L. P. (2015). Aprendizagem Matemática on-line: quando tensões geram conflitos. *Educação Matemática Pesquisa, 17*(5).

Souto, D. L. P., & Borba, M. C. (2015). Movimentos, estagnações, tensões e transformações na aprendizagem da matemática on-line. *Anais do VI SIPEM*. Retrieved from http://www.sbembrasil.org.br/visipem/anais/story_html5.html

Stahl, G., Koschmann, T., & Suthers, D. (2006). Computer-supported collaborative learning. In R. K. Sawyer (Ed.), *Cambridge handbook of the learning sciences* (pp. 196–223). Cambridge, UK: Cambridge University Press.

Tchounikine, P. (2009). *Précis de recherche de Ingénierie des EIAH*. Retrieved from hal.archives-ouvertes.fr/docs/00/41/36/94/PDF/PrecisV1.pdf.

Tikhomirov, O. K. (1981). The psychological consequences of computerization. In J. V. Wertsch (Ed.), *The concept of activity in Soviet psychology*. (James V. Wertsch, Transl). New York: M. E. Sharpe.

Trouche, L. (2005). Construction et conduit des instruments dans les apprentissages mathématiques: nécessité des orchestrations. *Recherches en Didactique des Mathématiques, 25*, 91–138.

Valente, J. A. (2005). *A espiral da espiral de aprendizagem: o processo de compreensão do papel das tecnologias de informação e comunicação*. Unpublished full professor thesis/ livre docência – Universidade Estadual de Campinas. Instituto de Artes, São Paulo.

Verillon, P., & Rabardel, P. (1995). Cognition and artifacts: A contribution to the study of though in relation to instrumented activity. *European Journal of Psychology of education, 10*(1), 77–101.

Chapter 8
Mathematics Teacher Education: Synthesis and Perspectives of Research Developed in Brazil

Adair Mendes Nacarato, Ana Cristina Ferreira, Cármen Lúcia Brancaglion Passos, and Márcia Cristina de Costa Trindade Cyrino

Abstract This chapter considers projects developed by the working group on teacher education of the Brazilian Society of Mathematics Education and discussed at the sixth International Seminar of Mathematics Education. Twenty-six projects were presented in four areas of focus: teacher education in the initial stages of elementary education; knowledge necessary for teaching; narratives and texts enunciated by (future) teachers; education practices and practices of teacher educators. Initially we present briefly the projects, and, at the end of each focus area, we produce a synthesis looking for interconnections between the projects and international literature. We identified a multiplicity of focus points and methodological approaches, but also some fragilities and gaps. Our analysis indicates that some themes have been recurrent since the working group was founded in 2000 although with some theoretical references updated and expanded; others are emerging, such as concept study, and others remain absent, such as teacher professionalisation processes. Finally, we point out some particularities of the Brazilian studies that reveal involvement between the Brazilian scientific community and trends that mark the important practices of teacher education and the education of teacher educator.

A. M. Nacarato (✉)
Universidade São Francisco, Itatiba, Brazil

A. C. Ferreira
Universidade Federal de Ouro Preto, Ouro Preto, Brazil

C. L. B. Passos
Universidade Federal de São Carlos, São Carlos, Brazil

M. C. d. C. T. Cyrino
Universidade Estadual de Londrina, Londrina, Brazil

A. J. Ribeiro et al. (eds.), *Mathematics Education in Brazil*,
https://doi.org/10.1007/978-3-319-93455-6_8

8.1 Introduction

The research scenarios regarding the mathematics teacher education in Brazil and in the world are composed of several themes of study, namely: nature and structure of knowledge needed for teachers, organisation of teaching in teacher education, teacher education programmes, professional development of teachers, stories of teachers, professional identity, teacher learning, intervening factors in the processes of teacher education, and articulation between the university and the school for the teacher education, among others.

Teacher education in Brazil is a field where research occurs in the majority of postgraduate programmes linked to areas of education or teaching in CAPES.[1] Research topics include pre-service education (undergraduate courses in mathematics and education[2]) and continuous education in formal and non-formal spaces (in-service). Although the dichotomy between pre-service and in-service education is a point of debate on the understanding that education is a continuum, this division still occurs, mostly because of Brazilian public policies. The Ministry of Education published, in 2015, a document indicating, for the first time, that both should be articulated. In Article 2, section IX, it reads: "the articulation between initial and continuing education, as well as between the different levels and modalities of education" (Brazil, 2015, p. 5). However, to date the official documents do not indicate how this articulation should occur. Among the current public policies, Programa Institucional de Bolsas de Iniciação à Docência (institutional scholarship programme for initiation into professorship—PIBID), implemented in 2008, has promoted this articulation. This programme has promoted the relationship between undergraduate students and teachers at elementary, middle, and high school level (6–17 years old) in public schools; the future teachers accompany the work of the teachers who take on the role of teacher educator; both teachers and students receive a research scholarship. This programme, present in almost all universities, has generated many different research projects in different areas of educational knowledge.

Viñao (2008) postulated that the discipline in which the teacher teaches determines his identity and professionalisation. Thus, it is necessary to look at the teacher who teaches mathematics[3] seeking to understand his development process, the

[1] Coordenação de Aperfeiçoamento de Pessoal de Nível Superior (Coordination of Improvement of Higher Level Personnel). Capes, a foundation of the Ministry of Education (MEC), plays a key role in the expansion and consolidation of stricto sensu (master's and doctorate) postgraduate courses in all states of the federation. In 2007, it also began to work in the teacher education of basic education.

[2] An undergraduate degree in education (*pedagogia*) in Brazil is the course that enables teachers to exercise teaching in early childhood education programmes, from first to fifth grade (6–10 years old), and in youth and adult education programmes.

[3] We highlight that the expression "teachers that teach mathematics" refers to all teachers who, in one way or another, work with mathematics, from the elementary education level to higher education.

construction of his identity, and his professionalisation. Sociedade Brasileira de Educação Matemática—SBEM (Brazilian Society of Mathematics Education) has advanced as a professional field of investigation, since its creation, in 1988. This required math educators to organise themselves into working groups (WG), aiming at greater clarity between the research projects and the fields of action of its members. Among the different groups created since 2000, one of them was teacher education—WG07. Since then, the researchers in this working group have been dedicated to discussing and disseminating, in several academic spaces, the research projects that are carried out by its participants to intervene in public policies, and to articulate and identify emerging themes.

The main thematic investigations, in these last years, were: the teacher as a producer of knowledge (emphasis on research on teacher and professional knowledge and skills); the teacher as an agent of his/her own development; the teacher and research (developed by teachers on their practice or with the teacher educator); collaborative work and/or research produced in collaborative groups; specific disciplines for pre-service education (especially related to teaching practice and supervised internship); education of the teacher's educator; the role of mathematical knowledge in the teaching of initial and continuous development of teachers; the initial and continuous development of teachers who teach mathematics stemming from public policy actions (PIBID and OBEDUC[4]), its importance and contributions to future teachers' perceptions of professional practices and the process of teaching and learning as a collective and collaborative construction of knowledge. Researchers in WG07 are constantly concerned about following public policy movements and the analyses carried out in the area of teacher education. The group has thus been active in the organisation of national forums on undergraduate studies in mathematics and in the publication of research results, in addition to engaging in partnerships in other events organised by the Brazilian Society of Mathematics Education and other Brazilian associations.

In the last 15 years, with each edition of the Seminário Internacional de Pesquisa em Educação Matemática—SIPEM (International Mathematics Education Research Seminar), the number of researchers in WG07 has been increasing and, in the same proportion, the number of studies discussed in meetings. This fact led us to focus, in this article, on the analysis of the studies debated in SIPEM VI, held in 2015. These studies reflect, to a certain extent, the direction of the research in this field of investigation.

[4]The Education Observatory Program (Observatório da Educação—OBEDUC), a result of a partnership between Capes, the National Institute of Education Studies (INEP), and the Secretariat for Continuing Education, Literacy, Diversity and Inclusion (SECADI), was established in 2006 aiming to articulate postgraduate courses, undergraduate courses, and basic education, as well as to stimulate academic production.

Table 8.1 Studies in Group 1: studies that focus on initial and continuing development of teachers teaching the initial years of elementary education

Author(s)	Title
Carneiro (2015)	Teacher education of the early years in an education course: contributions of mathematics courses
Fanizzi and Santos (2015)	Public policies and continuing teacher education in the initial years of mathematics: an experience of the Municipal Secretary of Education of São Paulo
Giongo, Rehfeldt, and Quartieri (2015)	Continuing education and elementary school teachersschool and university partnership activities
Giusti and Justo (2015)	Intersection between collaborative and cooperative processes in continuing education of elementary school teachers
Marco, Lopes, and Sousa (2015)	Training project in the math teacher's educational activity
Nacarato and Grando (2015)	Research of and about mathematics teachers
Passos and Souza (2015)	Initial training/continuing training of elementary teacher: computer technology and mathematics
Santos (2015)	Teachers who teach mathematics: relevant dialogues between mathematics teacher and pedagogue

8.2 A Look at WG07 Research in SIPEM VI

WG07 included 29 projects approved and published in SIPEM VI.[5] In this chapter, we analyse the 26 projects presented and discussed in the event. In this analysis, we follow the organisation established by the coordinators of the WG to present the event by themes. These themes were: teacher education in the initial years of elementary education; knowledge necessary for teaching, narratives and texts enunciated by (future) teachers, and formative practices and practices of teacher's educators (challenges and possibilities in the mathematics teacher education).

We read and analysed the articles entirely, identifying their research focus, the theoretical and methodological perspectives and the results. This analysis allowed us to divide the set of projects into four groups. Nevertheless, the research focuses are the main aspects that differentiate these groups, since the same groups present several points of intersection, both in theoretical and in methodological aspects.

We present the organisation of each group. Group 1, with eight projects, focuses on the initial and continuing development of teachers in the initial years of elementary education (Table 8.1).

Group 2 is comprised of seven studies that focus on the knowledge necessary for teaching (Table 8.2).

Group 3 is comprised of four studies that focus on narratives and texts enunciated by (future) teachers (Table 8.3).

[5] Available at: http://www.sbembrasil.org.br/visipem/anais/story.html.

Table 8.2 Studies in Group 2: studies that focus on knowledge of teachers

Author(s)	Title
Galleguillos, Montes, and Ribeiro (2015)	Situations in mathematics investigations and the specialised knowledge of the teacher
Giraldo, Rangel, Quintaneiro, and Matos (2015)	Elementary mathematics and concept study: establishing relationships
Moretti and Radford (2015)	Culturally meant concept's history and the organisation of mathematics teaching activity
Oliveira and Cyrino (2015)	Learning to respect the proportional reasoning in a mathematics teacher community of practice
Powell and Pazuch (2015)	Tasks, collaboration, and the geometric justifications of teachers in VMTcG environment
Ribeiro and Oliveira (2015)	Mathematical knowledge of teachers and equation learning: a study about the lesson planning for basic education
Silva and Wrobel (2015)	Teacher's knowledge from the pre-service mathematics teacher standpoint

Table 8.3 Studies in Group 3: studies focusing on narratives and texts produced by teachers in pre-service and in in-service

Author(s)	Title
Marquesin and Nacarato (2015)	My time in school: narratives of future teachers
Santana and Barbosa (2015)	Collaborative work with math teachers: an analysis of the conflicts between/in the texts enunciated by the participants
Souza and Silva (2015)	Effects that induce and produce a learning tool in the education process of a teacher and her students
Taxa-Amaro, Perovano, Bortoldi, and Santana (2015)	Narratives of teacher educators crisscrossing the process of mathematical literacy

Group 4 is comprised of articles that focus on training practices and teacher's educators (difficulties and possibilities in the mathematics teacher education training of mathematics teachers) (Table 8.4).

8.3 What Has Been Researched in the Area of Working Group 07?

In this section, we identify theoretical and methodological approximations in the Brazilian and international context evident in the studies reviewed. We seek to describe the themes and theoretical-methodological approaches of these studies, as well as to position them, in some measure, within the national and international scenarios of mathematics teaching.

Table 8.4 Studies in Group 4: Studies focusing on education practices and on teacher educators (difficulties and possibilities in the development of math teachers)

Author(s)	Title
Côco and Silva (2015)	Supervised training and learning teaching: actions and reflections of mathematics graduates
Costa and Lucena (2015)	Formative practices as links between teachers and teacher educators in self-formations
Cyrino (2015)	Teachers' professional identity development in communities of practice: practice elements
Ferreira, Coura, and Barbosa (2015)	An analysis of research in Minas Gerais on the bachelors in mathematics: considering theses and dissertations defended between 2001 and 2012
Oliveira and Fiorentini (2015)	Knowledge and practices of math teacher educators: the role and the place of a special teaching methodology for mathematics in the bachelors' course
Silva and Cedro (2015)	Mathematics teacher education: the supervised internship as one of the spaces for the constitution of the teaching praxis
Pardim and Pereira (2015)	Interactions between the members of the university-school in a collaborative project

8.3.1 *Initial and Continuing Development for Teachers Teaching in the Early Years of Elementary School*

Initial and/or continuous development of teachers that teach mathematics in the initial years of schooling has been considered in research in all the editions of SIPEM. In this sixth edition, nine studies were presented. Carneiro's (2015) research was the only study that dealt specifically with initial teacher development by bringing dilemmas and perspectives from narratives of undergraduate education students who evaluated contributions from mathematics courses in an education programme. The theoretical background of the paper discussed studies by Brazilian researchers on the complexity of teacher education in the initial years (Gatti & Barreto, 2009; Nacarato, Mengali, & Passos, 2009). The other studies discuss contexts of continuous training of teachers.

The article of Passos and Souza (2015) discusses the initial development of teachers promoted by an extension course (a hybrid online and face-to-face course with teachers in their initial years of studies and teachers who had already graduated) where information technology is integrated in the process of learning and teaching mathematics. To the theoretical approaches related to teacher education (for example, Shulman, 1986), the authors incorporated the TPACK model (technological pedagogical content knowledge), developed by the researchers Mishra and Koelhler (2006), adding other categories to the typology. For these authors, technology knowledge is always changing and it includes the teachers' knowledge of digital and regular technology, the ability to operate this technology, and the capacity to learn and to adapt to technological advancements.

Five studies in this group refer to the importance of collaborative work in the continuing development of teachers. The study by Santos (2015) presents collaborative actions developed in the mathematics education laboratory of a university as practices that promote professional development for a sixth grade (11 years old) teacher who accompanies a fifth grade (10 years old) teacher and who discusses contents and practices of teaching mathematics. The research of Giusti and Justo (2015) refers to a partnership created between the researcher and the school in a perspective of intersectional collaboration and cooperation (Kemczinski, Marek, Hounsell, & Gasparini, 2007) as an intersectional analogy of collective work between the participants in the education programme.

The other three studies stem from a university–school partnership instituted through the Educational Observatory (OBEDUC) public policy. Giongo et al. (2015) present information from studies about teacher development. Marco et al. (2015) base themselves primarily on studies of the activity theory, founded on Leontiev's study (1978, 1983, 2001). Nacarato and Grando's research (2015), conducted with elementary school teachers, takes as reference studies related to literature that call for the legitimacy of research by elementary school teachers, especially that of Kenneth Zeichner (Geraldi, Fiorentini, & Pereira, 1998) and that focuses on mathematical literacy practices. The analysis follows from class narratives produced by the teachers. It is noteworthy that, in part of these studies, there is concern with the elaboration of activities to be developed in the classroom and discussed in the group; in others, the authors start from the written productions of the teachers. Opportunities for protagonism have been provided through these spaces of dialogue on the knowledge of content so that the group can find answers to their dilemmas. The school is recognised as a space of development and production of knowledge. The study by Fanizzi and Santos (2015), discussing São Paulo[6] Municipal Secretary's public policies on continuing development of mathematics teachers in the initial years of elementary, noticed that what is cited in the official documents is not actually carried out. In a way, this research corroborates the thesis that working group 07 advocated regarding the effectiveness of continuing education: that which offers the teachers a space to feel free, safe, and reliant, a space where their "voice" is indeed considered, a voice formed by the recontextualisations that the teachers make of the curriculum, of mathematics teaching, and of the student in the beginning of his schooling. The results of the studies that investigated the continuing development of teachers indicate that frequent meetings with the participation of teachers from different areas of knowledge, who propose to prepare, develop, and assess educational proposals with different methodological strategies (based on the emergent themes of their teaching practice and in a dialectical movement with theory), have allowed for the production and problematisation of knowledge and caused a movement of rupture with the current curriculum. The actions produced in these movements allowed for collective learning to be shared and for teachers to become protagonists of their development, feeling more confident and secure to innovate in their educational practices.

[6] São Paulo city is the capital of the São Paulo State. The population of the municipality is 10,886,518 habitants. It is the most populous city in Brazil and South America. http://www.cidadedesaopaulo.com/sp/br/a-cidade-de-sao-paulo.

These results show that it is necessary to invest in teacher development proposals by incorporating studies on collective and shared education, thus breaking with approaches founded on technical rationality, which prioritise only mathematical concepts in the perspective of knowledge to be applied by teachers in their practices.

The potentialities of collaborative spaces for teacher education have been highlighted in world literature, both in the field of education in general, and in mathematics education. The results of these studies are close to the discussions established by authors such as Wenger (1998), Cochran-Smith and Lytle (2009), and Jaworski (2008). What these authors have in common is the defence of teacher development in groups or communities (of research, learning, or practice) that allow for the production of situated knowledge, stemming from the needs of the teachers. Cochran-Smith and Lytle (2009, pp. 41–2) consider that "in most forms of practitioner inquiry, the role of the local community is critical, since this is the context in which knowledge is constructed and used, and it is also the context in which knowledge is initially made public and opened up to the scrutiny of others". In the same perspective, Jaworski (2008, p. 310) adopts the concept of community investigation, basing himself on the studies by Jean Lave e Ettiène Wenger, defending research as a tool: "Inquiry is first of all a tool used by participants in a community of practice in consideration and development of the practice, that of mathematics learning and teaching in classrooms. Inquiry mediates between the activity of the classroom and the developmental goals of participants". Robutti et al. (2016, p. 653) defend that "we consider teachers who are working together collaborating for some specific aims, which could be directed towards: improving students' learning; improving their professional role in the school; learning to use new resources (e.g., technological tools); creating a professional network within the school or region; and discussing institutional reforms and demands around the curriculum, the national evaluations system, etc.". We understand that Brazilian research also seeks these goals by forming groups or communities of teachers to work collaboratively. Another aspect that deserves attention is the analysis of public policies. PIBID and Plano Nacional de Formação de Professores—PARFOR (national plan for elementary teachers' education) are programmes aimed at the initial and continuing education of teachers that have been developed in large scale in Brazil in the last decade. Although OBEDUC is mainly focused on research, providing a link between university and elementary school, in the studies presented in SIPEM, teacher education was also an important issue.

Numerous studies in the field of mathematics education have discussed the results of those public policies (e.g. Munhoz & Giongo, 2014; Nacarato, 2016; Sousa, 2010; Tinti, 2012), and there is already some consensus as to its potential as a central space. By articulating school with university, bringing together teachers, future teachers, and researchers, educational actions have a greater impact. Of course, the challenges and the steps in building this joint articulation are numerous, but it is a path of great promise.

8.3.2 Knowledge of Teachers Who Teach Mathematics

Studies regarding the knowledge of teachers in the field of mathematics education have gained ground in the last decades (e.g. Ball, Thames, & Phelps, 2008; Blömeke & Delaney, 2012; Carrillo, Climent Rodríguez, Contreras González, & Montes Navarro, 2015; Clark, 2012; Hossin, Mendrick, & Adler, 2013; Lerman, 2000; Shulman, 1986; Tatto, Lerman, & Novotna, 2010). In national and international literature, the majority of studies take the cognitivist/structuralist theoretical perspective, compared to sociocultural and sociological perspectives. The frameworks of Shulman and Ball and their collaborators are the most widely used (Potari & Ponte, 2016). Among the studies presented at SIPEM VI, seven involved investigations approaching teacher knowledge (Galleguillos et al., 2015; Giraldo et al., 2015; Moretti & Radford, 2015; Oliveira & Cyrino, 2015; Powell & Pazuch, 2015; Ribeiro & Oliveira, 2015; Silva & Wrobel, 2015). Two of these studies did not involve processes of teacher education (Moretti & Radford, 2015; Silva & Wrobel, 2015). Moretti and Radford (2015) described a theoretical study aiming to investigate possible implications for the development of teachers, of the relationship between phylogenesis and ontogenesis in the organisation of mathematics teaching. The authors sought to establish relationships between an educational approach to mathematical concepts and the history of the concept and its cultural significance in order to understand the potential of the history of the concept as a support for the elaboration of situations that potentially trigger learning. The research is based on the cultural-historical perspective (Kopnin, 1978; Leontiev, 1983; Vygotsky, 2002) and elements of the cultural theory of objectification (Radford, 2006, 2013). Silva and Wrobel (2015) investigated what knowledge future math teachers point to "when they write about a good teacher and about basic knowledge necessary for the practice of teaching" (p. 1). The theoretical-methodological references used are Shulman (1986), Tardif (2002), and Fiorentini (2005).

In the other five studies, the analysis focused on teacher knowledge mobilised or constructed in processes of continuing education/development of the teachers. These investigations qualify teacher education as a complex, unfinished process, and indicate the need to consider teachers and future teachers as (co) responsible for their professional education process. Thus, in these processes, the teachers in formation developed work in groups (collaborative groups, communities of practice, groups of studies) and took their classroom practices and theoretical studies as the basis for discussions. In most cases, the themes discussed were negotiated with the teachers in training in order to meet the demands of their professional practice.

Galleguillos et al. (2015) study the mathematical production of teachers working in groups in the resolution of an exercise that has an open solution, aiming to find the flat figure of maximum area with fixed perimeter. Depending on the conditions established for this figure (which were not detailed in the description of the exercise), the problem could have different solutions. The analysis was based on group discussions and on different ways of solving the problem that generated opportunities for teachers to develop their mathematical knowledge. The research theoretically supports the notion of specialised knowledge to teach mathematics, by Carrillo et al. (2015).

Considering the ideas of Felix Klein on double discontinuity and elementarisation as a process of historical translation, and recent research literature on teacher knowledge (e.g. Ball et al., 2008; Shulman, 1986) and teacher education associated to the investigation of concepts proposed by Davis et al. (e.g. Davis & Renert, 2012), Giraldo et al. (2015) have designed a collective study model that was used as a strategy in the continuing training of teachers connected with experience of the classroom practice of these teachers. The theme of the collective study was rational numbers.

Powell and Pazuch (2015) studied how teachers who teach mathematics solve a task about quadrilaterals and their bisectors in the virtual math teams with GeoGebra (Stahl, 2009), and how they construct geometric justifications using relations and properties. The authors emphasise that the process of geometric justifications occurred in a collaborative and investigative way among those involved in the education process.

Oliveira and Cyrino (2015) analysed teachers participating in the 'Community of Practice of Teachers who Learn and Teach Mathematics'—CoP-PAEM, in a study of proportional reasoning (Lamon, 2012; Lesh, Post, & Behr, 1988). The actions of the teachers (in initial and continuing education) in solving, discussing and reflecting on problems involving proportionality were analysed from the perspective of the social theory of learning (Wenger, 1998). Ribeiro and Oliveira (2015) studied the knowledge of teachers on the concept of equation, namely knowledge of content, of students and of teaching, based on Ball et al. (2008).

The education contexts in which the studies of Oliveira and Cyrino (2015) and Ribeiro and Oliveira (2015) were developed included the participation of future math teachers and researchers who were part of projects linked to the 'Observatory of Education' and funded by CAPES. Among the actions developed by the teachers in formation, we highlight: problem solving, discussion and analysis of strategies and mathematical contents involved in problem solving, elaboration of lesson plans, the use of dynamic geometry software, and the study of theoretical texts. In the course of these actions, teachers in formation were able to engage in individual or collective discussions and reflections.

The results and conclusions of the studies involving education processes highlighted the importance of these actions, characterising them as key elements for the identification and analysis of knowledge constructed and mobilised by teachers.

It should be noted that in these actions, teachers in training had the opportunity to reveal and reflect on aspects that make up their professional practice, reflect on what they produced (problem solving, lesson plans), give meaning to mistakes (validating or not the conclusions presented by colleagues), and (re) signify/ (re) construct their own teaching knowledge.

The work groups in the education projects were created as a learning space through social interactions based on respect, trust, and the commitment of the participants. The responsibility, with respect to one's own learning and that of his/her colleagues, was fostered in the groups, enabling for the mobilisation/construction of knowledge for teaching.

Nevertheless, one of the challenges that occurs in Brazilian and international projects that involve working groups (collaborative groups, communities of practice, study groups) is how to identify and relate the learning of the participants to the work developed in these groups. Robutti et al. (2016, p. 682) also point to such difficulties: "However, learning is not easy to observe. Although many of the studies report developing or changing practices that claim that learning took place, sometimes with quotations from the teachers involved, we have no consistent clarity about the ways in which learning has occurred or on the issues that have been involved for teachers. In particular, we cannot overwhelmingly claim that the reported learning is due to collaboration".

Regarding results that don't involve teacher development, Moretti and Radford (2015) conclude that the study of the history of mathematics allows for the educator to understand the limits and qualitative changes of social practices related to the historical and cultural production of concepts and to organise teaching within a cultural-historical perspective and an epistemology of knowledge. Silva and Wrobel (2015) affirm that the teachers in formation can understand the different dimensions of teaching knowledge and of being a teacher, namely: of academic knowledge (mathematical knowledge), of the relationship between academic knowledge and know-how (knowing to teach) and of subjective knowledge, involving choice and identification with the profession (actions necessary for "being" a teacher).

It is noteworthy that most of the investigations analysed in this section, in addition to discussing mathematical knowledge mobilised by teachers in education process, considered the articulation that they established between such knowledge and their teaching practice, through reflections about their actions in the classroom. In the international literature on the topic, most studies still have as their central focus on the mathematical knowledge of the teacher rather than on knowledge of how to teach those mathematical contents (Ponte & Chapman, 2008, 2015; Potari & Ponte, 2016). Brazilian studies stand out for analysing the reflections promoted by teachers in education process (in initial and continuing education) and by teacher educators and researchers on teaching practice in contexts that aim to build relationships between university and school. This is probably because many researchers are linked to postgraduate programmes in education and mathematics education.

8.3.3 Narratives and Texts of (Future) Teachers

Writing and/or the production of texts by the teacher has gained space in the practices of teacher development. There is a variety of forms of oral or written production (diaries, stories, reflective texts, autobiographies, narrative practices, pictures, among other things). This production has been taken as sources of data for research in mathematics education with the aim of identifying and analysing: stories of development or professional paths, beliefs, conceptions, and/or representations of school mathematics and its teaching. Some of these discursive forms have been present in the study group since the SIPEM IV in a sporadic form; in edition VI, they gained more emphasis, with the creation of a section specifically for the discussion of studies that take the texts of teachers (autobiographical narratives, narratives, pictures, or mathematical texts that have an educational component) as object of analysis.

In this group of four studies, two of them (Marquesin & Nacarato, 2015; Taxa-Amaro et al., 2015) were undertaken with teachers or future teachers in the initial years of elementary education, and one of them (Souza & Silva, 2015) with a mathematics teacher. These three studies availed themselves of the qualitative approach with the use of narratives as sources of data. They have in common the assumption that, through oral, written, or pictorial (drawings) narrative, the teachers can narrate their professional experiences and development trajectories, unveiling the signs of schooling or important events and teachers; explain beliefs, representations, and/or conceptions about mathematics and its teaching created in these trajectories.

They can also be taken as education practices. This was the case of the research by Marquesin and Nacarato (2015), in which the first researcher used autobiographical narratives of future teachers to organise her discipline, Foundations and Methodology of Mathematics Teaching, aiming to break with conceptions constructed during student life. Likewise, in research by Taxa-Amaro et al. (2015), the authors started with the drawings of teacher educators who were participants in a national study of teachers education on literacy (National Pact for Literacy in the Right Age—PNAIC) to develop continuing education practices. Teacher educators problematised conceptions of the mathematics classroom during the education project.

Souza and Silva (2015), based on the sexualisation of a narrative interview with a mathematics teacher, analysed the evaluative practices and possible implications of these practices in the construction of the subjects involved: teacher and students. Based on Michel Foucault's ideas, the authors emphasised the evaluation practice as a pedagogical tool of subjectification of the teacher that induces his/her development as an evaluator.

Santana and Barbosa's work (2015) was developed in a continuing education project characterised by collaborative work within OBEDUC with the aim of producing educational materials for the teaching of mathematics. The researchers analysed texts produced by elementary and university teachers within the perspective of recontextualisation proposed by Basil Bernstein, and they looked to identify the (conceptual and pedagogical) conflicts that occurred in this process when texts are dislocated from one context to another (from the university to the school and from the school context to a confrontation with other texts—reverse recontextualisation).

The results of those works show that the materials produced by teachers can constitute practices of initial or continuing teacher education allowing for: a problematisation of mathematics teaching; mapping mathematics teaching practices in different historical periods; an identification of beliefs, conceptions, and representations on mathematics and its teaching, as well as the modes of promoting ruptures; an analysis of the ways in which teachers conceive their practices; contribute to teacher education from a reflection-action-reflection perspective. Teachers' written material, in addition to constituting data for research, is also an investigation tool in and for the development of teachers, as a self-development practice.

The role of narrative production in teacher education and research is also presented in the first line of this article in Carneiro's text (2015), which studies the formative process of teachers, and in that of Passos e Souza (2015) which investigates mathematical education using computer technology.

The written and oral narratives have been used as possibilities of hearing the teacher, as ways of highlighting the teacher's voice, one that is almost always silenced, and, through it, identifying the knowledge, professional development, and work conditions of teachers. Robutti et al. (2016, p. 676), for example, analyse the presence of narratives in research with teachers: "Related to the use of narrative in an educational context, Ponte and Chapman (2008) recognises that narrative could be considered as: a tool for collecting data; an object for analysis in studying teaching; a basis for, or tool in teacher professional development or teacher education; and as a basis for reflective thinking. Thus, we see the hybrid nature of narrative as a tool both for research and for development in learning and teaching".

In this sense, there are similarities with Brazilian studies, in which narratives are taken as data sources (in the form of texts written by teachers, such as reflective records, practice narratives, development narratives, autobiographies, or in the form of narrative interviews), or as a research methodology, with contributions, for example, in the works on narrative research by Jean Clandinin and Michael Connelly. Other relevant authors are Jorge Larrosa, António Bolívar, and Walter Benjamin. Given the strong emphasis on the reflective processes present in the narratives, they have been considered sources of (self) development.

If, on the one hand, the use of narratives has been presented as a potentiator of the protagonism of the teacher, on the other, it is still a field that is being built with consequent weaknesses in the analytical processes. The difficulty lies in crossing the voices of teachers with those of researchers and with framework taken as a reference, whether for categorical analysis or narrative analysis. Without a doubt, this is a methodological perspective that needs more attention.

8.3.4 Development Practices and Difficulties and Possibilities in the Development of Mathematics Teachers

The theme "Education practices and practices of teacher educator" has been discussed since the SIPEM I (see Ferreira et al., 2000). Although there has not been a significant increase in the quantity of projects presented in the previous editions of

SIPEM VI, concerns about education practices and the practices of teacher educators have persisted over the years.[7]

In SIPEM VI, the education practices and practices of the teacher educator were highlighted in four papers (Côco & Silva, 2015; Costa & Lucena, 2015; Oliveira & Fiorentini, 2015; Silva & Cedro, 2015). In the first three works, one common aspect is the notion of practice as a privileged field of learning how to teach. In the latter, the focus is on an experiment aimed at understanding the complexity of the constitution of the teaching praxis.

While Côco and Silva (2015) and Silva and Cedro (2015) approach the internship as immersion in practice, Costa and Lucena (2015) investigate a process of continuing education in which educator and teacher come together in the classroom. Oliveira and Fiorentini (2015), on the other hand, focus on education discipline (and other types of discipline, similar in purpose) and its role in the mathematics teacher education from the perspective of the teacher educator who taught that discipline.

Although Côco and Silva (2015) and Silva and Cedro (2015) developed their studies in the context of supervised training, their methodological paths are diverse. The first authors perform a documentary analysis of three training reports (observation, participation, and regency) and show learning about planning, school management, teacher–student relationship, as well as the understanding that mathematical knowledge does not happen in an isolated way, but is integrated with other knowledge. Silva and Cedro (2015), analyse an educational experiment in which they planned, in a coordinated and intentional way, sets of teaching activities to be applied in a high school class.

Costa and Lucena (2015) investigated a continuing education project in ribeirinha[8] schools in the city of Belém (Pará), where the educators entered the classroom with a second grade (8-year-old children) teacher to develop multiplication lessons, favouring what the authors called "self-formation" or "self-creation" for all those involved in creating the project.

Oliveira and Fiorentini (2015) strive to understand what has been done in disciplines dedicated to the practice of teaching (generally called Mathematics Didactic, Methodology of Mathematics, etc.) in some undergraduate mathematics courses in Campinas[9] (São Paulo) and Rio de Janeiro[10] (Rio de Janeiro).

Three of these studies (Côco & Silva, 2015; Costa & Lucena, 2015; Oliveira & Fiorentini, 2015), especially the last, found a certain mismatch between the

[7] In SIPEM II, we find four papers in which the subject matter is dealt with in a relevant way. In SIPEM III, we have nine papers. In SIPEM IV, four papers were presented, and in SIPEM V, two papers.

[8] In Brazil, and especially in the Amazon region, riverside schools in small townships or riverside communities are, according to the National Curricula Directives for Basic Education, in the domain of Rural Education (Brasil, 2013).

[9] Campinas is a city in the state of São Paulo, located 100 km from its capital. It is the centre of a metropolitan region constituted by 19 surrounding cities. Its population is approximately one million inhabitants.

[10] Rio de Janeiro is the capital of the state of the same name, with a population of approximately 6.3 million.

disciplines related to the practice of teaching and the teacher training. Even when the disciplines are concomitant, there is a lack of articulation between these activities. These are old, but still recurrent issues, in the teaching of mathematics in the country.

Another group of studies presented in the SIPEM VI collected studies that discussed "Difficulties and possibilities in mathematics teacher education". The studies in this group do not have common aspects as well defined as those in the previous group, but this is due to the difference in dynamics of the article presentation.

Two studies (Cyrino, 2015; Pardim & Pereira, 2015) approach teacher learning through practices in which collaboration is valued. In the first study, teacher learning is studied in a collaborative project (OBEDUC) involving the university and the school. In the second, the development of professional identity of mathematics teachers is studied in four practice communities (Wenger, 1998).

Pardim and Pereira (2015) present an analysis of interactions that occurred during the OBEDUC project that showed a favourable analysis of the continuing teacher training. Based on Fiorentini studies (2013), the authors analysed the reflexive movements of teachers and students participating in cycles of collaborative study. They collected information through reflexive registration methods, narrative interviews, autobiographies, and videos. They concluded that the project contributed to a collaborative space that favoured interactions between participants in the project, as well as important moments of reflection on teaching practices.

Cyrino (2015), in turn, through an intervention research project, analysed the elements of four study groups constituted as communities of practice (Wenger, 1998) of mathematics teachers who offered opportunities for learning and identity development of its members. Through the production and collection of audio records of the training sessions, written records of the teachers, and field notes of the education sessions produced by the teacher educator, the researcher analysed the discussions in the study group, identifying the participants' learning and the elements that offered opportunities for such learning. The results showed that this practice was marked by the search for an approximation between university and school, and especially, for the opportunity for teachers and future teachers to work collectively in organising proposals and materials for the classroom, sharing experiences, studying mathematical concepts, and actively participating in their training processes.

The third study (Ferreira et al., 2015) analysed the studies produced in the period from 2001 to 2012 in the postgraduate programmes of the state of Minas Gerais.[11] This analysis show that, in recent years, there has been a significant increase in research on mathematics teacher education and a significant part of this production happens in the scope of professional master's programmes.[12] In addition, it was

[11] Minas Gerais is one of the four largest Brazilian states, both in population (about 21 million inhabitants) and in income (third richest state in the country in 2016).

[12] In Brazil, there are two types of master's programmes, one academic and one professional. The academic master's degree "focuses on developing human resources with a view to strengthening research on teaching in the country. It aims to deepen the scientific education and integrate disci-

observed that new trends are gaining ground in this scenario (e.g. studies developed in the context of distance education or indigenous education, studies that analyse and problematise the mathematical knowledge offered to the future teacher). However, important curricular changes proposed by the Brazilian government in 2001 and 2002 for undergraduate courses—and especially for teacher education courses—did not focus on any of the studies analysed, what may mean that reforms take a long time to implement.

It is observed that, despite the variety of focuses and methodological contributions, there is a tendency to value the teacher practice as an important axis in teacher education, to value collective work that integrates school teachers or future teachers and university teachers (researchers), and to value the search for alternatives to overcome obstacles in order to improve the education processes. There are interrelations between the axes analysed here, showing that there are multiple theoretical and methodological possibilities used in the teacher education and professional development processes. For example, the work of Pardim and Pereira (2015) brings elements that were previously discussed: the role of the collaborative group in teacher education.

8.4 Final Remarks

What does this set of studies synthesised here reveal? What are the perspectives and challenges to be faced?

It can be said that there is a broad and dynamic process of teacher education that has culminated in a diversity of research foci as well as in multiple methodological approaches to production and data analysis.

Some of the themes that have been recurrent in Brazilian research in the field of teacher education in the last 20 years are: teacher education processes; knowledge required for teaching; teacher learning; professional development; and the formation of professional identity. Many of them have endured new theoretical perspectives, resulting, on the one hand, from the singularity of the Brazilian context (public policies of education, conditions of teaching work, reforms in undergraduate courses, etc.), and on the other, from the appropriation of international theorists that broaden previous discussions. In this sense, we can highlight the works of Ball et al. (2008), Mishra and Koelhler (2006) and Carrillo et al. (2015), amplifying the discussions of Lee Shulman and Maurice Tardif that marked the first Brazilian studies on the subject of teacher knowledge.

---

plinary and teaching knowledge and research results in the area of teaching to teaching practices". The professional degree focuses on "the application of knowledge, that is, in applied research and in the development of products and processes that are implemented in real teaching conditions [...] The final project should necessarily include a detailed description of this experience of which the educational product developed is an integral part" (document from the education area, available at: www.capes.gov.br/images/stories/download/avaliacaotrienal/Docs_de_area/Ensino_doc_area_e_comiss%C3%A3o_block.pdf).

Other theoretical approaches are emerging in Brazilian research, especially in *concept study*, proposed by Brent Davis and Moshe Renert. Giraldo, Rangel, Quintaneiro, and Matos (2015, p. 5), based on these authors, consider it "a model of continuing education, structured as a collective and collaborative study, focused on content, in which groups of teachers share the knowledge that emerges of their own practice, aiming to question and (re) build their knowledge of mathematics for teaching."

There is a significant number of studies of programmes where education occurs in groups of teachers "working together". We observe that both nationally and internationally there is a movement to investigate what kind of learning is promoted in these groups and how this learning relates to the type of work developed in them (Robutti et al., 2016). It is worth noting how much the research model has focused on research with the teacher and not for or about the teacher, as highlighted by Nacarato and Grando (2015). In groups or communities of practice, teachers take on the role of teacher educator, of professional and curricular development, breaking with the paradigm of technical rationality. However, these activities rarely mention the tensions that exist in groups. Many situations in these studies point to important processes with teachers, and, in some of these, teachers are placed in positions of vulnerability, exposing their insecurities and difficulties. Not the vulnerability that weakens and paralyses, but the one that allows one, for a moment, to suspend certainties and convictions (Oliveira & Cyrino, 2011). In these cases, the relationship between the participants and the educator can play a determining role in the search for a sense of agency. To this end, it is fundamental to investigate the role of the teacher educator and the tasks that are proposed in these education spaces. What is the role of the teacher educator in cultivating spaces in which those involved can organise, discuss, and reflect on proposals that can be implemented in the classroom and that promote the development of their professional identity (Cyrino, 2016; Oliveira & Cyrino, 2011)?

There are multiple sources for data production: interviews, class records, teachers' writings, field diaries, audio/video recording of classes and work groups, and use of oral or written narratives. The studies involving narratives have served as a source of investigation for the researcher and as a research tool in and for the development of teachers. However, as the use of this type of source is recent, answers to the following questions remain open in Brazilian research: how can narratives impact on the development of the teacher educator and of the teachers in education process? How to identify these impacts?

Most of the studies, besides establishing dialogues with the international and national literature, also show the national context, the working conditions of the teacher, and the role of public policies in the development and action of the teacher, although, in this event, there were no research projects discussing the process of teacher professionalisation in this scenario.

The projects analysed reflect a syntony between public policies—such as PIBID and OBEDUC—and teacher education and development practices, connecting, in several cases, initial and continuing education, in-service teachers and future teachers, as well as researchers.

One of the challenges Brazilian researchers face is the development of large-scale studies on teacher education, such as "Mathematics Teaching in the 21st Century" (MT21) (Schmidt, Blömeke, & Tatto, 2011) and TEDS-M (Tatto et al., 2010, 2012), which includes the participation of several countries.

In summary, the studies presented in SBEM's Working Group 07 in SIPEM VI reveal an intense involvement between the Brazilian scientific community in the area of mathematics education and national and international trends that mark the important practices of teacher education and the education of teacher educator. Such involvement does not always mean agreeing with the political decisions that influence the development of initial and continuing education, but it implies the construction of reflections about such a scenario and careful analysis of its impact on society. Often, as pointed out by Nacarato (2016), collaborative work spaces allow teachers to create resistance tactics against public policies in favour of quality education for their students.

References

Ball, D., Thames, M. H., & Phelps, G. (2008). Content knowledge for teaching: What makes it special? *Journal of Teacher Education, 59*(5), 389–407.

Blömeke, S., & Delaney, S. (2012). Assessment of teacher knowledge across countries: A review of the state of research. *ZDM-International Journal on Mathematics Education, 44*, 223–247.

Brasil. Ministério da Educação. Conselho Nacional de Educação. (2015). Resolução n. 2, de 1° de julho de 2015. Define as Diretrizes Curriculares Nacionais para a formação inicial em nível superior e para a formação continuada.

Brasil. Ministério da Educação. Secretaria de Educação Básica. Diretoria de Currículos e Educação Integral. (2013). *Diretrizes Curriculares Nacionais da Educação Básica*. Brasília: MEC, SEB, DICEI.

Carneiro, R. F. (2015). A formação de professores dos anos iniciais em um curso de pedagogia: contribuições das disciplinas de matemática. In *VI Seminário Internacional de Pesquisa em Educação Matemática* (pp. 1–11). Pirenópolis: SBEM.

Carrillo, J., Climent Rodríguez, N., Contreras González, L. C., & Montes Navarro, M. (2015). *Un marco teórico para el Conocimiento Especializado del Profesor de Matemáticas*. Huelva: Universidad de Huelva Publicaciones.

Clark, K. M. (2012). History of mathematics: Illuminating understanding of school mathematics concepts for prospective mathematics teachers. *Educational Studies in Mathematics, 81*, 67–84.

Cochran-Smith, M., & Lytle, S. (2009). *Inquiry as stance: Practitioner research for the next generation*. New York, London: Teachers College Press.

Côco, D., & Silva, S. A. F. (2015). Estágio supervisionado e aprendizagem da docência: ações e reflexões de licenciandos de matemática. In *VI Seminário Internacional de Pesquisa em Educação Matemática* (pp. 1–12). Pirenópolis: SBEM.

Costa, L. F. M., & Lucena, I. C. R. (2015). Práticas formativas como elos entre professores e formadores em autoformações. In *VI Seminário Internacional de Pesquisa em Educação Matemática* (pp. 1–12). Pirenópolis: SBEM.

Cyrino, M. C. C. T. (2015). Desenvolvimento da Identidade Profissional de Professores em Comunidades de Prática: Elementos da Prática. In *VI Seminário Internacional de Pesquisa em Educação Matemática* (pp. 1–11). Pirenópolis: SBEM.

Cyrino, M. C. C. T. (2016). Mathematics Teachers' Professional Identity Development in Communities of practice: Reifications of proportional reasoning teaching. *BOLEMA, 54*(30), 165–187.

Davis, B., & Renert, M. (2012). Profound understanding of emergent mathematics: Broadening the construct of teachers' disciplinary knowledge. *Educational Studies in Mathematics, 29*(3), 37–43.

Fanizzi, S., & Santos, V. M. (2015). Políticas públicas de formação continuada de professores dos anos iniciais em matemática: uma experiência da Secretaria Municipal de Educação de São Paulo. In *VI Seminário Internacional de Pesquisa em Educação Matemática* (pp. 1–11). Pirenópolis: SBEM.

Ferreira, A. C., Lopes, C. A. E., Fiorentini, D., Jaramillo, D., Melo, G. F. A., Carvalho, V. E, & Santos-Wagner, V. M. (2000). Estado da arte e da pesquisa brasileira sobre a formação de professores que ensinam matemática: Uma primeira aproximação. In SBEM. *Livro de resumos do I SIPEM* (pp. 264–271). Serra Negra: SBEM.

Ferreira, A. C., Coura, F. C. F., & Barbosa, C. P. (2015). Uma Análise da Produção Mineira sobre a Licenciatura em Matemática: Considerando as Teses e Dissertações Defendidas entre 2001 e 2012. In *VI Seminário Internacional de Pesquisa em Educação Matemática* (pp. 1–12). Pirenópolis: SBEM.

Fiorentini, D. (2005). A Formação matemática e didático-pedagógica nas disciplinas da Licenciatura em Matemática. *Revista de Educação, Campinas, SP, 18*(1), 107–115.

Fiorentini, D. (2013). Pesquisar Práticas Colaborativas ou Pesquisar Colaborativamente? In M. C. Borba & J. L. Araújo (orgs.). *Pesquisa Qualitativa em Educação Matemática.* (5° ed, pp. 53–85). Belo Horizonte: Autêntica.

Galleguillos, J., Montes, M., & Ribeiro, C. M. (2015). Situaciones de investigación matemática y el conocimiento especializado del professor. In *VI Seminário Internacional de Pesquisa em Educação Matemática* (pp. 1–13). Pirenópolis: SBEM.

Gatti, B. A., & Barreto, E. S. S. (2009). *Professores do Brasil: impasses e desafios*. Brasília: UNESCO.

Geraldi, C. M. G., Fiorentini, D., & Pereira, E. M. A. (Orgs.). (1998). *Cartografias do trabalho docente: Professor(a)-pesquisador(a).* Campinas, São Paulo: Mercado de Letras.

Giongo, I. M., Rehfeldt, M. J. H., & Quartieri, T. (2015). Formação continuada e professores da escola básica: movimentos de parceria universidade-escola. In *VI Seminário Internacional de Pesquisa em Educação Matemática* (pp. 1–10). Pirenópolis: SBEM.

Giusti, N. M. R., & Justo, J. C. R. (2015). Processos colaborativos e cooperativos interseccionados na formação em serviço de professores dos anos iniciais. In *VI Seminário Internacional de Pesquisa em Educação Matemática* (pp. 1–11). Pirenópolis: SBEM.

Giraldo, V., Rangel, L., Quintaneiro, W., & Matos, D. (2015). Matemática elementar e investigação de conceito: estabelecendo relações. In *VI Seminário Internacional de Pesquisa em Educação Matemática* (pp. 1–13). Pirenópolis: SBEM.

Hossin, S., Mendrick, H., & Adler, J. (2013). Troubling "understanding mathematics in-depth": Its role in the identity work of student-teachers in England. *Educational Studies in Mathematics, 84*, 32–48.

Jaworski, B. (2008). Building and sustaining inquiry communities in mathematics teaching development. In K. Krainer & T. Wood (Eds.), *Participants in mathematics teacher education: Individuals, teams, communities and networks. The international handbook on mathematics teacher education* (Vol. 3, pp. 309–330). Rotterdam, The Netherlands: Sense Publisher.

Kemczinski, A., Marek, J., Hounsell, M. S., & Gasparini, I. (2007). Colaboração e cooperação—Pertinência, concorrência ou complementaridade. *Revista Produção, 7*(3), 1–15.

Kopnin, P. V. (1978). *A dialética como lógica e teoria do conhecimento.* Rio de Janeiro: Editora Civilização Brasileira.

Lamon, S. J. (2012). *Teaching fractions and ratios for understanding: Essential content knowledge and instructional strategies for teachers* (3rd ed.). New York: Routledge.

Leontiev, A. N. (1978). *O desenvolvimento do psiquismo*. Lisboa: Livros Horizonte.
Leontiev, A. N. (1983). *Actividad, Conciencia, Personalidad*. Ciudad de La Habana: Editorial Pueblo Y Educación.
Leontiev, A. N. (2001). Os princípios psicológicos da brincadeira pré-escolar. In L. S. Vygotsky et al. (Eds.), *Linguagem, desenvolvimento e aprendizagem* (pp. 119–142). São Paulo: Ícone.
Lerman, S. (2000). The social turn in mathematics education research. In J. Boaler (Ed.), *Multiple perspectives on mathematics teaching and learning* (pp. 20–44). Westport, CT: Ablex.
Lesh, R., Post, T., & Behr, M. (1988). Proportional reasoning. In J. Hiebert & M. Behr (Eds.), *Number concepts and operations in the middle grades* (pp. 93–118). Reston, VA: Lawrence Erlbaum & National Council of Teachers of Mathematics.
Marco, F. F., Lopes, A. R. L. V., & Sousa, M. C. (2015). Projeto Formativo na Atividade Pedagógica do Professor que Ensina Matemática. In *VI Seminário Internacional de Pesquisa em Educação Matemática* (pp. 1–11). Pirenópolis: SBEM.
Marquesin, D. F. B., & Nacarato, A. M. (2015). Minha Época de Escola: Narrativas de Futuras Professoras. In *VI Seminário Internacional de Pesquisa em Educação Matemática* (pp. 1–12). Pirenópolis: SBEM.
Mishra, P., & Koelhler, M. J. (2006). Technological pedagogical content knowledge: A framework for teacher knowledge. *Teachers College Record, 108*(6), 1017–1054.
Moretti, V. D., & Radford, L. (2015). História do conceito culturalmente significada e a organização da atividade de ensino de matemática. In *VI Seminário Internacional de Pesquisa em Educação Matemática* (pp. 1–12). Pirenópolis: SBEM.
Munhoz, A. V., & Giongo, I. M. (2014). *Observatório da Educação I: tendências no ensino da matemática*. Porto Alegre: Evangraf.
Nacarato, A. M., Mengali, B. L., & Passos, C. L. B. (2009). *A matemática nos anos iniciais do ensino fundamental: Tecendo fios do ensinar e do aprender*. Belo Horizonte: Autêntica.
Nacarato, A. M., & Grando, R. C. (2015). A Pesquisa com e pelas professoras que ensinam Matemática. In *VI Seminário Internacional de Pesquisa em Educação Matemática* (pp. 1–12). Pirenópolis: SBEM.
Nacarato, A. (2016). A parceria universidade-escola: Utopia ou possibilidade de formação continuada no âmbito das políticas públicas? *Revista Brasileira de Educação, 21*(66), 699–716.
Oliveira, H., & Cyrino, M. C. C. T. (2011). A formação inicial de professores de Matemática em Portugal e no Brasil: Narrativas de vulnerabilidade e agência. *Interacções, 7*, 104–113.
Oliveira, L. M. C. P., & Cyrino, M. C. C. T. (2015). Aprendizagens a respeito do raciocínio proporcional em uma comunidade de prática de professores matemática. In *VI Seminário Internacional de Pesquisa em Educação Matemática* (pp. 1–12). Pirenópolis: SBEM.
Oliveira, A. T. C. C., & Fiorentini, D. (2015). Saberes e práticas de formadores de professores de matemática: O papel e o lugar da didática especial de matemática no curso de licenciatura. In *VI Seminário Internacional de Pesquisa em Educação Matemática* (pp. 1–12). Pirenópolis: SBEM.
Pardim, J. F. S., & Pereira, P. S. (2015). Interações entre os integrantes da universidade-escola em um projeto colaborativo. In *VI Seminário Internacional de Pesquisa em Educação Matemática* (pp. 1–10). Pirenópolis: SBEM.
Passos, C. L. B., & Souza, A. P. G. (2015). Formação inicial/continuada de professores dos anos iniciais: tecnologias informáticas e matemática. In *VI Seminário Internacional de Pesquisa em Educação Matemática* (pp. 1–11). Pirenópolis: SBEM.
Ponte, J. P., & Chapman, O. (2008). Preservice mathematics teachers' knowledge and development. In L. D. English (Ed.), *Handbook of international research in mathematics education: Directions for the 21st century* (2nd ed., pp. 225–263). New York: Routledge.
Ponte, J. P., & Chapman, O. (2015). Prospective mathematics teachers' learning and knowledge for teaching. In L. English & D. Kirshner (Eds.), *Handbook of international research in mathematics education* (3rd ed., pp. 275–296). New York: Routledge.
Potari, D., & da Ponte, J. P. (2016). Current research on prospective secondary mathematics teachers' knowledge. In M. E. Strutchens, L. Losano, D. Potari, R. Huang, J. P. da Ponte,

M. C. C. T. Cyrino, K. R. Leatham, B. E. Peterson, & R. M. Zbiek (Eds.), *The mathematics education of prospective secondary teachers around the world* (1st ed., pp. 5–15). Cham, Switzerland: Springer International Publishing.

Powell, A. B., & Pazuch, V. (2015). Tarefas, Colaboração e Justificações Geométricas de Professores no Ambiente VMTcG1. In *VI Seminário Internacional de Pesquisa em Educação Matemática* (pp. 1–13). Pirenópolis: SBEM.

Radford, L. (2006). Elementos de una teoría cultural de la objetivación. *Revista Latinoamericana de Investigación en Matemática Educativa, Special Issue on Semiotics, Culture and Mathematical Thinking*, pp. 103–129.

Radford, L. (2013). Three key concepts of the theory of objectification: Knowledge, knowing, and learning. *Journal of Research in Mathematics Education, 2*(1), 7–44.

Ribeiro, A. J., & Oliveira, F. A. P. V. S. (2015). Conhecimentos Matemáticos dos Professores e o Ensino de Equações: Uma Investigação Acerca do Planejamento de Aulas para a Educação Básica. In *VI Seminário Internacional de Pesquisa em Educação Matemática* (pp. 1–13). Pirenópolis: SBEM.

Robutti, O., Cusi, A., Clark-Wilson, A., Jaworski, B., Chapman, O., Esteley, C., et al. (2016). ICME international survey on teachers working and learning through collaboration: June 2016. *ZDM Mathematics Education, 48*, 651–690.

Santana, F. C. M., & Barbosa, J. C. (2015). Trabalho colaborativo com professores de matemática: uma análise dos conflitos entre/nos textos enunciados pelos participantes. In *VI Seminário Internacional de Pesquisa em Educação Matemática* (pp. 1–13). Pirenópolis: SBEM.

Santos, M. B. Q. C. P. (2015). Professores que ensinam matemática: Diálogos pertinentes entre a licenciada em matemática e a pedagoga. In *VI Seminário Internacional de Pesquisa em Educação Matemática* (pp. 1–12). Pirenópolis: SBEM.

Schmidt, W. H., Blömeke, S., & Tatto, M. T. (2011). *Teacher education matters: A study of middle school mathematics teacher preparation in six countries*. New York: Teacher College Press.

Shulman, L. S. (1986). Those who understand: Knowledge growth in teaching. *Educational Researcher, 15*, 4–14.

Silva, M. M., & Cedro, W. L. (2015). Formação de professores que ensinam matemática: O estágio supervisionado como um dos espaços de constituição da práxis docente. In *VI Seminário Internacional de Pesquisa em Educação Matemática* (pp. 1–12). Pirenópolis: SBEM.

Silva, S. F. A., & Wrobel, J. S. (2015). Saberes necessários aos docentes sob o olhar de professores de matemática em formação. In *VI Seminário Internacional de Pesquisa em Educação Matemática* (pp. 1–10). Pirenópolis: SBEM.

Sousa, M. C. (2010). O PIBID, área Matemática da Ufscar: contribuições para a formação docente. In XV ENDIPE—Encontro Nacional de Didática e Prática de Ensino, Belo Horizonte. *Anais*...Belo Horizonte: MG, 2010. Cd-Room.

Souza, D. M. X. B., & Silva, A. A. (2015). Efeitos que Induzem e Produzem uma Ferramenta Pedagógica na Formação de uma Professora e de seus Alunos. In *VI Seminário Internacional de Pesquisa em Educação Matemática* (pp. 1–11). Pirenópolis: SBEM.

Stahl, G. (Ed.). (2009). *Studying virtual math teams*. New York: Springer.

Tardif, M. (2002). *Saberes docentes e formação profissional*. Petrópolis, RJ: Vozes.

Tatto, M. T., Lerman, S., & Novotna, J. (2010). The organization of the mathematics preparation and development of teachers: A report from the ICMI study 15. *Journal of Mathematics Teacher Education, 13*(4), 313–324.

Tatto, M. T., Peck, R., Schwille, J., Bankov, K., Senk, S. L., Rodriguez, M., et al. (2012). *Policy, practice, and readiness to teach primary and secondary mathematics in 17 countries: Findings from the IEA Teacher Education and Development Study in Mathematics (TEDS-M)*. Amsterdam: International Association for the Evaluation of Educational Achievement (IEA).

Taxa-Amaro, F. O. S., Perovano, A. P. R., Bortoldi, R. A. M., & Santana, E. R. S. (2015). Narrativas Dos Professores-Formadores: Entrecruzando o processo de alfabetização matemática. In *VI Seminário Internacional de Pesquisa em Educação Matemática* (pp. 1–16). Pirenópolis: SBEM.

Tinti, D. S. (2012). *PIBID: um estudo sobre suas contribuições para o processo formativo de alunos de Licenciatura em Matemática da PUC-SP*. Dissertação (Mestrado em Educação Matemática). Pontifícia Universidade Católica, São Paulo.

Viñao, A. (2008). A história das disciplinas escolares. *Revista Brasileira de História da Educação, 18*, 173–215.

Vygotsky, L. S. (2002). *A formação social da mente* (6th ed.). São Paulo: Martins Fontes.

Wenger, E. (1998). *Communities of practice: Learning, meaning, and identity*. New York: Cambridge University Press.

Chapter 9
Assessment and Mathematics Education: Possibilities and Challenges of Brazilian Research

Maria Isabel Ramalho Ortigão, João Ricardo Viola dos Santos, and Jader Otávio Dalto

Abstract This chapter portrays Brazilian research that connects the fields of assessment and mathematics education. It reflects on what has been investigated in the scope of the working group on Assessment and Mathematics Education of the Brazilian Society of Mathematics Education since its conception in 2000. The research projects that have formed the basis of the group's work over the six editions of the International Mathematics Education Research Seminar (SIPEM) have been classified into the following categories: learning assessment, assessment as investigation practice, assessment and inequality processes and assessment and teacher training. Before detailing the approaches that characterise each category, an overview of the Brazilian assessment system is presented to situate the reader. More than providing solutions, this chapter attempts to consider research practices regarding mathematics education based on theoretical and methodological discussion and shared research experiences. It calls attention to some changes in school mathematics education, indicating some challenges associated with the integration and articulation between assessment, teaching and learning, the assessment criteria that teachers adopt, and how are they related to the development of learning and the articulations that can be established between internal and external assessment.

M. I. R. Ortigão (✉)
Universidade do Estado do Rio de Janeiro, Rio de Janeiro, Brazil

J. R. V. dos Santos
Universidade Federal de Mato Grosso do Sul, Campo Grande, Brazil

J. O. Dalto
Universidade Tecnológica Federal do Paraná, Cornélio Procópio, Brazil

A. J. Ribeiro et al. (eds.), *Mathematics Education in Brazil*,
https://doi.org/10.1007/978-3-319-93455-6_9

9.1 Beginning Discussion: Assessment as a Research Field in Education

In the last decades, assessment has gained more and more centrality in the field of education and, specifically, in mathematics education. There is a significant set of research conducted within a confluence of distinct theoretical and methodological perspectives and inscribed in different epistemological approaches. This diversity of perspectives has contributed to deepen and enrich the theoretical referential that grounds assessment research processes and practices, enabling a dynamic debate about the decisions that mediate the field development, as well as stimulating the development of investigation routes that help both thinking about education and mathematics education and problematising mathematics teaching, learning and assessment in the school.

Several countries identify educational assessment systems, conducted at national or regional level, as the main reason for inserting the topic of assessment in the centre of discussions about education. Since the early 1960s, a significant set of research on education assessment has shed light on the implications of school mechanisms of sociocultural discrimination and domination on social inequalities. Works by Coleman (1966), Anyon (1978), and Lee and Bryk (1989) in the United States, by Bourdieu and Passeron (1975) in France and by Franco and Ortigão (2007) and Ortigão, Franco, and Carvalho (2007) in Brazil are examples of this approach.

In a different perspective, other researchers in many countries have investigated the relationship between assessment and the educational reforms that occurred in the 1980s. The recent book edited by Nigel Brooke (2012)[1] brings a collection of texts that show strong similarities between these reforms—"as if they were following a prescription of educational policies, some kind of orchestration or at least a history of common origins" (Brooke, 2012). Brooke warns about the plurality of meanings that "educational reform" carries. For him, when understood in the sense of a change in educational policy to influence directions, a multiplicity of foci and very broad cases are observed. There would be no ways to create an organisation capable of dealing with so much variety.

On the other hand, there is quite strong criticism related to the forms of assessment that have been used, from the metrics used to monitor students' performance to methods employed to assess educational systems. Among this criticism, some studies that show that external assessments are guided by a market logic that justifies meritocratic practices, leading to a narrow and reduced understanding of the meaning of "quality", inducing a standardisation of curricula production, silencing differences and leading to a process of homogenisation of educational systems

[1] The book "Historical marks in Education Reform", organised by Nigel Brooke, presents a collection of articles that allows tracing an evolution of the main ideas behind educational reforms occurred in several countries in the last five decades, which, according to the author, changed the way of thinking education.

(Fernandes, 2009; Freitas, 2012; Ortigão & Pereira, 2016; Ravitch, 2011; Tura & Pereira, 2013, among others).

Educational policies in many countries, including Brazil, are presently influenced by assessment. The disclosure of external assessment results highlights school mathematics, showing the low performance achieved by students at different education levels. However, very little is said about unequal conditions such as the origin of students and their families, which are the basis of the construction of the inequality of the results.

It is in this scenario that, in the scope of Brazilian mathematics education, the first works aiming to investigate other possibilities for the assessment practices of teachers who teach mathematics in basic education arose. These studies indicate some strategies for teachers, such as the possibility of taking learning activities as assessment activities and placing them in the centre of the pedagogical process. Much of this research considers students' mistakes as a way of producing meanings that may be taken as a resource for mathematical discussions and learning (Borasi, 1996; Buriasco, 2004; Cury, 2007). There are also studies focusing on students' written productions, which highlight the role of context and language in the statements of the proposed activities, and enable reflections about students' ways of interpreting and solving mathematical problems, as well as their methods of elaborating strategies and using procedures (Dalto & Buriasco, 2009; Cooper & Harries, 2002, 2003; D'Ambrosio, Kastberg, & Lambdin, 2007).

Other investigations have analysed alternative possibilities for written assessment (Buriasco, 1995, 2000, 2004; Trevisan, 2013). In this direction, an assessment in phases can be used as a training process in which teachers are led to analyse and reflect on students' productions and to elaborate questions and comments that may help students to rethink their own solution processes.[2]

This chapter portrays Brazilian research involving connections between assessment and mathematics education fields. More specifically, the authors propose to reflect on what has been investigated in the framework of the Working Group Assessment and Mathematics Education of the Brazilian Society of Mathematics Education. From the analysis of the information presented in the six proceedings of the Seminário Internacional de Pesquisa em Educação Matemática—SIPEM (International Mathematics Education Research Seminar), the texts were grouped as follows: learning assessment, assessment as investigation practice, assessment and inequality processes and assessment and teacher training. This classification came from a mapping based on theoretical perspectives guiding the surveys and the main assessments foci on each of them. This analysis covers all papers presented in all editions of SIPEM, although our primary focus is to analyse those in the last edition (2015) of the proceedings. In sequence, given the centrality gained by

[2] For other information on test in phases, see works produced in the scope of GEPEMA—Grupo de Estudo e Pesquisa em Educação Matemática e Avaliação (Study and Research Group in Assessment and Mathematics Education), coordinated by Professor Regina Luzia Corio de Buriasco (Universidade Estadual de Londrina—UEL (Londrina State University), Londrina, Paraná, Brazil).

assessment policies in the present scenario, we situate the reader as to the Brazilian assessment system.

9.2 National Assessment System in Brazil

The establishment of an assessment system for Brazilian education begun in the late 1980s, with the recognition of the nonexistence of studies that would clearly show the educational services offered to the population. The general objective was to map how, when and who has access to quality education.

The notion of quality, as a new competitive strategy for education, marks the construction of assessment systems in several countries, including Brazil. This logic is based on the idea that quality is obtained through "measuring, comparing and assessing knowledge/learning patterns to achieve the measurement/comparison/assessment of people" (Lopes, 2015, p. 455).

The first Brazilian large-scale assessment experience started in 1990 with the creation of the *Sistema de Avaliação do Ensino Público (SAEP)* (Public Teaching Assessment System). At two consecutive moments (1990 and 1992), SAEP applied tests[3] and questionnaires to a sample of infant school students in state schools. In 1995, the *Sistema de Avaliação da Educação Básica—SAEB*[4] (Basic Education Assessment System), the first great reform of the assessment system took place, covering in its sampling plan not only the state network but the private education and involving Elementary, Middle and High School levels, in all state schools. Thus, students in the final years of each school level—5th and 9th grades of elementary and middle levels—were assessed through mathematics and reading tests. It was also in this year that Brazilian assessment incorporated the item response theory (IRT) as a methodological approach in the construction and analysis of information collection tools—tests and questionnaires.

Unlike the classical theory of measurements (Vianna, 1987), which takes into account the score of the individuals in the tests as a whole, the item response theory (IRT) method is assessed based on answers given to each item. In this way, IRT makes possible comparisons both between differentiated populations, if only subjected to tests with common items, and between individuals that integrate the same population and who are subject to totally different tests.

With this, it is possible to assess the results of tests applied to different types of schools, as well as student performance from one school year to the other. These advantages explain how the national assessment has advanced towards greater theoretical and methodological sophistication. Another advance is the use of assessment

[3] The first assessment involved test application (in Mathematics, Reading, Science, Geography and History) to students in the 3rd, 5th and 7th grades of elementary and middle school.

[4] As from this moment, SAEB is conducted by the *Instituto Nacional de Estudos e Pesquisas Educacionais Anísio Teixeira—INEP* (Anísio Teixeira National Institute for Educational Studies and Research), an authority linked to the Ministry of Education (MEC).

reference matrices, which direct what will be assessed in each area of knowledge from the assessment system. These matrices are organised to describe the skills and abilities in articulation to the content that each student must master. The main justification for matrix elaboration is the need to establish evidence from consensual parameters.

Since the mid-1990s, by means of large-scale assessment, an inductive process in Brazilian curricula started. Assessment gained centrality and began to be conceived as an instrument that would boost the quality of the education. In 2005, the Brazilian government extended the external assessment role with the creation of *Prova-Brasil* (Brazil-Test), a census system to assess Brazilian state schools. It thus expanded the scope of the results, offering information not only for the whole country but also for the states, cities and individual schools. *Prova-Brasil* made possible the creation of the *Índice de Desenvolvimento da Educação Básica—IDEB* (Basic Education Development Index), a measure of the ratio between the average score of students in the national exam and school pass rate.

The *Sistema Nacional de Avaliação da Educação Básica—SAEB* (National System for Basic Education Assessment)—includes two assessment systems: Avaliação Nacional da Educação Básica—ANEB (National Basic Education Assessment) and Avaliação Nacional do Rendimento Escolar—ANRESC (National School Performance Assessment), also known as Prova-Brasil. Recently, a change in SAEB took place, to cover the last year of the Brazilian literacy circle[5] (1st to 3rd years of Elementary School). This occurred with the creation of the Avaliação Nacional da Alfabetização—ANA (National Literacy Assessment), which focuses on assessing students from third grade of public Elementary Schools, extending the role of SAEB to all Brazilian Basic Education.

At the same time, many Brazilian states and/or municipalities have formed their own assessment systems. In general, these systems are conducted by private companies which took responsibility for organising and conducting the assessment, as well as reporting on the results obtained by each participating school.

Student's performance at the end of Basic Education is also estimated through the Exame Nacional do Ensino Médio—ENEM (National Final Year School Exam), also conducted by INEP, which is used in allocating access to state and federal universities.

Since 2000, Brazil has participated in the Programme for International Student Assessment (PISA), conducted by the Organisation for Economic Cooperation and Development (OECD), enabling a worldwide comparison of results.

In these 25 years of coexistence with large-scale evaluation, assessment has become a central element in the regulation of education. Such centralisation has been identified as responsible for producing a process of erasing difference. To Ball (2005), this process causes increased individualisation, including destruction of

[5] According to Mainardes (2015), different kinds of education organisation cycles have been created since the 1980s, including the literacy cycle (Ciclo de Alfabetização), which includes the first 3 years of elementary school. The author considers the creation of this cycle as a relevant proposal to ensure a better literacy, based on one fundamental principle: learning as a continuous process.

solidarities based on common professional identity. More than that, large-scale evaluation induces a homogenisation (Ortigão & Pereira, 2016) that, in the name of equal opportunities, impedes teachers and students from "creative insubordination" (D'Ambrosio & Lopes, 2015).

In the next section, we describe concisely the focus of research conducted by the Working Group Assessment and Mathematics Education.

9.3 Assessment as a Research Field in Mathematics Education in Brazil

The Working Group Assessment and Mathematics Education (WG08) was formed at the first *Seminário Internacional de Pesquisa em Educação Matemática—SIPEM I* (International Mathematics Education Research Seminar), which took place in the city of Serra Negra (São Paulo—Brazil), in November 2000. SBEM organises SIPEM every 3 years; the last was held in the city of Pirenópolis (Goiás—Brazil) in 2015. WG08 had the specific objective of gathering researchers and/or research groups to discuss theoretical and methodological issues related to the interrelationship between educational assessment and mathematics education. Table 9.1 summarises the number of contributions presented in the different editions of the event in the course of these 16 years.

The contributions published in WG08 in the first three SIPEM proceedings addressed the following research issues:

- The role of assessment in the classroom and its relations to school failure, teachers' perspective about assessment processes that are conducted in mathematics classrooms (SIPEM, 2000);
- Assessment practices in mathematics classrooms; assessments trends in the educational context; error potentialities in learning assessment; discussions on large-scale assessments (SIPEM, 2003);
- Analysis of the written production of basic education students, memories of undergraduates on their assessment processes in basic education (SIPEM, 2006).

Table 9.1 Number of works contributions in WG08—Assessment and Mathematics Education—SIPEM I-VI, by focus

	Number of works presented in each SIPEM edition					
Focus	2000	2003	2006	2009	2012	2015
Learning assessment	1	8	2	–	1	1
Assessment as investigation practice	–	2	4	6	3	2
Assessment and inequalities	1	–	–	–	1	2
Assessment and teacher training	1	–	1	–	3	4
TOTAL	3	10	7	6	8	9

In turn, papers published in the three last SIPEM proceedings approached the following themes:

- Analysis of students' and teachers' written production, analysis of students' errors, references of external assessment results for the practice of the teacher who teaches mathematics, the relationship between literacy and numeracy which assisted in building the new matrix on which the research of the *Indicador Nacional de Alfabetismo Funcional—Inaf* (National Indicator of Functional Literacy) is based (SIPEM, 2009);
- Analysis of written productions, test in phases as assessment tool in the realistic mathematics education approach, choice of textbooks (SIPEM, 2012);
- Assessment as investigation practice, error analysis, contributions of written production analysis in the development of assessment-related knowledge; differences between genders in mathematics average scores by means of large-scale assessments, school failure and elements that impact students' learning development, assessment practices of teachers who teach mathematics, exclusively assessment (SIPEM, 2015).

So far this item has presented the main themes discussed in order to build a historical line of possibilities and challenges in this research field—assessment and mathematics education—in the scope of WG08. Consequently, by means of an analysis of these contributions, we grouped the themes in four classes: (1) learning assessment, (2) assessment as investigation practice, (3) assessment and inequality processes and (4) assessment and teacher training.

9.3.1 Learning Assessment

Concerns with learning assessment in mathematics appear in research presented in WG08 since the first edition of the event. The initial question was whether another assessment form that could set an alternative to the predominant assessments would be possible. Experiments using different assessment tools were conducted and their potential analysed. In this perspective, investigations involving error diagnosis and students' written production analysis gained importance. In line with the education field at national[6] and international levels, assessment is understood, in the WG scope, as a constitutive and integrated part of the teaching-learning process, with the main function of helping students and teachers to learn and develop their learning. For the teacher, it is understood as a process that allows a diagnosis both of his/her students' development and his/her own educational planning:

> The assessment of the students' learning paths in the classroom context is only significant if articulated with learning and teaching. It will be hard to advance markedly in a domain so intrinsically pedagogical such as assessment without investigating its relations with

[6] See, for example: Esteban et al. (2010); Esteban (2012); Fernandes (2014).

learning, with teaching and with the existing dynamics and environments in the classrooms (Fernandes, 2009, p. 132).

We agree with Fernandes and advocate that an adequate integration between the three processes allows, or should allow, the monitoring of teaching and learning, using tasks that, simultaneously, aim to teach, learn, evaluate and contextualise the assessment, thereby merging assessment tasks and teaching objectives.

It is with this understanding of assessment that the study *Avaliação e Ensino na Educação Básica em Portugal e no Brasil: Relações com as Aprendizagens (AERA)* (Assessment and Teaching in Basic Education in Portugal and in Brazil: Relationships with Learning), an international cooperation between Universidade Federal do Pará (Federal University of Pará), in Brazil and Universidade de Évora (Évora University), in Portugal (Borralho & Lucena, 2015) was conducted.

AERA aims to understand the elements that impact upon student's learning development. It considers the classroom as a unit of analysis and involves information gathered from teachers and students in early elementary school grades (7–10 years of age) in Portuguese (Évora region) and Brazilian (Amazon region) schools. According to the authors, the following are considered: (a) teachers' teaching and assessment practices; (b) teachers' perceptions on teaching, assessment and learning, (c) the nature of the assessment tasks used in the classroom, (d) frequency, distribution and nature of the *feedback* used, and (e) students' participation in teaching and learning processes.

The study adopts methods of both qualitative and quantitative natures, making use of several data collection and analysis procedures such as questionnaires, interviews, class observation and assessment tasks, students' notebooks, among others. According to the researchers, the mixed nature of the research requires the employment of several data collection and analysis tools, which has enabled triangulation and construction of narratives for each observed teacher and class.

The research covers a *design* involving five "distinct, but strongly interdependent" (p. 6) stages, including: the construction of a conceptual, critical and analytical frame of reference (stage one) and the document analysis (stage two: analysis of regulations, curricula, guidelines of didactic-pedagogic nature, among others). The third stage involves the construction and application of a questionnaire to a sample of teachers, in order to gather a larger amount of data that allow

> Identifying and establishing relationships between aspects such as: (a) teaching, assessment and learning perception; (b) more frequent pedagogical dynamics and styles; (c) feedback distribution and use; and (d) students' participation in learning and assessment processes. (Borralho & Lucena, 2015, p. 6)

The fourth phase of the research involves class observation in participating schools (at least ten class hours for each teacher). This stage allows detailed description of action and interaction in teaching, learning and assessment activities and, according to the authors, "it is a unique opportunity for understanding a variety of relations between the above mentioned elements, taking the classroom, and not students or teachers, as the unit of analysis" (p. 6). Phase five consists of an intense and interpretative triangulation process of the data and is present throughout the development of research; it includes close contact with the participants, promoting reflections that contribute to data interpretation and analysis.

As partial results, especially from phase four of the research, it was possible to perceive a concern with teaching planning and organisation. For Borralho and Lucena (2015), teachers recognise the importance of planning for the development of classroom tasks, which are seen "as essential strategy so that students can work and progress towards the objectives proposed" (p. 8). Nevertheless, the author point to how, given the different working conditions in both countries, teachers guide their teaching differently, taking into account: (a) use of teaching materials with little or no use of information and communication technologies, (b) some relevance of the interaction between the students and discussions on the work done; (c) some appreciation of the importance of feedback.

Regarding assessment, generally, the research showed that it is most often used to assign ratings to students at the end of school terms, and "its presence was sporadic and punctual and not used to help students to learn" (p. 9), a traditional assessment approach invariably associated with a record, which will, somehow, support the assignment of a rating. For them, this is a limited vision of the assessment processes, that sees assessment, teaching and learning as decoupled and independent processes; a reductive vision of what assessment is, since "it does not foresee, for example, that through work on a given task, it should be possible to teach, learn and assess" (Borralho & Lucena, 2015, p. 9).

The analysis of the interviews with students[7] highlighted that, although they were very conscious of the importance of their participation in the learning process, they generally limited themselves to answer the questions asked by their teachers.

The work presented by Borralho and Lucena (2015) shows that educators need to give more attention to assessment. According to them, although teachers know that assessment is important to the learning and teaching processes, they seem not to understand it as integrated to these processes. Assessment is conducted, in general, using traditional means, focusing on merit and on the idea of accumulation of knowledge.

The analysis of the literature in the fields of education and mathematics education suggests an urgent need for a serious investment in teacher training, but not in a general training for assessment. It is urgent to contextualise and carry out assessment strategies that integrate teaching and assessment, answering to the difficulties teachers find in promoting alternative, transparent, demanding, diversified and significant tasks, from the educational and formative perspective. In other words, an assessment strongly concerned with students' learning. This is a challenge that must be faced!

[7] Student participation dynamics occurred in collective and individual work contexts. For example, task resolution could occur in pairs or small groups, but discussion of the different group's resolutions always happened in a large group (Borralho & Lucena, 2015, p. 10).

9.3.2 Assessment as Investigative Practice: Written Production Analysis

Assessment as investigative practice is based on works by Maria Teresa Esteban (Esteban, 2001, 2002), from the recognition of the existence of a multiplicity of paths taken by students to solve a particular issue or task. For the author, students are in constant process of knowledge elaboration; the teacher, therefore, must take an investigation posture. According to Esteban (2002), the teacher is evaluated as he evaluates; he is put in contact with a permanent movement of elaborating knowledge and lack of knowledge, and, as he investigates the peculiar paths of his pupils, he knows that he also confronts his own knowledge and lack of knowledge. In this perspective, assessment as an investigative practice is configured by the recognition of multiple knowledge, logic and values that permeate the fabric of knowledge. In this sense, assessment is being constituted as a process that questions the results presented, the paths travelled, the routes planned, the relationships established among people, knowledge, information, facts and contexts (Esteban, 2001, p. 11).

The research developed with such a theoretical and methodological approach in the scope of WG08 starts from discussions conducted by researcher Regina Buriasco[8] and her group. For them, assessment as investigative practice is an alternative through which you can seek information about how the subject (student or teacher) mobilises his repertoire in the development of knowledge (Buriasco et al., 2012). It requires, therefore, a change in the look commonly directed to assessment. A change towards disrupting the idea that mathematical results provided by students in their tasks are always accurate, precise, if not immutable.

More recently, research by Buriasco and her group has been influenced by Hans Freudenthal's study and the realistic mathematics education movement, defending the idea of mathematics as a human activity. Buriasco considers that learning should be "actually designed from the mathematisation of situations that can enable students to 're-invent' mathematics" (Buriasco et al., 2012, p. 84).

It is in this perspective that Neves, Silva, and Baccarin (2015) conducted a study in which they analysed the written production of freshmen and graduates in mathematics, through applying a question in the ENEM (National High School Exam) involving the concept of similar triangles.[9] The question was applied to two groups of students, both composed by freshmen and graduates in mathematics, from two higher education institutions. The analysis of the students' written production enabled a classification in the following categories: 1—does not present a written record (in blank); 2—presents incorrect numerical answers, without calculation

[8] Regina Luzia C. de Buriasco, from Universidade Estadual de Londrina (UEL) (Londrina State University, Paraná, Brazil) was the first coordinator of WG08 and remained in charge for four consecutive mandates, from 2000 to 2012. She coordinates the GEPEMA—Grupo de Estudo e Pesquisa em Educação Matemática e Avaliação (Study and Research Group in Assessment and Mathematics Education).

[9] The study is part of an inter-institutional project involving students at the beginning and end of mathematics and education courses at two universities in the south of the country.

records or discussions; 3—presents ratings and records of calculation that do not lead to the correct answer; 4—presents written records that show a mastering of ideas on similar triangles, but does not provide a satisfactory answer to the problem; 5—presents written records that evidence mastering of ideas on similarity and a satisfactory answer to the proposed problem. For the authors, most freshmen productions were found in categories 1 and 2 and few graduate productions could be classified as 5.

The analysis potential of the written production in mathematics education is also the focus of studies developed by Dalto, Araman, and Barbosa (2015) in a university extension project involving undergraduate students in public schools, aiming at school diagnosis of mathematics learning. Initially, a questionnaire was applied to undergraduates to obtain information about their conceptions of assessment. The analysis of this initial stage indicated a traditional perception of assessment, conceived as a measurement to classify those who "know" and those who "don't know". As of this phase, undergraduates were asked to analyse public school students' written production and reflect about the paths students chose to solve the proposed questions. For Dalto et al. (2015), there was a significant change in understanding assessment: after the written production analysis activity, the participants began to realise that errors are learning opportunities, a condition that could enable the teacher to rethink his/her teaching practice, with a view to overcoming the still current hit-error dichotomy in school activities.

For the authors, students' written production analysis practice allowed a change in teachers' attitude as to the teaching-learning process, increasing and resignifying knowledge related to assessment and teaching practice. A perspective that, according to Buriasco et al. (2012, p. 14),

> can provide a more refined look on the involvement of students with mathematics when they deal with the proposed tasks, as well as help to perceive the multiple processes students may have developed, recognise that resolutions of school type are not the only possible, as it accepts diversity, explores the particular ways students develop their justifications, explanations and arguments, that is, their way of giving answers in their own words.

Analysing written productions may be a practice of researchers with research intentions and objectives, as well as of teachers in the classroom who want to know their students' meaning production processes. From this point of view, these works contribute to and may offer changes in school mathematics education. Learning assessment, considered an investigative practice focused on the analysis of written productions, is a promising alternative to the pedagogical practice of teachers who teach mathematics. Promising, because it offers different ways of producing meaning (of students and teachers) in relation to the themes/contents discussed; provides evidence of learning (of students and teachers); enables the teacher to rethink and modify work planning and class dynamics. These, among others, are arguments that motivate us to indicate possibilities of transformation in the assessment practices of mathematics teachers (Buriasco, 2004; Kazemi & Megan, 2004).

9.3.3 Assessment and Inequality Processes

Issues related to inequalities in access to school and in student performance have been present in educational research for at least four decades.[10] An important study conducted in the United States at the end of the 1970s by Jean Anyon showed the association between students' social profile and the math curriculum in each of the schools (Anyon, 1978). In this study, the author tried to discuss the pedagogical work carried out in five North-American schools (the choice was based on the social profile of the students enrolled), through lesson observations (that occurred along a period of 1 year in fifth grade classes) and interviews with students and teachers. Anyon concludes that math curricula, pedagogical practices and tasks proposed in the classroom are associated to the different social profiles of students, contributing to the maintenance, reproduction and production of social and cultural inequalities.

In Brazil, mainly from national assessments data, it has been possible to investigate the relationship between school characteristics, socioeconomic and cultural factors and students' academic achievement. In general, Brazilian research has suggested that schools are different not only in relation to students' socioeconomic and cultural profile, but also due to internal processes developed in each school (Franco & Ortigão, 2007).

In the scope of WG08 concerns with school failure and the learning processes between different groups (boys and girls, for instance) are present since the 2012 edition.[11] A notable example, in this sense, is the research by Ortigão, Aguilar-Junior, and Zucula (2015), which involved the analysis of the PISA 2012 data, in order to identify factors associated with school failure. For the authors, despite having been considered a positive practice, the practice of holding back students who fail a school year is now questioned by research and educational policies, mainly due to its negative social consequences. It is recognised as a complex social phenomenon produced through an interaction of characteristics of schools and school policies and practices, along with profiles of the students and their families. It is also directly responsible for the age-grade gap, for school dropout and removal of the students of their age group (Leon & Menezes-Filho, 2002; Ortigão & Aguiar, 2013). According to Ortigão et al. (2015), high failure rates place Brazil in the condition of school failure champion in basic education, among the 41 countries of Latin America and the Caribbean (Unesco, 2011). Considering high school only, school census data (Brasil, 2015) indicate that school failure in Brazil reached its highest rate in 12 years, reaching 13.1% in 2011—5% more than in the previous year. The authors

[10] According to Forquin (1995), several studies conducted in the United States, France and England, with a view to verify the effects of socioeconomic factors, family and the school in relation to access and school performance, had the merit of drawing the attention of governments and society in general to the effect of social inequalities, such as access conditions and performance.

[11] In the previous SIPEM editions, we identified only one work (presented in SIPEM I) focusing on failure in mathematics, but without a discussion that would allow to relate it to social inequality processes.

also warn that in the Brazilian public network, national failure rate is 14.1%, against 6.1% recorded in private schools.

In the course of the study, an indicator of the number of students repeating the school year was built from the students' answers to a contextual questionnaire, specifically to the question about how many times the student had failed. Answers were grouped into two categories: "yes", if the student had already repeated a school year at least once throughout his/her school progression; and "no" if he or she had never failed. The study's analytical approach was based on two approaches: to describe the studied phenomenon by means of descriptive statistics and to explain it through the implementation of a logistic regression model for retention (Ortigão et al., 2015). The first analysis indicated a difference in math averages between students who claimed never to have failed and those who failed at least once along their school progression of almost 26 points (404.21 against 378.30, with standard deviation of 76.33 and 69.15, respectively). If the student claimed to have failed more than once throughout his school path, the difference increased 86 points compared to the average of the students who had never failed.

Logistic regression model results indicate that failure is affected by characteristics of students and their families such as declared colour, gender, socioeconomic status, family support to studies, family education and future professional/academic expectations, among others. These results are similar to those obtained in other Brazilian studies that show school failure as a problem that affects more deeply the students of low socioeconomic status, whose families are less likely to purchase and mobilise the kind of cultural resources and social support that cause a positive impact on the education of these students. Such results highlight two challenges for schools: (1) the need for policies able to foster both a closer relationship between the family and the school, and the mobilisation of family social support and (2) the need for the school to provide social support to students. School-based social support is associated, at first, with two basic dimensions. On the one hand, it concerns the collective attitude of teachers towards teaching and learning processes. On the other hand, it is associated with the "school climate".[12] Schools that promote a cooperation climate create favourable conditions for reducing the impact of socioeconomic background on school results and the chances of retention. Thus, one of the most important messages of this study is that factors internal to the school can contribute to reducing the chances of failure among male and female students of different races and socioeconomic levels. These factors cover, mainly, school social support focused on material conditions that favour the pedagogical process as well as on the organisational conditions that make the school a learning environment.

Studies that investigate gender relations gain prominence in the educational setting. Jo Boaler and collaborators (Boaler, 1993, 1994), for example, conducted a longitudinal study in two schools in the United Kingdom and observed that teachers'

[12] School climate is a recent expression in educational literature. Associated to the school environment it may be related to various dimensions of school organisation: relations between teachers and management team; relations between teachers, management team and community; relations between teachers and students and school curriculum, among others.

attitudes may favour or not the different performance between girls and boys. Schools where teachers encourage students (boys and girls) to solve problems and engage with mathematics learning tend to minimise differences in gender.

The international investigation developed in the scope of PISA 2000—considered the most comprehensive study on gender and educational performance in all the 32 countries that participated in the assessment—indicates that boys had higher performance than girls in mathematics; however, in many countries, unlike what happens in Brazil, this difference is already zero.

In Brazil, Andrade, Franco, and Carvalho (2003) and Franco and Ortigão (2007) found similar results using, respectively, 1999 and 2001 SAEB data. Both studies were conducted by means of multilevel models (Lee & Bryk, 1989) controlled by retention, socioeconomic status and work simultaneous to study. In both studies, the authors identify that boys have outperformed girls who study in the same schools. However, the study by Andrade et al. (2003) showed that in the schools where socioeconomic status of students is higher and there is a good "academic climate" the difference is quite small. Nonetheless, in schools where students have a low socioeconomic profile and an unfavourable academic climate, the opposite occurs: girls have outperformed boys who study in the same schools. These results impose on the school and education the challenge of thinking about mathematics education considering the diversity of Brazilian classrooms.

The work by Chagas, Cantão, and Leinke (2015), presented in SIPEM VI, investigated differences between boys' and girls' average scores in mathematics, based empirically on data from two relevant assessment means, ENEM and PISA. In this study, they investigated the mean scores in mathematics obtained by girls and boys both in the general scale and in the four subscales considered in the assessments,[13] also making use of a specific modelling to determine the distances between the averages calculated. The study allowed the authors to note that boys obtained better averages than girls in mathematics, both in the global scale and in the subscales, confirming previous national and international studies that investigate performance between genders. They also observed that distances between average scores, obtained based on Cohen's statistics (1988), were positive. For the authors,

> When analysing Cohen's distance, we see that in all cases the value is positive, that is, the performance of boys is always greater than that of girls. The intensity of this performance difference depends on test format and structure and of the responding public. In the case of PISA, it is a controlled sample of adolescent boys and girls, while in ENEM all candidates present at the test were considered. (Chagas et al., 2015, pp. 6–7)

For the authors, it is possible that the difference between genders relates to the still high inequality in social and economic relations in our country. For them, "this performance inequality between genders in the area of mathematics may affect with discouragement female participation in exact sciences" (p. 6).

[13] Both PISA and ENEM allow that averages be obtained in global scale (mathematics) and subscales (mathematics subareas). PISA calls subareas: quantity, uncertainty, space and shape, and change and relationships. ENEM uses the following names: numbers, statistics and probability, geometry and algebra.

The results of this study suggest two aspects that need to be carefully analysed. On the one hand, quantitative results present consonance with the criteria of educational literature, which has recorded the trend of differences that favour boys when assessment focuses on the area of mathematics. Nonetheless, in spite of this convergence, the above-mentioned research by Andrade et al. (2003) shows that in the lower layers of the socioeconomic levels, girls do better in math. Our hypothesis is that girls, especially those belonging to lower socioeconomic level families, stay longer in school. Therefore, we believe that the difference is associated with a greater quantity of girls, which affects average calculation.

On the other hand, results observed in large-scale assessments, such as that conducted by Chagas et al. (2015), allow an enlarged and panoramic view of the studied phenomenon. Therefore, based on these results, it is not possible to make statements about the students' choice for professional careers. After all, as shown by Boaler's studies, girls, when encouraged by their teachers, have results equal to those of boys. PISA also shows that, in many countries, this difference is practically null.

Below, we present the research involving relations between assessment and teacher training.

9.3.4 Assessment and Teacher Training

Educational literature has long highlighted that assessment processes conducted in classrooms, schools or educational systems strongly depend on the conceptions we hold of learning. Fernandes (2009) warns that a hundred years ago the tests or examinations already had characteristics similar to those still found today, such as questions focused on memorising routines or on establishing correspondence between statements. For him, "these characteristics simply match what was then considered important to learn and thought to be the ways in which students learned" (p. 31). In this section, we present research that seeks to articulate assessment and teacher training that were the subject of discussion at the SIPEM VI.

A contribution from Viola dos Santos (2015)[14] explored the advantages of creating a space for basic education mathematics teachers where they can debate the assessments used in their professional practice. The study adopted an approach inspired by the working group perspective described in Lins (2012), and involved ten public and private school mathematics teachers, as well as students (mathematics and mathematics education postgraduates) from a Brazilian public university. The author argues that participation in the project enabled teachers and students to reflect about the ways they understand and deal with assessment in their classes, as

[14] This work is the result of a research project between the postgraduate programme in mathematics education of the *Universidade Federal de Mato Grosso do Sul* (Federal University of Mato Grosso do Sul) and postgraduate programme in science teaching and mathematics education of the *Universidade Estadual de Londrina* (State University of Londrina) with CNPq support (Universal Notice, 14/2012).

well as to reflect about their experiences in evaluative processes, contributes to training teachers, in particular by allowing participants

> /.../ to leave behind the vision of assessment as an *old camera* which records a static, unique moment, to "take" a picture and use, instead, digital cameras, which provide a procedural look, modifiable with various angles to be looked at, with various tools for various "pictures" to be constituted (p. 4).

For the author, the working group as a formative space enables participants to learn, unlearn, share classroom stories, suggest activities, speak of their anguish, barriers and clearly define their accomplishments and achievements—thus becoming a space for discussions and movements provided by teachers' demands.

Adopting another theoretical-methodological perspective, the study by Teles (2015) analysed teachers' written production about mistakes made by students when solving activities that involve subtraction. The author specifically used students' written production in an activity involving a composition problem with one of the parts unknown. The activities followed Vergnaud's classification (1986) and were proposed to 40 fifth grade elementary school teachers, in order to identify elements related to pedagogical content knowledge (Shulmam, 2005) from the analysis of students' productions. As a result, Teles (2015) found that teachers related errors to the student's knowledge or ignorance of some aspect of the calculation (subtraction), to the choice of an inappropriate procedure to execute the algorithm or to little knowledge of decimal numbering system. With regard to content pedagogical knowledge, explicit when teachers presented strategies to overcome the students' difficulties, Teles made several indications (2015). On the one hand, teachers suggested the need to strengthen teaching of subtraction, providing more repetition and training moments—a technicist perspective of teaching, still present in the Brazilian classrooms, as verified in the study of Borralho and Lucena (2015). On the other hand, they suggested the use of concrete material to support understanding of operations and drive the students to reflect on the error committed.

A third theoretical-methodological approach emerged in the survey conducted by Santos and Gontijo (2015). In order to investigate teachers' perceptions[15] of assessment, they conducted a survey by applying a questionnaire to a sample of 39 teachers teaching mathematics in high school, randomly selected from public schools located in the Distrito Federal (Federal District)—Brazil.

The questionnaire was built especially for the survey, with open questions that enabled respondents to express what they thought about assessment, such as: "to evaluate is…", "Assessment is useful for you, because …" and "Assessment is useful to your students, because…"

[15] For the authors, perception delimits all that we are able to feel and understand, corresponding thus to a selective ordering of the stimuli, creating a dichotomy between what we perceive and what we do not realise. Most of the human sensitivity, including internal sensations, is linked to the unconscious. Another part however, even also participating in the sensory, comes to our knowledge in articulated way, that is, in organised forms. This organised and articulated construction is nothing more than perception, which includes the intellectual being.

In general, teachers who answered the questionnaire said that assessment is synonymous with evidence and serves to "check the content assimilated" (p. 7). They also understood the utility of assessment as a "learning progress indicator" (p. 8). For the student, assessment is useful, as it "shows the points they need further study or the points where they did well" (p. 9). For the authors, teachers carry a vision of assessment as a measure, pointwise and external to the process of teaching and learning—a vision contrary to what has been discussed regarding the assessment of learning (Fernandes, 2009, Esteban, 2003, among others).

Menezes and Santos (2015) explored yet another aspect of assessment in their investigation of the "assessment agreement", based on the ideas of a didactic contract by Brousseau and of didactic transposition by Chevallard (1999). Data collection involved videos made in classes (6th grade) of an elementary school teacher during a school year. The methodological strategy of the research consisted of three steps: textbook analysis; observation of a teacher's lessons during a school year; written tests analysis.

There is, for the authors, a set of rules constructed specifically for the assessment, which they call "assessment agreement". This contract presents the teacher's expectations as to the knowledge to be assessed, how the questions will be presented and how students should provide the answers. Moreover, the teacher gives hints of what is expected and of ways to obtain "success" in the assessment (Menezes & Santos, 2015, p. 11). Despite being a case study, the authors believe that each teacher conducts an assessment contract in accordance with his/her conceptions of teaching and learning and also with his/her relationship to the knowledge in question.

These four examples of research, conducted in the interrelation assessment—teacher training, show that training must be well related to investigation and, surely, with practices. We consider that training processes must include alternative approaches that go beyond placing teachers as listeners to what tutoring teachers have to say. Teacher training, in our view, must be done *with* teachers and not *for* them. This research also shows the need to discuss and problematise teachers' pedagogical practice, specifically their assessment practice. Many studies in mathematics education literature indicate possibilities of using different assessment tools in the classroom; many other studies indicate the possibilities of assessment training for teachers' and students' teaching and learning processes. Conducting discussions with teachers on assessment in the classroom necessarily implies holding discussions about the school, which leads to discussions on performance and (initial and continuing) training processes of teachers who teach mathematics.

9.4 Final Considerations

Studies focusing on learning assessment have received significant attention from researchers in Brazil as well as in other countries. An analysis of works presented in the different SIPEM editions demonstrates the presence of different methodological approaches and different theoretical perspectives. It is interesting to observe that research focusing on learning assessment has changed through time, with the current conception of assessment as that of a research practice. Themes and approaches in the course of SIPEM editions have widened, including: analysis of teaching materials, collaborative research (with the teacher) and quantitative research approaches, involving specific statistics modelling and assessment as an investigative practice.

Assessment in the scope of mathematics education has been constituted as a research field, given a significant increase in research conducted and research projects that coalesce researchers from different parts of the country at different levels of education (master's and Ph.D.), as well as some international collaboration, such as the project AERA.

It is undeniable that basic education teachers, as well as university lecturers, need to transform their assessment practices and make them part of the teaching and learning processes. It is necessary to discuss public policy—in which large-scale assessments fall—and build some ways of subversion when they attack the educational process as children's and adolescents' rights. It is urgent that the discussion concerning assessment be present in initial and continuing training courses for mathematics teachers. It is also important to create research fields focused on classrooms and schools and give us the opportunity to understand what happens in classrooms, in schools, in the realm of learning assessment. We should investigate questions, such as:

- How do teachers integrate and articulate assessment, teaching and learning?
- What assessment criteria do teachers adopt, and how are they related to the development of learning?
- Which assessment tools and strategies do teachers prioritise?
- What effects does assessment have on students' motivation and development of learning? What type of analysis do teachers make of the results of the external assessments? How do you use them? What articulations do they establish between internal assessment and external assessment? How do teachers and students construct/understand school assessment policy?

We believe, as researchers and educators, that our efforts need to move in the direction of finding answers and promoting the emergence of new questions in the midst of our investigations.

References

Andrade, M., Franco, C., & Carvalho, J. B. P. (2003, jan.–jun.). Gênero e Desempenho em Matemática ao Final do Ensino Médio: quais as relações? *Estudos em Avaliação Educacional, 27*.

Anyon, J. (1978). Social class and the hidden curriculum at work. *Journal of Education, Boston, 162*(1), 67–92.

Ball, S. (2005). Profissionalismo, gerencialismo e performatividade. *Cadernos de Pesquisa, São Paulo, 35*(126), 539–564. https://doi.org/10.1590/S0100-15742005000300002

Boaler, J. (1993). Encouraging the transfer of "school" mathematics to the "real world" through the integration of process and content, context and culture. *Educational Studies in Mathematics, 25*, 314–373.
Boaler, J. (1994). When do girls prefer football to fashion? An analysis of female underachievement in relation to "realistic" mathematic contexts. *British Educational Research Journal, 20*, 551–564.
Borasi, R. (1996). *Reconceiving mathematics instruction: A focus on errors*. Norwood: Ablex Publishing Corporation.
Borralho, A., & Lucena, I. (2015). *Avaliação e ensino na educação básica em Portugal e no Brasil: relações com as aprendizagens (AERA)*. Trabalho apresentado no VI SIPEM. Pirenópolis. Retrieved from http://www.sbembrasil.org.br/visipem/anais/story_content/external_files/AVALIA%C3%87%C3%83O%20E%20ENSINO%20NA%20EDUCA%C3%87%C3%83O%20B%C3%81SICA%20EM%20PORTUGAL%20E%20NO%20BRASIL.pdf
Bourdieu, P., & Passeron, J.-C. (1975). *A reprodução. Elementos para uma teoria do sistema de ensino*. Rio de Janeiro: Francisco Alves.
Brasil. (2015). *MEC/INEP. SAEB/PROVA BRASIL 2011—primeiros resultados*. Retrieved from http://download.inep.gov.br/educacao_basica/prova_brasil_saeb/resultados/2012/Saeb_2011_primeiros_resultados_site_Inep.pdf
Brooke, N. (Org.). (2012). *Marcos Históricos na Reforma da Educação*. Belo Horizonte: Fino Traço.
Buriasco, R. L. C. d. (1995). Sobre a resolução de problemas. *Nosso Fazer, 1*(5), 1.
Buriasco, R. L. C. d. (2000, jul–dez). Algumas considerações sobre avaliação educacional. *Estudos em Avaliação Educacional, 22*, 155–178.
Buriasco, R. L. C. d. (2004). Análise da produção escrita: a busca do conhecimento escondido. In J. P. Romanowski, P. L. O. Martines, & S. A. Junqueira (Orgs.), *Conhecimento local e conhecimento universal: a aula e os campos do conhecimento* (pp. 243–251). Curitiba: Champagnat.
Buriasco, R. L. C., Ferreira, P. E. A., & Ciani, A. B. (2012). Avaliação como Prática de Investigação (alguns apontamentos). Bolema, 22(33), 69–96 Retrieved from http://www.redalyc.org/pdf/2912/291221900005.pdf
Chagas, E. A., Cantão, R. F., & Leinke, M. U. (2015). *Gênero e desempenho em matemática nas provas do ENEM e do PISA 2012*. Trabalho apresentado no VI SIPEM. Retrieved from http://www.sbembrasil.org.br/visipem/anais/story_content/external_files/G%C3%8ANERO%20E%20DESEMPENHO%20EM%20MATEM%C3%81TICA%20NAS%20PROVAS%20DO%20ENEM%20E%20DO%20PISA.pdf
Chevallard, Y. (1999). L'analyse des pratiques enseignantes en Théorie Anthropologie Didactique. *Recherches en Didactiques des Mathématiques*, 221–266.
Coleman, J. S. (1966). *Report on equality of education opportunity*. US Government Printing Office for Department of Health, Education and Welfare.
Cooper, B., & Harries, T. (2003). Children's use of realistic considerations in problem solving: Some English evidence. *Journal of Mathematical Behavior, 22*(4), 449–463 Retrieved from http://www.elsevier.com/wps/find/journaldescription.cws_home/620200/description#description
Cury, H. N. (2007). *O que podemos aprender com as respostas dos alunos* (Vol. 1, 1st ed.). Belo Horizonte: Autêntica.
D'Ambrosio, B., Kastberg, S., & Lambdin, D. (2007). Designed to differentiate: What is NAEP measuring? In P. Kloosterman & F. Lester Jr. (Eds.), *Results and interpretations of the 2003 mathematics assessments of the National Assessment of Education Progress*. Reston, VA: NCTM.
D'Ambrosio, B., & Lopes, C. E. (Eds.). (2015). *Creative insubordination in Brazilian mathematics education research*. Ohio: Lulu Press, Inc.
Dalto, O., Araman, E. M. O., & Barbosa, L. N. S. C. (2015). *Avaliação como saber docente*. Trabalho apresentado no VI SIPEM. Retrieved from http://www.sbembrasil.org.br/visipem/anais/story_content/external_files/AVALIACAO%20COMO%20SABER%20DOCENTE.pdf

Dalto, J. O., & Buriasco, R. L. C. (2009). Problema proposto ou problema resolvido: qual a diferença? *Educação e Pesquisa (USP), 35*, 449–461.
Esteban, M. T. (2001). A avaliação no processo ensino/aprendizagem: os desafios postos pelas múltiplas faces do cotidiano. Apresentado na 24° Reunião Anual da Associação Nacional de Pós-Graduação e Pesquisa em Educação, Caxambu. *Intelectuais, conhecimento e espaço público*. Retrieved from http://www.anped.org.br/24/T0682499156560.doc
Esteban, M. T. (2002). *O que sabe quem erra? Reflexões sobre avaliação e fracasso escolar* (3rd ed.). Rio de Janeiro: DP&A.
Esteban, M. T. (org.). (2003). Escola, currículo e avaliação. São Paulo: Cortez.
Esteban, M. T. (2012, set.–dez.). Considerações sobre a política de avaliação da alfabetização: pensando a partir do cotidiano escolar. *Revista Brasileira de Educação, 17*(51).
Esteban, M. T.; Afonso, A. (Orgs.). (2010). Olhares e interfaces: reflexões críticas sobre a avaliação. São Paulo: Cortez.
Fernandes, D. (2009). *Avaliar para Aprender/Fundamentos, práticas e políticas*. São Paulo: Editora UNESP.
Fernandes, C. O. (Org.). (2014). *Avaliação das aprendizagens: sua relação com o papel social da escola*. São Paulo: Cortez.
Forquin, J.-C. (1995). As Abordagens Sociológicas do Currículo: orientações teóricas e perspectivas de pesquisa. *Educação e Realidade, 21*(1), 187–198.
Franco, C., & Ortigão, M. I. R. (2007). Eficácia escolar em Brasil: Investigando prácticas y políticas escolares moderadoras de desigualdades educacionales. In S. Cueto (Org.), *Educación y brechas de equidad em América Latina, Tomo I*. (pp. 223–249). Santiago, Chile: Fondo de Investigaciones Educativas/PREAL.
Freitas, L. C. (2012). Os reformadores empresariais da educação: da desmoralização do magistério à destruição do sistema público de educação. *Educação & Sociedade, 119*, 379–404 Retrieved from http://www.scielo.br/pdf/es/v33n119/a04v33n119.pdf
Kazemi, E., & Megan, L. F. (2004). Teacher learning in mathematics: Using student work to promote collective inquiry. *Journal of Mathematics Teacher Education, 7*, 203–235.
Lee, V., & Bryk, A. (1989). A multilevel model of the social distribution of high school achievement. *Sociology of Education, 62*, 172–192.
Leon, & Menezes-Filho. (2002). Reprovação, Avanço e Evasão Escolar no Brasil. *Pesquisa e Planejamento Econômico (PPE), 32*(3), 417–451.
Lins, R. C. O. (2012). Modelo dos Campos Semânticos: estabelecimentos e notas de teorizações. In C. L. Angelo, et al. (Eds.), *Modelo dos Campos Semânticos e Educação Matemática: 20 anos de história. Vol. 1* (pp. 10–20, 1st ed.). São Paulo: Midiograf.
Lopes, A. C. (2015). Por um currículo sem fundamentos. *Linhas Críticas, 21*(45), 445–466 Retrieved from http://periodicos.unb.br/index.php/linhascriticas/article/view/16735
Mainardes, J. (2015). Avaliação da aprendizagem na Alfabetização. In M. C. S. Cruz & R. E. S. R. Borba (Orgs.). *Ciclo de Palestras, Vol. 1* (pp. 109–125). Recife: Ed. Universitária da UFPE. Retrieved from https://drive.google.com/file/d/0BxwtaLINTVLEYXJ0YjBvMV9ibGc/view
Menezes, M. B., & Santos, M. C. (2015). *Contrato de Avaliação: Uma Análise dos Efeitos da Transposição e do Contrato*. Trabalho apresentado no VI SIPEM. Retrieved from http://www.sbembrasil.org.br/visipem/anais/story_content/external_files/CONTRATO%20DE%20AVALIA%C3%87%C3%83O_UMA%20AN%C3%81LISE%20DOS%20EFEITOS%20DA%20TRANSPOSI%C3%87%C3%83O%20E%20DO%20CONTRATO%20DID%C3%81TICOS.pdf
Neves, R. S. P., Silva, J. C. S., & Baccarin, S. A. O. (2015). *A produção escrita de estudantes de licenciatura em matemática em questão do ENEM*. Trabalho apresentado no VI SIPEM. Pirenópolis. Disponível em: http://www.sbembrasil.org.br/visipem/anais/story_content/external_files/A%20PRODU%C3%87%C3%83O%20ESCRITA%20DE%20ESTUDANTES%20DA%20LICENCIATURA%20EM%20MATEM%C3%81TICA%20

EM%20QUEST%C3%83O%20DO%20EXAME%20NACIONAL%20DO%20ENSINO%20M%C3%89DIO.pdf. Acesso em: 20/06/2016.

Ortigão, M. I. R., & Aguiar, G. (2013). Repetência escolar nos anos iniciais do ensino fundamental: evidências a partir dos dados da Prova Brasil 2009. *Revista Brasileira de Estudos Pedagógicos, 94*, 364–389.

Ortigão, M. I. R., Aguilar-Junior, C. A., & Zucula, A. F. (2015). *Analisando a repetência escolar a partir dos dados do PISA 2012*. Trabalho apresentado no VI SIPEM. Retrieved from http://www.sbembrasil.org.br/visipem/anais/story_content/external_files/ANALISANDO%20A%20REPET%C3%8ANCIA%20ESCOLAR%20A%20PARTIR%20DOS%20DADOS%20DO%20PISA%202012.pdf

Ortigão, M. I. R., Franco, C., & Carvalho, J. B. P. (2007). A distribuição social do currículo de matemática: quem tem acesso a tratamento da informação? *Educação Matemática Pesquisa, 9*(2), 249–273 Retrieved from http://revistas.pucsp.br/index.php/emp/article/viewFile/904/597.

Ortigão, M. I. R., & Pereira T. V. (2016). Homogeneização curricular e o sistema de avaliação nacional brasileiro: o caso do estado do Rio de Janeiro. In *Educação Sociedade & Cultura*. (Edição especial: Avaliação das escolas: políticas, perspectivas e práticas), Retrieved from http://www.fpce.up.pt/ciie/sites/default/files/ESC47Maria.pdf

Ravitch, D. (2011). *Vida e morte no grande sistema escolar americano – como os testes padronizados e o modelo de mercado ameaçam a Educação*. Porto Alegre: Sulina.

Santos, V. S., & Gontijo, C. H. (2015). *Percepções de Docentes de Matemática de Ensino Médio em Relação ao Processo de Avaliação da Aprendizagem*. Trabalho apresentado no VI SIPEM. Retrieved from http://www.sbembrasil.org.br/visipem/anais/story_content/external_files/PERCEP%C3%87%C3%95ES%20DE%20DOCENTES%20DE%20MATEM%C3%81TICA%20DE%20ENSINO%20M%C3%89DIO%20EM%20RELA%C3%87%C3%83O%20AO%20PROCESSO%20DE%20AVALIA%C3%87%C3%83O%20DA%20APRENDIZAGEM.pdf

Shulmam, L. S. (2005). Conocimiento y Ensenanza: Fundamentos de la nueva reforma. Profesorado. *Revista de currículum y formación del profesorado, 9*, 2.

Sipem. *Seminário Internacional de Pesquisa em Educação Matemática*. Retrieved from http://www.sbembrasil.org.br/sbembrasil/index.php/anais/sipem

Sipem. (2000). I Seminário Internacional de Pesquisa em Educação Matemática. Retrieved from http://www.sbembrasil.org.br/files/sipemI.pdf

Sipem. (2003). II Seminário Internacional de Pesquisa em Educação Matemática. Retrieved from http://www.sbembrasil.org.br/files/sipemII.pdf

Sipem. (2006). III Seminário Internacional de Pesquisa em Educação Matemática. Retrieved from http://www.sbembrasil.org.br/files/sipemIII.pdf

Sipem. (2009). IV Seminário Internacional de Pesquisa em Educação Matemática. Retrieved from http://www.sbembrasil.org.br/files/sipemIV.pdf

Sipem. (2012). V Seminário Internacional de Pesquisa em Educação Matemática. Retrieved from http://www.sbembrasil.org.br/files/v_sipem/

Sipem. (2015). VI Seminário Internacional de Pesquisa em Educação Matemática. Retrieved from http://www.sbembrasil.org.br/visipem/anais/story_html5.html

Teles, R. A. M. (2015). *Repetir, Refletir ou Omitir? O que dizem professores sobre erros de alunos no algoritmo da subtração*. Trabalho apresentado no VI SIPEM. Retrieved from http://www.sbembrasil.org.br/visipem/anais/story_content/external_files/REPETIR,%20REFLETIR%20OU%20OMITIR.pdf

TREVISAN, A.L. (2013). Prova em fases e um repensar da prática avaliativa em Matemática. 2013. 168 f. Tese (Doutorado em Ensino de Ciências e Educação Matemática) – Universidade Estadual de Londrina, Londrina.

Tura, M. L. R., & Pereira, T. V. (2013). Políticas curriculares, sistemas de avaliação e conhecimentos escolares. In M. L. R. Tura & M. M. A. Garcia (Orgs.), *Currículo, políticas e ação docente* (pp. 111–125). Rio de Janeiro: EdUERJ.

Unesco. (2011). Llegar a los marginados. In *Informe de Seguimiento de la Educación Para Todos en el mundo*. Retrieved from www.unesco.org/publishing

Vergnaud, G. (1986). Psicologia do desenvolvimento cognitivo e didáticas da matemática. Um exemplo: as estruturas aditivas. *Análise Psicológica., 1*(V), 75–90.

Vianna, H. M. (1987). *Testes em Educação*. São Paulo: IBRASA.

Viola Dos Santos, J. R. (2015). *Discussões De Professores De Matemática A Respeito Da Avaliação Em Um Grupo De Trabalho*. Trabalho apresentado no VI SIPEM. Retrieved from: http://www.sbembrasil.org.br/visipem/anais/story_content/external_files/DISCUSS%C3%95ES%20DE%20PROFESSORES%20DE%20MATEM%C3%81TICA%20A%20RESPEITO%20DA%20AVALIA%C3%87%C3%83O%20EM%20UM%20GRUPO%20DE%20TRABALHO.pdf

Chapter 10
Cognitive and Linguistic Processes in Brazilian Mathematics Education: Theoretical Considerations and Educational Implications

Airton Carrião, Sintria Labres Lautert, and Alina Galvão Spinillo

Abstract The aim of this chapter is to contribute to the debate about the trends, objects of study, and theoretical and methodological assumptions that have marked and constituted the current identity of working group Cognitive and Linguistic Processes in Mathematics Education of the Brazilian Society of Mathematics Education. The considerations presented bring together a range of investigations that explore the cognitive and linguistic aspects involved in the teaching and learning of mathematics in different learning contexts and different levels of schooling. The work developed by the group has been characterised by investigations conducted on language and communication in the classroom and their sociocultural aspects, alongside studies into the cognitive processes involved in mathematical reasoning. The discussion about the theoretical-methodological questions underlying the reflections on the cognitive and linguistic processes in the Brazilian scenario has been divided into two parts. In the first, we present a historical review of the main trends considered during the first 10 years of working group's history. In the second, explore the more recent objects of study. We describe how this development indicates a convergence of Brazilian researchers with different theoretical and methodological affiliations, as they search for theoretical models that can explain the role of language, cognition and cultural aspects in the teaching and learning of mathematics. The advances made in terms of knowledge of cognitive and linguistic processes in mathematics education within Brazil are also presented.

A. Carrião (✉)
Universidade Federal de Minas Gerais, Belo Horizonte, Brazil

S. L. Lautert · A. G. Spinillo
Universidade Federal de Pernambuco, Recife, Brazil

A. J. Ribeiro et al. (eds.), *Mathematics Education in Brazil*,
https://doi.org/10.1007/978-3-319-93455-6_10

10.1 Introduction

This chapter presents and discusses the proposals, ideas, and theoretical assumptions that underlie the research conducted by the Working Group Cognitive and Linguistic Processes in Mathematics Education (Processos Cognitivos e Linguísticos em Educação Matemática), which has been part of the Brazilian Society of Mathematics Education (SBEM) for the last 16 years. The chapter also presents the changes that have taken place over this period. Before considering the issues that have generated reflections on the cognitive and linguistic processes in the area of mathematics education in Brazil, it is important to consider the objectives that led to the creation of the Working Group (WG09).

This working group aims to discuss and disseminate research on the cognitive and linguistic aspects involved in the teaching and learning of mathematics at different levels of schooling and in different contexts. Three factors have characterised these studies: (1) language as the means of formation and expression of mathematical reasoning; (2) communication in the classroom; and (3) the cognitive processes involved in the mathematical reasoning. Thus, both the role of language in the production of meaning and communication in the classroom, and the description of the reasoning processes (limits and possibilities) involved in mathematical knowledge are topics that have permeated the production of the working group.

In an attempt to give visibility to the diversity of proposals, objects of study and theoretical-methodological assumptions that mark and constitute the current identity of WG09, we have chosen to present the most frequent and relevant questions that can be seen in the work produced by the group. To do so, we have presented the investigations that demonstrate the extensiveness and variety of subjects considered, and the interdisciplinary nature of the work conducted by the group, as it brings together different areas of knowledge such as cognitive psychology and mathematics education. The studies of this working group have been arranged into two groups. The first offers a historical review of the first 10 years of the working group, presenting the main trends that characterised the first meetings of the group. The second discusses the research presented in the last two SIPEMs, which took place in Petrópolis (2012) and Pirenópolis (2015). These studies depict two thematic and methodological tendencies that remain the focus of the investigations, namely: (1) research which emphasises cognitive processes from a developmental perspective and (2) research emphasising the language and cultural aspects in the teaching and learning processes of mathematics. This organisation reflects the thematic and methodological trends that have characterised the work of WG09 since its establishment.

10.2 The Research Conducted in the First Years of WG09

The first meetings were marked by discussions about how students at different levels of education (elementary, middle and high school, university education and youth and adult education—EJA)[1] were taught mathematics and about their mathematical learning processes. The aim of these first meetings was to study teacher–student interactions, student–student interactions, as well as how students interacted with the mathematical knowledge in which argumentative and metacognitive strategies were identified as support for learning situations in the classroom context. The solving of problems involving various concepts (e.g. percentage, function, graphs, and fraction) and different fields of the mathematical knowledge (e.g. geometry, combinatorics, and arithmetic) were also investigated. These studies, according to Rabello, Lins, and Da Rocha Falcão (2000), focused on issues of great theoretical and practical interest in the field of mathematics education:

1. Language, cognition, and communication—emphasising the role of language, embodied language (according to Lakoff & Johnson, 1999), representation (according to Duval, 1993, 1995), and discourse genres (according to Bakhtin & Voloshinov, 1992), and highlighting the differences between natural language and the language of mathematics (Pires, Curi, Rabelllo, Pavanello, & Valente, 2003) and the boundaries between the use of rationality in daily life and formal logic;
2. Cognitive and linguistic processes in the classroom, focusing on the interface between theory and practice in relation to the development of cognitive and language skills considered relevant to the teaching and learning of mathematics.
3. Vygotsky's social-historical-cultural perspective and Piaget's constructivism, addressing some of the theoretical debates about conceptualisation in mathematics and processes such as internalisation, scaffolding, and interaction.

In summary, the results of the studies presented at the first meetings pointed to the following facts: (1) students used concepts acquired in extracurricular social contexts and different strategies to grasp the mathematical concepts presented in the classroom; (2) the mathematical language used in the classroom, particularly notation and the use of symbols, was of little significance for the students; and (3) the dialogues between teacher and student often inhibited the expression of the ways of reasoning of the students.

Some of the research studies presented have gone on to gain international recognition, especially those which focused on the study of cognitive and linguistic

[1] The educational system in Brazil is divided into basic education and higher education. Basic education corresponds to: nursery or early-years education (0–5 years old), (2) elementary school (elementary and middle school, from 6 to 14 years), and high school (15–17 years). Higher education corresponds to university education, undergraduate and postgraduate courses (Masters and PhD). This educational system also includes youth and adult education (EJA), distance learning and educational technologies, technological education and vocational training, special education and indigenous education.

aspects in the learning of mathematical concepts by blind students (Healy & Fernandes, 2008), and on the role of language, corporeality, and technology in supporting the analysis of the process of teaching and learning mathematics (Frant, 2009). Research on the tacit and explicit knowledge involved in the learning of mathematics in the school context (Frade & Borges, 2006); and discussions about the learning of mathematics in the context of youth and adult education (Fonseca, 2010) have also received international attention.

Another feature of the work conducted by WG09 during this period was the concern with how students with atypical profiles learn mathematics, such as blind learners and children with epilepsy. There was also a great interest in how the processes of teaching and learning take place in the classroom, in investigating how teachers understand the learning process and the exchanges between the students and the teacher. In other words, there was an interest in how the teacher's role was defined in relation to their teaching practice, the transfer of learning, and the classroom as a "community of practice" (Garnica, Soares, & Buriasco, 2006). These studies seek to construct a knowledge of flexible bases, one that no longer focuses on the idealised student. Instead, this knowledge brings into play a teacher who respects and integrates different ways of learning and of communicating in the classroom. The discussions derived from these studies later appeared in national and international publications (David & Watson, 2008; Frade & Falcão, 2007; Frade & Tatsis, 2009; Frade, Winbourne, & Braga, 2009; Hazin, Da Rocha Falcão, & Leitão, 2006).

Despite the thematic, theoretical, and methodological variations that characterised the research conducted by members of WG09 during this period, it is possible to see a general common ground in relation to the teaching and learning of mathematics. This involves cognitive and linguistic processes that, whether investigated separately or jointly, are interconnected and become inseparable in the classroom.

10.3 The Most Recent Trends in the Research Conducted by WG09

In this part, we consider the propositions and theoretical assumptions present in the theoretical-methodological production of the V and VI Brazilian Symposium on Mathematics Education, in which twelve and three papers were presented, respectively. First, we offer an overview of the investigations presented at these two seminars. Then we discuss a selected set of studies that show the heterogeneity of the research conducted by the group. These studies also illustrate two recurring themes of the investigations since the creation of the GW9, namely: research that emphasises cognitive processes from a development perspective, and research that emphasises the language and the cultural aspects in the teaching and learning processes of mathematics.

The fifth seminar illustrates the diversity of the objects of investigation. The vast majority of the investigations focused on language and cultural aspects of the processes of reasoning and of learning mathematics. These aspects were investigated both from the perspective of those performing the activity individually (questions, tasks, or problems) and in the interactions which occur during a mathematics class (e.g. Carrião, 2012; David & Tomaz, 2012a; Frant, 2012; Torisu, 2012). There were also three studies on the processes involved in the teaching and learning of specific contents: rational numbers in fractional form (Freire & Lima, 2012), division (Lautert & Spinillo, 2012), and geometry (Viana, 2012). In addition, we identified two studies which looked into students' ability to solve mathematical problems and their relationship with the activity of playing chess (Lopes & Magina, 2012) and creativity in mathematics (Oliveira, Albuquerque, & Gontijo, 2012). There was also a paper on the production of meanings and the decision-making ability of consumers (Kistemann Jr., 2012). Two studies drew attention to the education of blind and deaf students, with regard to the material and semiotic tools for the acquisition of mathematical knowledge (Fernandes, Healy, & Serino, 2012) and the configuration of the mathematics classroom when deaf students are present and interpreters are required (Borges & Nogueira, 2012).[2]

Although these investigations focus mainly on elementary school students, diversity is maintained by also conducting studies that consider high school and undergraduate students. The three studies presented at SIPEM VI focus on the solving of problems involving the multiplicative conceptual field in elementary school. Two of them focus on the teachers' conceptions (e.g. Merlini, Santos, Teixeira, & Magina, 2015; Spinillo, Lautert, Santos, & Silva, 2015) and one on the students' conceptions (e.g. Magina, Spinillo, & Melo, 2015).

In summary, in recent years, despite the thematic, theoretical, and methodological variability that has characterised the work developed by WG09 since its inception, the studies that have been carried out can be grouped together according to their main theme:

1. The cognitive aspects involved in solving problems relating to specific mathematical concepts (Freire & Lima, 2012; Lautert & Spinillo, 2012; Lopes & Magina, 2012; Magina et al., 2015; Oliveira et al., 2012; Viana, 2012) and the views of the teachers (e.g. Merlini et al., 2015; Spinillo et al., 2015);
2. Communication in the classroom which includes the participation of deaf and blind students (Borges & Nogueira, 2012; Fernandes et al., 2012) and the cultural aspects involved in the teaching and learning of mathematics (e.g. Carrião, 2012; David & Tomaz, 2012a, b; Frant, 2012; Kistemann Jr., 2012; Torisu, 2012).

In the next section, we discuss the advances made with regard to the knowledge about the cognitive and linguistic processes in mathematics education. We consider the studies that focus on the cognitive aspects in a developmental perspective, and the studies which emphasise the language and the cultural aspects involved in the reasoning processes. The division of the latest research by WG09 in two groups shows how Brazilian researchers from different theoretical and methodological affiliations have come together in their search for theoretical models that can explain the role of language, cognition, and of cultural aspects in the teaching and learning of mathematics.

10.3.1 Studies That Highlight the Cognitive Processes in a Developmental Perspective

According to Spinillo and Lautert (2006), cognitive psychology has distanced itself from more general theories that explain cognitive development in a broad manner, such as the Piagetian theory. Instead, cognitive psychology has become more closely identified with more specific theories, such as the theory of conceptual fields, proposed by Vergnaud, which deals specifically with mathematical knowledge. This has created a scenario that has led to the understanding of the multiple connections between psychology and education, and between theory and school practice. In the last few years, much of the research on cognitive processes that has been presented in the Cognitive and Linguistic Processes in Mathematics Education Working Group has adopted a developmental perspective. Based mainly on the theory of conceptual fields (Vergnaud, 1983, 1988, 1990, 2003), these investigations have examined problem solving in the field of multiplicative structures. Problems of this nature have been widely documented, both in the national and international literature and are considered a challenge for both those who teach and for those who learn them. Following the international trend,[2] the issues addressed in WG09 emphasise problem solving, drawing attention to the importance of learning how to solve problems in order to learn mathematics, and to the importance of learning mathematics in order to solve problems (Charnay, 2001; Onuchic & Alevato, 2004).

Problem solving has been investigated, particularly, aiming at: (1) understanding how students conceive certain concepts considered to be complex, such as fractions, proportions, probability, divisions, multiplications, and the solving of problems involving combinatorial reasoning (e.g. Freire & Lima, 2012; Lautert & Spinillo, 2012; Lopes & Magina, 2012; Merlini et al., 2015; Spinillo et al., 2015); (2) analysing the nature of different types of problems since the way of reasoning adopted is related to the type of problem being solved (e.g. Magina et al., 2015).

The research conducted in the last few years has taken an interdisciplinary approach, involving psychology and mathematics education. To illustrate this interdisciplinary nature of the group, we will briefly present some studies that look into the solving of problems in the field of multiplicative structures. They are

[2] The solving of problems continues to be a relevant topic in the field of mathematics education. It has been the topic of numerous international conferences, such as the 13th International Congress on Mathematical Education (ICME) in 2016, the 40th Conference of the International Group for the Psychology of Mathematics Education (PME) also in 2016, and of many of the National Council of Teachers of Mathematics (NCTM) documents.

grouped as follows: (1) *diagnostic research*, which seeks to describe how reasoning is organised and how it develops in students from different age groups and at different levels of schooling, when these are working with a particular concept; analysing the strategies adopted; (2) *intervention research*, which seeks to examine ways to develop mathematical reasoning; and (3) research which looks into *the teacher's conception* when dealing with problem-situations in the field of multiplicative structures.

A study by Magina et al. (2015) is an example of *a diagnostic study*. The authors analysed the performance and the strategies adopted by 3rd, 4th, and 5th graders of elementary school in the resolution of Cartesian product problems. The results showed that the students used different strategies, expressing different levels of combinatorial reasoning, although their performance, in terms of the process employed, did not vary throughout the different school years. The most elementary of these levels corresponds to the inability to understand the relationships between the variables, and the use of inappropriate operations. The strategies at this elementary level were the ones adopted more frequently. The less elementary strategies employed by the students were characterised by the establishment of fixed combinations based on term-to-term correspondences. Slightly more sophisticated strategies, but still not yet entirely appropriate, were those in which a flexible combination of the elements was observed. Such strategies, however, did not exhaust all the possible combinations between the elements from the two elementary sets. Appropriate strategies were rare. These were considered to be the strategies that, as stated by Moro and Soares (2006), resulted in combinatorial solutions derived from one-to-many correspondences, and that made use of all possible combinations, involving all the elements from both elementary sets. These results confirm Vergnaud's (1983, 1988) claim that conceptualisation in mathematics develops gradually and over a long period since the most sophisticated solutions, although rare, were found among the 5th graders. Moreover, the results of this study point to both the limitations and to the potential of the reasoning ability of elementary school students. These results have educational implications that can be translated into didactic actions that can support the development of these concepts, as well as helping to overcome the difficulties identified.

As an instance of *intervention research,* we have selected the study conducted by Lautert and Spinillo (2012). In this study, the authors investigated the effect of a specific intervention on the understanding of division in children who had difficulties understanding this concept. The research involved an experimental group (EG) and a control group (CG), and all the participants in the groups were required to take a pre-test and a post-test. The intervention, which was offered to the EG only, consisted of problem-solving activities, and discussed the forms of resolution adopted by the children. The intervention also explained two invariant principles of division considered as determinants of the difficulties experienced by students during the early years of elementary school: co-variation relations between the terms of division when the dividend is kept constant, and how to deal with the remainder. The results showed that the EG students reached more advanced levels of understanding than those of the CG, overcoming the difficulties they had before the intervention. According to the authors, this progress was due to the fact that the intervention allowed participants to perform a metacognitive activity—that is, to think about their ways of thinking when solving problems—and to reflect on the invariant principles of the concept of division.

Finally, as examples of studies that investigate *the teacher's conception* with regard to mathematical concepts, we focus on two studies that have analysed the notions held by teachers about the resolution of problems involving both division and multiplication. Most research is based on the assumption that the understanding of a concept manifests itself not only in problem-solving situations, but also in situations where the individual is asked to formulate problems. Research conducted in Brazil (Chica, 2001; Itacarambi, 2010), and abroad (English, 1997; Lowrie, 2002; Zunino, 1995) assume this perspective regarding the formulation of problems by students, showing that students' ability to formulate problems is related to their understanding of the concept involved in the solving of the problems they formulate. If we apply this reasoning to the teachers, we can see that the posing of problems by teachers needs further investigation. If problem solving is considered a didactic strategy, and if the problems presented relate to the content to be taught, we can conclude that knowledge about problem solving is a didactic knowledge of the content that needs to be mastered by those who teach mathematics.

As a result, research in Brazil (Cunha, 2015; Souza, 2015) and abroad (Crespo, 2003; Leung & Silver, 1997) have looked at how teachers of mathematics formulate problems. The results of these studies led Spinillo et al. (2015) to investigate the formulation of problems by elementary school teachers. Participants in this study were asked to formulate, in writing, problems whose resolution required the use of multiplication and/or division. It was observed that the vast majority of the problems involved simple proportion. With regard to problems involving division in particular, most of them involved the idea of partition. Thus, it was concluded that simple-proportion problems are considered prototypical, and that teachers seem to have a limited notion about problem-situations that involve multiplication. Such limitation may have repercussions on how teachers deal with the textbooks that they adopt, especially, on their ability to analyse critically the problems proposed in these books, and on their capacity to make adjustments that they consider necessary in order to develop the knowledge of mathematics of their students.

Another study also investigated elementary school teachers was conducted by Merlini et al. (2015). In this study, about the introductory teaching of multiplication, the authors analysed how the participants, during a teaching training, reflected about their own didactical actions when teaching multiplication to 4th graders. A semi-structured interview was held shortly after the researchers observed classes the participants ministered. The way the teachers reflected about their own didactical practices were stimulated according to Schön's model (2000): reflection-on-action (reflection on the actions carried out in the classroom), reflection-in-action (reflection made when the action is being carried out), and reflection-for-action (reflection on the outcomes of the actions performed). The main result was that after applying the three types of reflection, teachers tended to change their unsuccessful didactical actions and could adopt ones that were more efficient. For the authors, teachers showed a progress in their understanding and evaluation of their own didactical actions, being able to improve their teaching practice. Probably, this was because the teachers were aware of the discontinuity between the addition and multiplication conceptual fields, and of the need to explore situations that require students to apply forms of reasoning that were appropriate to the field of multiplication.

Taken together, the investigations mentioned above have relevant educational implications. A first implication is that teachers need to be fully aware of the limitations and difficulties experienced by students in the field of multiplicative structures. Diagnostic studies contribute to this knowledge. Once these difficulties are known, it is possible to think of ways of teaching that will help students overcome these difficulties, and that will promote increasingly more sophisticated forms of reasoning. In this sense, intervention research can serve as a basis for the creation of more effective and appropriate didactic situations that will allow the development of the mathematical reasoning. It is necessary, however, to adapt these types of research to the school context. Such adaptation, although complex, is possible.

A second implication is that the teacher's knowledge and conceptions are equally crucial to creating an effective learning environment. Therefore, it is necessary to make teachers more knowledgeable, both in relation to the knowledge that the students possess, and in relation to methodological issues to be considered by the teacher, such as the ability to formulate problems and the ability to reflect about their actions in the classroom (Gonzales, 1994; Tardif, 2002).

It is important to mention that the studies carried out within this cognitive approach are characterised by continuity, complementarity, and depth. More than theoretical or methodological tensions and disagreements, the results of these investigations have gradually added new information to this field of knowledge, both with regard to the operative invariants mobilised by the students, as well as the teacher's conception of the conceptual field of multiplicative structures. Moreover, in spite of the theoretical cohesion between them, these studies present a wide diversity of methodological resources, and allow investigating different aspects of the same phenomenon. A final, but equally important, point about these cognitive-based studies is that they have implications for mathematics education, and belong therefore to a field of knowledge of international recognition: the psychology of mathematics education.

10.3.2 Studies Which Focus on the Linguistic and the Cultural Aspects Involved in the Processes of Teaching and Learning of Mathematics

To consider language in the classroom during a mathematics class means to discuss its different aspects and uses. These can range from the individual level (expressing the understanding of knowledge) to the social level (involving the different socially constructed discourses and interactions, and how they affect the learning process). In this focus, the studies discussed in the Working Group Cognitive and Linguistic Processes in Mathematical Education also adopted different theoretical and methodological perspectives. Most frequently adopted is the social-historical-cultural perspective (Vygotsky and Leontiev), which calls attention to the fact that all human activity is *mediated* by semiotic instruments, and that these instruments are constructed during the interactions and discursive practices developed in the social context, in this case, the mathematics class. These issues have been widely discussed in the international literature, with reference to the role attributed to language and the cultural aspects in the practices involving the learning of mathematics.

The latest studies presented in the working group (Carrião, 2012; Frant, 2012; Kistemann Jr., 2012) draw attention to the fact that language in the investigation of learning processes not only has the role of *mediator* between the student and the mathematical knowledge, but it is also the sole form of access to students' way of thinking. In other words, according to the authors, it is through the language used by the teacher and by the students that we can obtain information about how they understand a particular topic.

Another language aspect considered in the investigations is the social norms of conduct that determine what can be said, how and where. This stems from the fact that every social group has its own repertoire of forms of discourse in communication (Carrião, 2012). Kistemann Jr. (2012) draws attention to the fact that we aspire to be socially accepted, and this will determine our discursive choices. This view of language as social practice means that the study of language cannot be restricted to the analysis of isolated phrases or words. For someone to understand the use of words in a given context, it is necessary to interact with the social group that makes use of those words. Thus, in order to analyse the discourse used in the classroom during a mathematics class, we must consider the various elements present in the interactions that take place, such as gestures, intonation, written material, and drawings. This is because the mental processes emerge, mainly, through the internalisation of the social discourse (Carrião, 2012; Frant, 2012).

In order to examine the role of language and the cultural aspects involved in the acquisition of mathematical knowledge, researchers have adopted some theoretical references, namely: (1) the semantic field theory and critical mathematics education (Lins, 1994; Skovsmose, 2005); (2) the historical-cultural perspective of the activity (Engeström, 1987; Leontiev, 1978); (3) the conceptual mapping theory (Lakoff & Johnson, 1999) and the idea of metaphor (Sfard, 2008); and (4) Vygostky's sociocultural perspective and Bakhtin and Voloshinov's (1992) discourse analysis.

Research that adopt these theoretical perspectives often involve the observation and systematic analysis of the classroom activities and of the teaching material used by the teacher. Our focus on this topic is in the discussion of three papers: Carrião (2012), David and Tomaz (2012a, b), and Frant (2012). These studies carried out in WG09 address recurrent questions regarding the role of language and the cultural aspects present in the processes of teaching and learning of mathematics.

The recourses of analysis used, as in any research, are determined by the research question and by the theoretical perspective adopted. Frant (2012) uses the idea of conceptual mapping to verify the use of compression and metonymy in the student's text and in the textbook. David and Tomaz (2012a, b), on the other hand, focus on the rules that govern the activities and the actions of the individuals, analysing the way students participate in the classroom. Carrião (2012), also investigating classroom discourse, uses strategies of discourse analysis to look for language features brought from other fields, in this case nominalisations.

A number of studies in WG09 share the idea that social aspects are involved in the learning process and that the development of thought is determined by language, in a specific way and by the cultural context of the individual's social group.

David and Tomaz (2012a, b) and Frant (2012) contrast two different perspectives of learning. Frant (2012) stands in favour of the one that she calls participatory metaphor, and according to which mathematics is a discourse and learning mathematics is to participate in the development of this discourse. The author highlights the role of everyday language in the construction of knowledge. For her, from the moment that the cognoscente subject takes hold of the text, everything that is said by others acquires meaning, they create objects and produce knowledge. David and Tomaz (2012a, b), on the other hand, adopt the perspective of expansive learning. From this perspective, it is assumed that learners construct a new object and motive for its use, and that they put it into practice in a process of expansive transitions of the actions of an activity.

Both the studies by Frant (2012) and Carrião (2012) point to questions related to language elements (e.g. nominalisation, compression, and metonymy) in the classroom discourse. These studies show that students' ability to use such elements will determine how they participate in the classroom activities, as well as influence their understanding of the concepts presented. The authors also draw attention to the fact that these language elements are used in our daily life although we are not always aware of this fact. In order to understand the use of words in the classroom dialogue, participants need to agree on the meanings attributed to these words, in particular with regard to compression and metonymy. Metonymy occurs when one entity is used to refer to another, and its main function is referential. Compression is a process that shortens the distance between mental spaces, juxtaposing two spaces and creating a third. Without this agreement, the classroom dialogue can become difficult (Frant, 2012).

Nominalisation is the linguistic process that turns verbs into nouns, indicating the resulting action or event, instead of the verb. Carrião (2012) shows that nominalisation is a feature of the mathematics class discourse, being a direct reflection of the discourse of the field of academic mathematics, as shown by other studies (e.g. Burton & Morgan, 2000; Meaney, 2005). For Carrião (2012), a grasp of the mathematics class discourse, nominalisations in particular, can reconfigure the meaning-construction process. This will not only lead to better school performance, but also, and most importantly, to the development of a specific form of thought that is valued socially.

David and Tomaz (2012a, b), on the other hand, discuss the complexity of the activities performed during a mathematics class. By analysing the mobility of the components of these activities, they explain the complexity of their structure and the superposition of the rules that govern the activities in the classroom and the actions of the subjects, identifying tensions and potential learning. The authors also point out that in order not to miss learning opportunities, we must focus on the main object of the activity. Moreover, they point out that the changes associated to momentary overlapping rules are an aspect that seems characteristic of mathematics activities in school. David and Tomaz (2012a, b) bring to the analysis of the micro level of the classroom, the perspective of Engeström's third generation activity theory (1987, 2001), which is interdisciplinary. This approach connects different realities. It analyses situations ranging from mathematical situations produced in the classroom to everyday situations, being therefore a methodology that can relate the micro and the macro context.

Research developed according to this approach has revealed important educational aspects. These aspects need to be problematised and discussed with the teachers when pondering the role they attribute to language, and to the cultural aspects present in the teaching and learning processes. Among such aspects, the following can be highlighted: (1) the need to develop pedagogical practices in which the tensions brought about by the different ways of interpreting mathematical records are confronted, rather than ignored; (2) the need for teachers and authors of textbooks and curricula to use an approach based on research on cognitive and linguistic processes, and not only on mathematical logic; (3) the need for students to have a good command of the mathematics class discourse, in particular nominalisations, and of the use of symbols, because not mastering these can result in exclusion, or at least limited classroom participation.

The studies that consider the role of language and of the cultural aspects in the teaching and learning processes, although presenting different theoretical-methodological approaches, share a common ground. For example, all such studies consider the student as a participant in a community, and analyse the learning process in a contextualised way, looking not only at the nearest environment, the classroom, but also at the wider social context. This contributes to the construction of a learning that is consistent and, to a certain extent, coherent. In addition, this diversity of research has contributed to the area of mathematics education, bringing different references from other areas and adapting them to the interpretation of several questions in this field, such as learning and communication in the school context.

10.4 Final Remarks

A retrospective look at the investigations about the cognitive and linguistic processes in mathematics education in Brazil shows that they reflect an ongoing search for possible answers about the role of language, of cognition, and of the cultural aspects present in the teaching and learning of mathematics. WG meetings represent an opportunity to open dialogues between different theoretical frameworks from different areas of knowledge and the use of a variety of methodological approaches. The diversity of the studies also contributes to establishing a dialogue between these theoretical frameworks and methodologies and the cultural reality of Brazil, and, when necessary, adapt them to this reality.

Such diversity has also been recognised on the international scene (e.g. English & Sriraman, 2005; Hoffmann, 2006; Moschkovich, 2010; Sfard, 2012), among researchers interested in the role of language in teaching and learning in the area of mathematics education. For Hoffmann (2006), this diversity can be related to the fact that there is no unifying theory that can encompass the different ways in which semiotic questions have been addressed in research. This indicates the need for more investigations that can avoid an undue separation between the internal and external processes of learning mathematics of the individual, from its historical and semiotic aspects.

The theoretical choices made with regard to the linguistic, cognitive, and cultural aspects explored make it possible to organise different explanatory models about the cognitive functioning of individuals and its relation to the mathematical knowledge. The research presented by WG09, and which constitutes its identity, portrays the interests within this field in the Brazilian scenario.

It should be stressed that the research developed in Brazil provides data that support theoretical questions posed by different scholars (e.g. Bakhtin, Engeström, Lakoff, Leontiev, Nemirovsky, Vergnaud, Piaget, and Vygotsky). These investigations have also contributed to the discussion of various topics. Among the several contributions of the studies area presented in the group, we highlight the following:

1. The role of gestures in language with reference to the coordination of semiotic tools to create and communicate mathematical meanings, and to the coordination of material tools and semiotics in the construction of learning environments that can contribute to the acquisition of mathematical knowledge by students with special needs. Such discussions highlight the cognitive importance of the body in mathematical thinking, a topic explored by several researchers in the international arena. The studies with blind students developed in Brazil (Fernandes & Healy, 2010; Healy & Fernandes, 2008, 2011) have revealed the role that gestures can play in the construction of mathematical knowledge. For Healy and Fernandes (2008, p. 2), "[...] If we can identify the differences and similarities in the mathematical practices of those whose knowledge of the world is mediated through different channels, perhaps we gain more robust understanding of the relationships between experience and cognition more generally".
2. The resolution of mathematical problems as a way to access both students' and teachers' conceptions, and as a way to relate problem solving to specific mathematical concepts (e.g. Lautert, Spinillo, & Correa, 2012; Spinillo & Lautert, 2006; Spinillo & Silva, 2010; Spinillo, Silva, & Lautert, 2016).
3. The analysis of classroom practices, in particular the multiple perspectives of the different participants in this complex social context, has led Brazilian researchers to make use of different methodological perspectives. Some researchers have adopted the theory of activity, taking Leontiev and Engeström as reference. Tomaz and David (2011) bring to the analysis of the micro level of the mathematics

classroom Engeström's interdisciplinary perspective, in an attempt to understand the complexity of the structure of the activities in the classroom, identifying the tensions and potential learning.

4. The language currently used in mathematics classes and in mathematics schoolbooks is the result of several research works that seek to understand their role in the communication and learning processes. Research by Frant (2012) and Carrião (2012) brings this discussion to the daily activities of the classroom, analysing how elements such as metonymy and nominalisation, which characterise this language, affect communication, the participation in the activities, and student's learning. In these works, we can see this language being used in the social interactions of the classroom.

To conclude, we bring the statement made by Rabello et al. (2000, p. 346) at SIPEM I. With reference to two approaches relevant to the theory and practice that have accompanied all the reflections produced over the last 16 years, they said "[...] At the same time that theory makes practice meaningful, practice makes theory significant: practice reveals to what extent the theoretical assumptions that support it can provide answers to the questions that emerge in the classroom".

References

Bakhtin, M., & Voloshinov, V. (1992). *Marxismo e Filosofia da Linguagem* (6th ed., M. Lahud & Y. L. Vieira, Trans.). São Paulo: Ed. Hucitec. (Original work published 1929).

Borges, F. A., & Nogueira, C. M. I. (2012). Aulas de matemática para alunos surdos inclusos no ensino fundamental. In *Anais do V Seminário Internacional de Pesquisa em Educação Matemática* (pp. 1–20), Petrópolis, Rio de Janeiro.

Burton, L., & Morgan, C. (2000). Mathematicians writing. *Journal for Research in Mathematics Education, 31*(4), 429–453.

Carrião, A. (2012). A nominalização como marca do discurso na aula de matemática. In *Anais do V Seminário Internacional de Pesquisa em Educação Matemática* (pp. 1–18), Petrópolis, Rio de Janeiro.

Charnay, R. (2001). Aprendendo (com) a resolução de problemas. In C. Parra & I. Saiz (Orgs.), *Didática da matemática: reflexões pedagógicas* (pp. 36–47). Porto Alegre: Artmed.

Chica, C. (2001). Por que formular problemas? In K. Smole & M. Diniz (Orgs.), *Ler, escrever e resolver problemas: habilidades básicas para aprender matemática* (pp. 151–173). Porto Alegre: Artmed.

Crespo, S. (2003). Learning to pose mathematical problems: Exploring changes in preservice teachers' practices. *Educational Studies in Mathematics, 52*, 243–270.

Cunha, M. J. G. (2015). *Elaboração de problemas combinatórios por professores de matemática do ensino médio.* Dissertação de Mestrado, Programa de Pós-graduação em Educação Matemática e Tecnológica da Universidade Federal de Pernambuco, Brazil.

David, M. M., & Tomaz, V. S. (2012a). Perspectiva de análise micro da estrutura da atividade matemática em sala de aula. In *Anais do V Seminário Internacional de Pesquisa em Educação Matemática* (pp. 1–20), Petrópolis, Rio de Janeiro.

David, M. M., & Tomaz, V. S. (2012b). The role of visual representations for structuring classroom mathematical activity. *Educational Studies in Mathematics, 80*, 413–431.

David, M. M. M. S., & Watson, A. (2008). Participating in what? Using situated cognition theory to illuminate differences in classroom practices. In A. Watson & P. Winbourne (Orgs.), *New directions for situated cognition in mathematics education* (pp. 31–57). New York: Springer.
Duval, R. (1993). Registres de représentation sémiotique et fonctionnement cognitif de la pensée. In *Annales de Didactique et de Sciences cognitives, IREM de Starsbourg* (Vol. 5, pp. 37–65).
Duval, R. (1995). *Sémiósis et pensée humaine:registres sémiotiques et apprentissages intellectuels*. Suisse: Peter Lang.
Engeström, Y. (1987). *Learning by expanding: An activity-theoretical approach to developmental research*. Helsinki: Orienta-Konsultit.
English, L. (1997). The development of fifth-grade children's problem-posing abilities. *Educational Studies in Mathematics, 34*, 183–217.
English, L., & Sriraman, B. (2005). Theories of mathematics education. In H. L. Chick & J. L. Vincent (Eds.), *Proceedings of the 29th Conference of the International. Group for the Psychology of Mathematics Education* (Vol. 1, pp. 170–202). Melbourne: PME.
Fernandes, S. H. A. A., & Healy, L. (2010). Embodied Mathematics: Relationships between doing and imagining in the activities of a blind learner. In *Proceedings of the 34th Conference of the International Group for the Psychology of Mathematics Education* (Vol. 2, pp. 345–352), Belo Horizonte, Brazil.
Fernandes, S. H. A. A., Healy, H., & Serino, A. P. (2012). Das relações entre figuras para relações em um espaço matematizável: as percepções de alunos cegos sobre transformações geométricas. In *Anais do V Seminário Internacional de Pesquisa em Educação Matemática* (pp. 1–19), Petrópolis, Rio de Janeiro.
Fonseca, M. C. F. R. (2010). Adult education and ethnomathematics: Appropriating results, methods, and principles. *ZDM Mathematics Education, 42*(3), 361–369.
Frade, C., & Borges, O. (2006). The tacit-explicit dimension of the learning of mathematics: An investigation report. *International Journal of Science and Mathematics Education, 4*, 293–317.
Frade, C., & Falcão, J. (2007). Exploring connections between tacit knowing and situated learning perspectives in the context of mathematics education. In A. Watson & P. Winbourne (Orgs.), *New directions for situated cognition in mathematics education* (Vol. 1, pp. 203–231). Norwell: Springer.
Frade, C., & Tatsis, K. (2009). Learning, participation and local school mathematics practice. *The Montana Mathematics Enthusiast, 6*, 96–112.
Frade, C., Winbourne, P., & Braga, S. A. M. (2009). A mathematics-science community of practice: Crossing boundaries. *For the Learning of Mathematics, 29*, 14–22.
Frant, J. B. (2009). As time goes by: Technology, embodiment and Cartesian graphics. In *Proceedings of the 33th Conference of the International Group for the Psychology of Mathematics Education* (Vol. 1, p. 343), Tessaloniki, Greece.
Frant, J. B. (2012). Linguagem, compressão e algumas implicações para a matemática escolar. Perspectiva de análise micro da estrutura da atividade matemática em sala de aula. In *Anais do V Seminário Internacional de Pesquisa em Educação Matemática* (pp. 1–10), Petrópolis, Rio de Janeiro.
Freire, P. C., & Lima, R. N. (2012). O subconstruto parte-todo: uma análise com os três mundos da matemática. In *Anais do V Seminário Internacional de Pesquisa em Educação Matemática* (pp. 1–20), Petrópolis, Rio de Janeiro.
Garnica, A. V. M., Soares, M. T. C., & Buriasco, R. L. C. (2006). *Anais do III Seminário Internacional de Pesquisa em Educação Matemática*. Águas de Lindoia: Sociedade Brasileira de Educação Matemática, São Paulo.
Gonzales, N. A. (1994). Problem posing: A neglected component in mathematics courses for prospective elementary and middle school teachers'. *School Science and Mathematics, 94*(2), 78–84.
Hazin, I., Da Rocha Falcão, J. T., & Leitão, S. (2006). Mathematical impairment among epileptic children. In *Proceedings of the 34th Conference of the International Group for the Psychology of Mathematics Education* (Vol. 3, pp. 249–256), Prague, Czech Republic.

Healy, L., & Fernandes, S. H. A. A. (2008). The role of gestures in the mathematical practices of blind learners. In *Proceedings of the Joint Meeting of PME 32 and PME-NA XXX* (Vol. 3, pp. 137–144). Morelia, México: Cinvestav-UMSNH.

Healy, L., & Fernandes, S. H. A. A. (2011). The role of gestures in the mathematical practices of those who do not see with their eyes. *Educational Studies in Mathematics, 77*(2), 157–174.

Hoffmann, M. H. G. (2006). What is a "semiotic perspective", and what could it be? Some comments on the contributions to this special issue. *Educational Studies in Mathematics, 61*, 279–291.

Itacarambi, R. R. (2010). *Resolução de problemas nos anos iniciais do ensino fundamental*. São Paulo: Editora Livraria da Física.

Kistemann, M. A., Jr. (2012). A produção de significados e a tomada de decisão de indivíduos-consumidores. In *Anais do V Seminário Internacional de Pesquisa em Educação Matemática* (pp. 1–13), Petrópolis, Rio de Janeiro.

Lakoff, G., & Johnson, M. (1999). *Philosophy in the flesh: The embodied mind and its challenge to western thought*. New York: Basic Books.

Lautert, S. L., & Spinillo, A. G. (2012). Os princípios invariantes da divisão como foco de um estudo de intervenção com crianças. In *Anais do V Seminário Internacional de Pesquisa em Educação Matemática* (pp. 1–13), Petrópolis, Rio de Janeiro.

Lautert, S. L., Spinillo, A., & Correa, J. (2012). Children's difficulties with division: An intervention study. *Journal of Medicine and Medical Sciences, 1*, 447–456.

Leontiev, A. N. (1978). *Activity, consciousness, personality*. Englewood Cliffs, NJ: Prentice-Hall.

Leung, S. S., & Silver, E. A. (1997). The role of task format, mathematics knowledge, and creative thinking on the arithmetic problem posing of prospective elementary school teachers. *Mathematics Education Research Journal, 9*(1), 5–24.

Lins, R. C. (1994). O modelo teórico dos campos semânticos: uma análise epistemológica da álgebra e do pensamento algébrico. *Revista Dynamis, 1*(7), 29–39.

Lopes, A. C., & Magina, S. (2012). O xadrez e o estudante: uma relação que pode dar certo na resolução de problema matemáticos. In *Anais do V Seminário Internacional de Pesquisa em Educação Matemática* (pp. 1–20), Petrópolis, Rio de Janeiro.

Lowrie, T. (2002). Young children posing problems: The influence of teacher intervention on the type of problems children pose. *Mathematics Education Research Journal, 14*(2), 87–98.

Magina, S. M. P., Spinillo, A. G., & Melo, L. M. (2015). As estratégias de estudantes dos anos iniciais na resolução de problema combinatório. In *Anais do VI Seminário Internacional de Pesquisa em Educação Matemática* (pp. 1–12), Pirenópolis, Goiás.

Meaney, T. (2005). Mathematics as text. In A. Chronaki & I. M. Christiansen (Orgs.) *Challenging perspectives on mathematics classroom communication*. Connecticut: Information Age Publishing.

Merlini, V., Santos, A., Teixeira, A. C., & Magina, S. M. P. (2015). Processo formativo centrado na escola: as reflexões da professora Maria. In *Anais do VI Seminário Internacional de Pesquisa em Educação Matemática* (p. 13), Pirenópolis, Goiás.

Moro, M. L. F., & Soares, M. T. C. (2006). Níveis de raciocínio combinatório e produto cartesiano na escola fundamental. *Revista Educação Matemática Pesquisa, 8*(1), 99–124.

Moschkovich, J. N. (2010). *Language and mathematics education: Multiple perspectives and directions for research*. Charlotte: Information Age Publishing.

Oliveira, D. L., Albuquerque, L. C., & Gontijo, C. H. (2012). Criatividade matemática: alguns elementos na divisão de quadrados. In *Anais do V Seminário Internacional de Pesquisa em Educação Matemática* (pp. 1–19), Petrópolis, Rio de Janeiro.

Onuchic, L. R., & Alevato, N. S. G. (2004). Novas reflexões sobre o ensino-aprendizagem de matemática através da resolução de problemas. In M. A. V. Bicudo & M. C. Borba (Orgs.), *Educação matemática: Pesquisa em movimento* (pp. 213–223). São Paulo: Cortez.

Pires, C. M. C., Curi, E., Rabelllo, M., Pavanello, R., & Valente, W. R. (2003). *Anais do II Seminário Internacional de Pesquisa em Educação Matemática*. Sociedade Brasileira de Educação Matemática. São Paulo: Santos.

Rabello, M., Lins, R., & Da Rocha Falcão. (2000). Processos cognitivos e linguísticos na Educação matemática. In T. M. Campos & C. M. C. Pires Anais do I Seminário Internacional de Pesquisa em Educação Matemática. *Sociedade Brasileira de Educação Matemática* (pp. 329–372). São Paulo: Serra Negra.

Schön, D. (2000). *Educando o Profissional Reflexivo: um novo design para o ensino e aprendizagem*. Porto Alegre: Artes Médicas Sul.

Sfard, A. (2008). *Thinking as communicating*. Cambridge University Press. Ebook.

Sfard, A. (2012). Introduction: Developing mathematical discourse—Some insights from communicational research. *International Journal of Educational Research, 51–52*, 1–9.

Skovsmose, O. (2005). Foregrounds and politics of learning obstacles. *For the Learning of Mathematics, 25*(1), 4–10.

Souza, E. R. R. de. (2015). *Estruturas multiplicativas: Concepção de professor de Ensino Fundamental.* Dissertação de Mestrado, Programa de Pós-Graduação em Educação Matemática, Universidade Estadual de Santa Cruz.

Spinillo, A. G., & Lautert, S. (2006). Exploring the role played by the remainder in the solution of division problems. In *Proceedings of the 30th Conference of the International Group for the Psychology of Mathematics Education* (Vol. 5, pp. 153–160).

Spinillo, A. G., Lautert, S. L., Santos, E. M., & Silva, J. F. G. (2015). Uma análise de problemas do campo multiplicativo elaborados por professores do Ensino Fundamental I. In *Anais do VI Seminário Internacional de Pesquisa em Educação Matemática* (pp. 1–9), Pirenópolis, Goiás.

Spinillo, A. G., & Silva, J. F. G. (2010). Making explicit the principles governing combinatorial reasoning: Does it help children to solve Cartesian product problems? In *Proceedings of the 34th Conference of the International Group for the Psychology of Mathematics Education* (Vol. 4, pp. 216–224).

Spinillo, A. G., Silva, J. F. G., & Lautert, S. L. (2016). Ensino e aprendizagem de conceitos matemáticos a partir da explicitação dos princípios invariantes. In J. A. de Castro Filho, M. Chagas, P. M. Barguil, D. L. Maia, & J. Lima (Orgs.), *Matemática, cultura e tecnologia: perspectivas internacionais* (pp. 35–48) Curitiba: CRV.

Tardif, M. (2002). *Saberes docentes e formação profissional*. Petrópolis, Rio de Janeiro: Editora Vozes.

Tomaz, V. S., & David, M. M. (2011). Classroom activity promoting students' learning about the use of drawings in geometry. In *Proceedings of the 35th Conference of the International Group for the Psychology of Mathematics Education* (Vol. 4, pp. 257–264).

Torisu, E. M. (2012). Acompanhamento extraclasse e fortalecimento das crenças de autoeficácia matemática. In *Anais do V Seminário Internacional de Pesquisa em Educação Matemática* (pp. 1–22), Petrópolis, Rio de Janeiro.

Vergnaud, G. (1983). Multiplicative structures. In R. Lesh & M. Landau (Eds.), *Acquisitions of mathematics concepts and procedures* (pp. 127–174). New York: Academic.

Vergnaud, G. (1988). Multiplicative structures. In H. Hiebert & M. Behr (Eds.), *Research agenda in mathematics education. Number concepts and operations in the middle grades* (pp. 141–161). Hillsdale, NJ: Lawrence Erlbaum.

Vergnaud, G. (1990). La théorie des champs conceptuels. *Recherches en Didactique des Mathématiques, Grenoble, 10*(23), 133–170.

Vergnaud, G. (2003). A gênese dos campos conceituais. In E. Grossi (Org.). *Por que ainda há quem não aprende? A teoria* (pp. 21–60). Ed. Vozes, RJ: Petrópolis.

Viana, O. A. (2012). A identificação de propriedades e a habilidade de planificação de figuras geométricas espaciais. In *Anais do V Seminário Internacional de Pesquisa em Educação Matemática* Petrópolis, Rio de Janeiro.

Zunino, D. L. (1995). *A matemática na escola: aqui e agora*. Porto Alegre: Artes Médicas.

Chapter 11
Research on Mathematical Modelling in Mathematics Education in Brazil: Overview and Considerations

Elizabeth Gomes Souza, Lourdes Maria Werle de Almeida, and Tiago Emanuel Klüber

Abstract This chapter aims at presenting an overview of the research about mathematical modelling and mathematics education in Brazil from the reports and studies developed within the scope of the International Seminars for Research in Mathematics Education, editions from 2003 to 2015. This diachronic choice aims to thematize the current studies on modelling, placing them in the historical perspective of themes and debates in the field of mathematical modelling in mathematics education throughout the different editions of the seminars. The analysis of the work discussed with the working group since its conception indicated that most research is exploratory with a strong tendency for qualitative methodologies to dominate. Brazilian studies in this area thematize aspects that emerge from the interior of modelling practices, viewed through the lens of several theories, some with a philosophical basis, some with theories of language and others with mathematics education. The chapter ends by considering future directions and suggests that what still needs to be consolidated is a more proficient and frequent dialogue between the different research groups established in the different Brazilian regions.

11.1 Introduction

Research in the field of education has been regularly present in the Brazilian scientific community since the late 1930s. According to Gouveia (1971), in Brazil, the recognition of the importance of scientific research in this area results, among other factors, from the installation, in 1937, of the National Institute for Educational

E. G. Souza (✉)
Universidade Federal do Pará, Belém, Brazil

L. M. W. de Almeida
Universidade Estadual de Londrina, Londrina, Brazil

T. E. Klüber
Universidade Estadual do Oeste do Paraná, Cascavel, Brazil

A. J. Ribeiro et al. (eds.), *Mathematics Education in Brazil*,
https://doi.org/10.1007/978-3-319-93455-6_11

Studies and Research *Anísio Teixeira,* aimed to develop research on the problems perceived in education to provide public administration with decision-making subsidies.

The new math movement, which occurred in the 1950s and 1960s, advanced significantly the emergence of the research in the area of mathematics education. The establishment of mathematics education as a professional and scientific field in Brazil also began from this movement that became stronger in the late 1970s and during the 1980s. The repercussions of this new math movement generated among mathematicians and mathematics educators favourable and unfavourable arguments regarding the problems it would bring to mathematics education. The Brazilian Society of Mathematics Education was established in this period, as well as the first post-graduate study programmes in mathematics education.

The configuration of mathematical modelling as a research field in mathematics education was identified as of the 1980s (Biembengut, 2009). Since then, it has intensified and can be found with a certain regularity in national and international scientific journals in the past decades.

Sometimes aligned, sometimes in disagreement, the studies characterized as knowledge development in mathematical modelling are based on mathematics, mathematics education, and education arguments and subsidies to respond to questions on the connections between modelling[1] and teaching, learning, curriculum, and teacher training, among other dimensions of mathematics education.

In general, the social spaces of education have been expanding in face of different fronts of knowledge production, creation, and recognition of identities and cultural and social practices. According to Candau (2000), "different educational ecosystems have been proposed as new space-time of knowledge production necessary for the formation of active citizenship in society" (p. 13).

In this context, the development of mathematical modelling as a field of knowledge is consolidated and recorded by research reports and the debates they initiate. In this sense, the VI International Seminar for Research in Mathematics Education (SIPEM VI) occupies a prominent position, considering that it aimed to bring together the national and international scientific communities to disseminate results, provoke reflections, and seek advances for research and practices in the field.

Regarding the development of mathematical modelling in Brazilian education, since the earliest experiments conducted by Professor Aristides Barreto at the Universidade Estadual de Campinas (UNICAMP), in the state of São Paulo, Brazil, in 1979, and the first courses for elementary school teachers conducted by Professor Rodney C. Bassanezi at the Faculdade de Filosofia, Ciências e Letras Guarapuava (FAFIG) in the, state of Paraná, Brazil, the area has moved forward—both with respect to the insertion of new researchers and study themes and the establishment of its own space for debate and sharing of research—and notably the National Conference on Modelling in Mathematics Education, which held its 10th edition in 2017.

[1] The term "modelling" should be considered in this article with reference to mathematical modelling in Mathematics Education.

With the presence of a group of researchers who share views and practices in mathematical modelling; of specific events for such sharing and the continuity of research; and of the strengthening increase of discussions on these practices and views in the scope of the SIPEMs, it is possible to affirm that the research community in mathematical modelling is consolidated.

Grounded upon this consolidation, this article aims at presenting an overview of the research about Brazilian mathematical modelling from the reports and studies developed within the scope of the International Seminars for Research in Mathematics Education, editions from 2003 to 2015. This diachronic choice aims to thematize the current studies on modelling, placing them in a historical perspective of themes and debates in the field of mathematical modelling in Mathematics Education throughout the different editions of the seminars.

From the experience of two of the authors of this article—one as a coordinator and vice-coordinator of the Working Group10 (WG10) between 2009 and 2015, and the other as an active participant of the WG10 also in the same period—and based on the fact that one of the authors has participated in all SIPEM editions since the creation of the WG10, we admitted the possibility of articulating the contents of the reports with the authors' memories. This process provides argumentative consistency to the chapter in such a way that we can undertake reflections on the itinerary of research on mathematical modelling in Brazilian mathematics education.

From the authors' point of view, the methodological and argumentative choice contemplates the proposal of this study, namely, to offer a comprehensive view of the research on mathematics education in Brazil, focusing on the investigations concerning the WG10 throughout the different international seminars' editions. In other words, the existence of the reports, which already expresses an initial and situated reflection of the research for each edition, the resumption of central aspects of the published works, the authors' experiences in the administrative instances, and the extensive involvement with the community guarantee the explicitness of a perspective that is relevant to the area.

It is worth mentioning that other studies discussing research on mathematical modelling have already been developed. Some of which have been published in the proceedings of the international seminars. Nevertheless, this is an original work, because it is does not focus exclusively on a specific edition of the event, as the studies by Bicudo and Klüber (2011), Klüber and Burak (2014), and Klüber, Tambarussi, Loureiro, Wichnoski, and Oliveira (2015), but presents interlocutions. It is also different from other works that focused on studies within research groups, such as those by de Almeida (2006) and Soares and Borba (2014). Finally, it does not present the same connotation of the study by Barbosa (2007), in which the author presented an essay on Brazilian mathematical modelling research proposing some criteria of scientificity as lenses to guide the area in the research.

However, these studies present indicatives of a research agenda whose consolidation we investigated in this chapter, taking into account the research presented in the different SIPEM editions.

11.2 A Summarized History of the Main Themes Approached by Working Group 10 of SIPEM (2003–2015)

Working Group 10, which focuses on research on mathematical modelling was created during the First International Seminar for Research in Mathematics Education (SIPEM) in 2000.

In 2003, during the second edition of the event[2], with the WG10 already established, nine articles attempted to identify boundaries that legitimized the presence of mathematical modelling in mathematics education. The repercussions of these discussions on research and practice of mathematical modelling in mathematics education in Brazil can be observed in the first joint publication by WG10, organized by Barbosa, Caldeira, and Araújo (2007).

In SIPEM III, in 2006, syntheses of the investigations discussed, as well as their possible contributions to the configuration of practices and research in mathematical modelling, were recorded in a report produced in the scope of WG10. Thus, we can refer in this chapter to the themes and discussions indicated in the reports of the last four editions of the SIPEM, related to the years 2006, 2009, 2012, and 2015, as well as the articles published in proceedings of each SIPEM.

At the SIPEM III (2006), WG10 had about 30 participants, including postgraduate students and professors/researchers. Fourteen research reports were discussed, which Barbosa, Araújo, and Caldeira (2006) subsumed in five thematic groups: (1) Epistemological issues and perspectives; (2) Educational practices focusing on student development; (3) Information and communication technologies; (4) Professional development of teachers; and (5) Reflections on research in mathematical modelling.

The research on aspects of epistemology and delimitation of modelling addressed key terms to understand modelling in mathematics education. These studies invited the research community to discuss themes such as mathematical reality and mathematical models.

Regarding educational practices, the emphasis of the studies seems to focus on the argumentation on the potential use of modelling to motivate students in mathematics classes.

In these studies, communication and information technologies focused on the creation of a virtual locus of discussions on the insertion of classroom modelling by teachers and researchers, identifying the formation of virtual centres that would provide the participation of a large number of individuals interested in such discussions, disregarding geographical barriers.

Concerning teacher training, the studies highlighted the need to insert mathematical modelling in teacher training courses aiming at familiarizing them with teaching practices based on modelling. During this SIPEM edition, teacher's lack of knowledge on modelling in teacher training courses became clear.

---

[2] The report produced from discussions occurred in this edition of the SIPEM is not available.

As for the reflections about research on mathematical modelling, although only one of the studies presented at the event explicitly referred to this theme, the report indicates that the characterization of research in mathematical modelling was still in its consolidating phase. Thus, the rapporteurs documented discussions conducted by WG10 coordination to evaluate works on relatively primary research questions, such as: "What elements comprise research on modelling in mathematics education?" or "What would differentiate a research report from an experience report?" These questions extended throughout the days of the event and were inserted in the axis of research in modelling, proposing that future studies should "pay strict attention to the methodology, explaining the concepts adopted and the consonance with the purpose of the research" (Barbosa et al., 2006, p. 9).

In 2009, during the SIPEM IV, about 20 participants discussed 11 research reports. This edition's report (Barbosa, Caldeira, & Araújo, 2009) indicates that "the approved studies [...] were organized around four themes [...]": (1) Research on mathematical modelling; (2) Students' practice in mathematical modelling; (3) Philosophical reflections and implications for modelling; and (4) Mathematical modelling and teacher training.

Analysis of the texts of the studies discussed in the IV SIPEM indicates that some problems identified in the previous edition of the event were resumed and that advances were observed especially concerning mathematical modelling practices at different levels of schooling.

The agenda of this SIPEM edition also focuses on meta-analysis about mathematical modelling studies presented at other events in the area of mathematics education. This survey identifies the need for the area to address its own research, continuing the discussions raised in the previous edition of the event.

Epistemological and philosophical aspects were also amplified and diversified in the SIPEM IV compared with the previous edition of the event. In 2009, some of these aspects were thematized due to the implications of adopting a specific epistemological basis for the configuration of the development of modelling activities in the field of mathematics education.

Regarding teacher training, some studies did not show the intention to hold a debate on teacher training in modelling, but indicated some possibilities that the use of modelling can generate for this training, such as critical reflection on the role that mathematics plays in society.

In contrast, the discussions on aspects linked to teacher training in modelling itself focused on the continuing education of teachers in modelling mediated by virtual environments. In this context, the study by de Almeida (2006) argues that teachers must be prepared to teach mathematics through modelling. To this end, the author indicates that teacher training programmes should provide the teachers with the opportunity to: "learn" about mathematical modelling; "learn" through mathematical modelling; "teach" using mathematical modelling.

In conclusion, the report of the IV SIPEM recommends that researchers in that area should focus on three aspects: discussion of teacher training considering that "the group listed issues not explored by the studies" (Barbosa et al., 2009, p. 6); the characterization of the studies developed with a focus on the relations and limits between research on modelling and teachers' practice; and the need for researchers to invest on generating meta-studies on research on modelling in mathematics education, in order to analyse and reflect about the field and its development.

In 2012, during SIPEM V, approximately 28 participants discussed 15 research reports. A significant number of these works brought argumentative and theoretically grounded elements regarding students' learning in mathematical modelling activities. This fact suggests a theoretical growth of thematic studies in the field.

Furthermore, this theme was inserted in the context of WG10 discussions at V SIPEM in the light of a diversity of theoretical perspectives, such as sociocultural, discursive, cognitive, and semiotic biases employed in the analysis of activities developed with students at different schooling levels. In this context, Peirce's (1970) semiotics and the social aspects pointed out by Bernstein (2000) are theoretical references that could bring elements to address the relevance of mathematical modelling for the mathematical formation of students.

Several researchers from different Brazilian states participated, in this edition of the event (Almeida, Araújo, & Bisognin, 2012). The issues addressed were organized into four groups: (1) Epistemological questions of mathematical modelling itself; (2) Teacher training in mathematical modelling; (3) Reflections on research in mathematical modelling; and (4) Articulation of mathematical modelling with other theories.

Concerning the epistemological aspects of modelling, elements that had been little addressed until then were discussed. One of them refers to the role played by the definition of premises and assumptions for the development of mathematical models (Bean, 2012). In contrast, the perspective called "analysis of models" was identified as a possibility to address a non-mathematical problem through mathematics. The very understanding of problem in mathematical modelling was discussed, especially when the context is a cybernetic reality.

With respect to the theme group "teacher training in mathematical modelling", we noticed the presence of studies, the analyses of which indicate the importance of modelling in teacher training in general, such as the discussion of its potential to initiate reflections on teaching practice. However, in this edition of the SIPEM, there was a significant number of studies that directed their objectives and conclusions to the understanding of formative processes related to teacher training on modelling.

The reported studies showed specificities of mathematical modelling, indicating that it may give rise to teaching practices that are different from those experienced by teachers in the classroom. These practices would shape the way teachers develop mathematical modelling activities, generating tensions in the choices of teaching strategies they would select.

Research in SIPEM V presented elements that had not yet been explored in previous editions. They were concerned with the understanding of teacher training on modelling, regarding preparation of curricular educational modelling materials for teachers. However, the discussion indicated that the people who organized those materials needed to be explicit on their intentions.

Just like in the previous editions, in this SIPEM an investigation on studies on mathematical modelling, a type of meta-analysis, was discussed. More specific than the study by Araújo (2010), this work focused on the analysis of the research objectives present in the articles published in the proceedings of a specific national modelling event in mathematics education, the National Conference on Mathematical Modelling. According to Araújo (2010), the research community is undergoing a movement of reflection about their own field of research. One of those debates was conducted by Araújo herself, indicating that the studies presented in the event under analysis can still be improved, both with regard to the methodology used and with the theoretical articulation established.

At SIPEM V, researchers showed the intention to explain the relations between research on modelling and its implementation in the school context—a thematic deemed necessary in the reports of the SIPEM III. In fact, Araújo and Campos (2015) suggested the need to indicate successful and unsuccessful practices with modelling in the research reports. This would be a possibility for those reports to reflect on school reality matters more accurately, thereby reducing the distance between the arguments presented in the modelling research reports and those arising from their inclusion in the school/academic context.

Regarding the recurrence of different theoretical contributions for the analysis and interpretation of mathematical modelling activities, more specifically with the purpose of paying attention to the students' learning process when involved in modelling activities, elements of semiotics were presented as indicators that the mathematical modelling favours the student's understanding of concepts and applications. Besides, the modelling potential to provide interactions that are source for learning was also evidenced in this edition of SIPEM. These interactions could also be strengthened with the use of information technology, such as computers and software programmes.

The themes discussed in the different editions, however, while indicating the consolidation of the modelling area in mathematics education, also give rise to new research, as presented in the analysis of the discussions in the ambit of the SIPEM VI, as we present in the next section.

11.3 The Research Discussed at the SIPEM VI

The VI International Seminar for Research in Mathematics Education (SIPEM VI) was held in November 2015 in the state of Goiás, Brazil, where approximately 25 teachers/researchers and graduate students discussed 12 research reports.

According to Klüber and Almeida (2015), the research topics addressed in the event can be organized into three theme groups: (1) Investigations of aspects concerning "doing and the theory of mathematical modelling" in the light of different theoretical frameworks; (2) Investigations on the use of curricular (educational) materials of mathematical modelling in mathematics education; and (3) Investigations on teacher training in mathematical modelling.

With regard to the discussions held at WG10 debate sessions, Klüber and Almeida (2015, p. 55) stated that:

> In practically all the studies, we inquire about the position of research in the field of modelling in mathematics education. We identified that the area resumes themes that have already been discussed at other times, but under a different perspective and at a more comprehensive level of analysis.

In the research group that discusses mathematical modelling in the light of some theoretical and philosophical perspective, we locate 8 of the 12 studies reported. One of them (Almeida, Tortola and Sousa, 2015) discusses epistemological questions regarding the formulation of hypotheses in mathematical modelling activities.

In this case, the authors have brought to light the philosopher Ludwig Wittgenstein, either to elucidate that different meanings can be traced from the identification of different contexts, or to present Wittgenstein's indications for hypotheses. The analysis Almeida, Tortola, and Sousa (2015) conducted, based on the theoretical assumptions about the development of an activity, fosters the idea that the formulation of hypotheses in activities of mathematical modelling cannot occur regardless of the characteristics of the phenomenon. However, the experience of individuals, either in relation to knowledge on the phenomenon, or with respect to mathematics, or regarding their experience with the development of modelling activities, favours "the mechanism of the working hypothesis", as considered by Wittgenstein.

In another paper, Brito, Oliveira, and Milani (2015) present a phenomenological analysis aiming at unveiling the meaning of simplification of reality in modelling activities involving geometry contents in school environments. The authors consider that if, on the one hand, simplifications and mathematical idealization in modelling are essential to facilitate the resolution of a problem or to adapt it to the students' level, on the other hand, they can mask and overlap the lived space, excluding the subjective aspects that comprise the meaning source of modelling.

By means of the analysis of modelling activities described in the seven editions of the International Conference on the Teaching of Mathematical Modelling and Applications (ICTMA), a phenomenological understanding identifies three categories: Idealized Spaces, Partially Idealized Spaces, and Spaces with Absence of Idealizations. By presenting three possible levels of simplification of the reality, the results of this study also show three levels of idealization of the lived space in the activities of modelling with geometry. In addition, the results show how the level of simplification shown in these activities can be correlated with the level of subjectivity required for solving the problem. This subjectivity is intrinsic to the modelling process in choosing the theme, formulating the problem, obtaining data, preparing the model, resolving and evaluating the solution, etc. The previous simplifications in modelling activities tend, therefore, to hide and mask the subjectivity of the mathematical modelling process itself.

The analysis of the first experience of students with mathematical modelling activities is the focus of another study. In this case, Canedo Jr and Istemann Jr (2015) investigated the participation and decision-making of students when involved with modelling. To analyse the actions of these students as a collective of human-with-media beings, the authors use Engeström Yrjö's activity theory. The outcomes indicate that the participation of media in modelling exceeds that of artefacts, because they not only mediate the actions of the individuals, but also participate in them and influence the modelling activity in all of its content. Another result of the same study shows that the presence of modelling in the basic education curriculum both demands and favours the development of student autonomy under teacher mediation with the use of media and problem-situations that invite students to develop creativity.

Another study was based on elements of critical mathematics education and aimed to characterize its influence in the negotiation space constituted by a group of students when they discussed the theme of their modelling project, aiming at deciding what to consider in a mathematical modelling activity. Supported by a qualitative methodological approach, the authors analysed the discussions of a group of undergraduate students while developing a modelling project. Araújo and Campos (2015) indicated four characteristics relevant to this influence: the constitution of the negotiating space itself; the mediation style used by the teacher; the consideration of social variables in modelling; and how these variables are incorporated into the theme of mathematical modelling.

Milani, Kato, and Cardoso (2015), based on the cognitive theory of multimedia learning, investigated the potentialities of the videos that accompany the textbook of high school for the development of mathematical modelling activities. The authors analyse the videos that accompany textbooks and reckon them as allies of the process of teaching and learning geometry through mathematical modelling. According to the authors, videos are potential generators of learning environments with reference to reality and semi-reality, exhibiting and instigating environmental, cultural, historical, social, and other situations that demand translations of reality through mathematics.

Developing research with higher education students, more specifically in the operational research discipline of a production engineering course, the authors discussed the unfolding of the process of problem determination in the development of mathematical modelling activities. Supported mainly by Deleuze (2011) and their characterization for meaning and making sense, Weingarten and Dalla Vecchia (2015) addressed the relevance of linking the determination and resolution of the problem to the multiplicity of possibilities for resolution rather than the presentation of a single solution. The authors concluded that the relationship between problem and meaning was provided by the sense that students attributed to mathematics, making it possible to present a solution to the non-mathematical problem.

Also from a Wittgensteinian perspective, Souza and Luna (2015), in a theoretical essay, presented a way to see mathematical modelling and proposed forms of practising mathematics in mathematical modelling activities. The foundation in Wittgenstein's theory, more specifically in one of his works well recognized by the scientific community, the book *Philosophical Investigations*, enables the authors to understand modelling as a way of organizing and dealing with empirical situations for the learning of mathematics. According to the authors, this understanding also considers ways of seeing mathematics, reality, and mathematical learning itself in modelling activities.

Almeida, Silva, and Veronez (2015) considered the routing onto which modelling activities are oriented by the search of the solution to a problem whose origin is outside mathematics, and refer to the possibilities and necessities of the production and action of signs for the understanding of the phenomenon and the mathematics used in this search. An approach based on the semiotic theory of Charles Sanders Peirce indicates that the functioning of the signs provides, and at the same time describes a continuous interaction between the phenomenon and the signs generated by the interpreter. The relationship thus constituted indicates that, in modelling activities, the initial questions, such as student instabilities, generate new instabilities that are overcome by semiosis over time, seeking understanding for the phenomenon and for mathematics. The authors argue that, although mathematical modelling activities can, from a given viewpoint, simplify the complexity intrinsic to the problems of reality, analysis of the articulations established by the students allows us to conjecture that the understanding of a problem and the understanding of mathematics are interdependent, so that one is being configured in the interpreter's mind to the extent that the other is on the same way.

With respect to the group of papers that investigate the use of educational curriculum materials of mathematical modelling in mathematics education, a study presented in this edition of the SIPEM (Silva, Barbosa, & Oliveira, 2015) addressed the production and use of curriculum materials for the implementation of modelling and presents an analysis of the modifications conducted by teachers when they use mathematical modelling curricular materials in their classrooms. The authors emphasized that some principles govern the teacher's choices in the selection of which material will be used and how this use will be conducted. The results the authors presented pointed out three principles that guided teachers in the recontextualization of educational curricular materials on mathematical modelling for pedagogical practices: discussion and reflection on the theme, curriculum content, and structure of curriculum material. These principles regulate the movement of texts from the field of pedagogical recontextualization to the field of reproduction.

Two studies presented at the VI SIPEM can be allocated in the group that investigated teacher training on mathematical modelling in mathematics education. One of them, Tambarussi and Klüber (2015) addressed the relevance of teacher training for the incorporation of mathematical modelling in school activities at different levels of schooling. In this case, the authors direct their reflections towards the training of teachers on modelling in the context of a teacher training public programme. The study shows the need for clarity regarding the peculiarities of teacher training on modelling and on what the guidelines for such training could be. In a sense, the authors problematize teacher training on mathematical modelling and plead the need for characteristics inherent in this training. They also clarify that attending long training on modelling is not sufficient for teachers if the principles of training are incommensurable to the principles of modelling.

In another study belonging to the same theme group, Sant'Ana and Sant'Ana (2015) investigate the relationship between the teachers' initial questions about a topic of interest to mathematical modelling and the tasks they planned from these questions. Based on questions posed under the perspective of Basil Bernstein, most of them of the open type, a specific group prepared a task with a strong framework initial phase (close to the rules imposed by the discourse in force in conventional teaching), which was weakened in later phases, culminating in a final phase that favoured communicative interactions between teacher and students in the classroom.

In the next section, from the considerations on the reported studies, we present some reflections aiming to outline an overview of the research on Brazilian mathematical modelling discussed in the different editions of the SIPEM.

11.4 About the Research Presented at Working Group 10 of SIPEM (2003–2015)

The creation of SIPEM, in the year 2000, aimed at promoting an exchange between national and international groups dedicated to research on different themes of the mathematics education area, thus constituting a space in which research could be discussed and shared and where advances and gaps could be evidenced. Discussions on mathematical modelling were included from the SIPEM II, in 2003.

The themes presented throughout the editions of the event, the dialogue between them, and the respective internal debate indicate a movement that seeks to provide more coherence and consistency to research conducted in the area.

Regarding the geographical coverage of the participants, analysing the reports, published texts, and syntheses herein presented, we observed that the different editions of the SIPEM have congregated only Brazilian researchers and were concentrated in some states of the federation, as shown in Fig. 11.1.

This characteristic of not strengthening relations with foreign researchers in the scope of the SIPEM does not represent the interactions established by the Brazilian scientific community, considering that there are international partnerships, but it seems to indicate fragility that is not restricted to the Working Group of Modelling but can be observed in other WGs structured in the SIPEM.

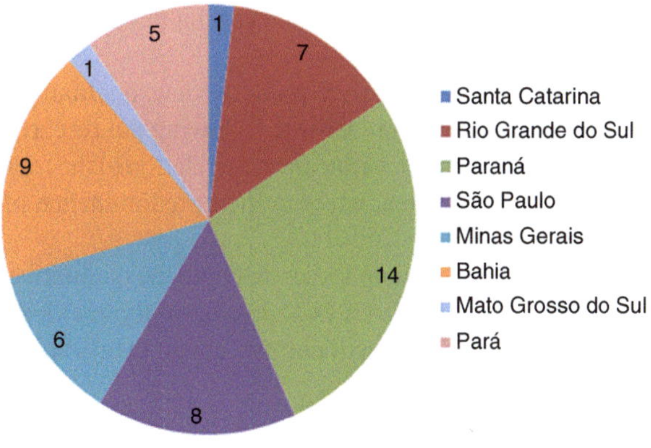

Fig. 11.1 Spatial distribution of research reports by Brazilian states

Table 11.1 Number of participants with studies accepted in the different editions of the SIPEM according to state of origin

Brazilian states	2006	2009	2012	2015	Total
Santa Catarina	0	1	0	0	1
Rio Grande do Sul	3	1	1	2	7
Paraná	3	2	4	5	14
São Paulo	3	2	3	0	8
Minas Gerais	1	1	2	2	6
Bahia	2	2	4	1	9
Mato Grosso do Sul	0	1	0	0	1
Pará	2	1	1	1	5

Although researches in mathematical modelling are distributed among the different Brazilian regions, in the south and southeast regions there is a higher concentration of research reported in the SIPEM. In fact, the states of Paraná, Bahia, São Paulo, Rio Grande do Sul, Minas Gerais, and Pará have consolidated research groups. The state of Paraná is the one with the highest number of studies and diversified groups, as can be seen in Table 11.1.

This is probably due to post-graduate courses at the doctoral level in the area of mathematics education created in the south and southeast in the last decade. Table 11.1 shows the distribution of the works presented in the SIPEM in the different Brazilian states.

The data in Table 11.1 shows certain regularity from the standpoint of spatial distribution and the number of studies reported and discussed at the different editions of the SIPEM: there were nine studies in first edition of the event in 2003, 14 in 2006, 11 in 2009 and 2015, and 15 in 2012.

An aspect related to research in mathematical modelling that has already been pointed out by Barbosa (2007) refers to the need to consolidate the sharing of Brazilian research in international journals and books.

In this sense, the study of Brito et al. (2015), which analysed articles of the ICTMA, signals this movement of international coverage of the topics discussed in SIPEM. However, even though the international authors that investigate mathematical modelling are inserted in the national research reports to discuss their themes, we have not yet been able to detect joint publications with international, especially Latin American, researchers.

From the methodological point of view, qualitative research prevails in the context of mathematical modelling, following the research profile in the scope of education and mathematics education. In relation to the research procedures/instruments adopted in the articles, we identified predominance of the use of observation and interviews.

Nevertheless, we emphasize that both the qualitative approach and the qualitative methods used in research lack details on the methodological ways chosen by the researcher, based on justifications about their choices according to the research objects in question.

The methodologies frequently used in this research modality, according to Martins (2004), include research methods by which reality is investigated and "analytical manipulations through which the researcher seeks to ensure favourable conditions for observing the phenomena" (Fernandes, 1959, p. 13), but they may require the use of modes of inference and interpretive arguments.

Furthermore, a characteristic of qualitative research is the possibility of considering individual experiences and the contexts in which they are experienced and reported, as well as the researcher's position while observing or interacting with these experiences. According to Bicudo (2014), its core "is the individuality and detailed description of what is perceived/observed" (Bicudo, 2014, p. 3).

During the last edition of SIPEM, this time of observation, which would be plausible or not for researchers to be able to make in-depth considerations about their object of study, was discussed in the context of research on classroom modelling practices of teachers (Klüber & Almeida, 2015). However, we understand that this analysis is important for research on modelling in general, not to list patterns of temporality, but to reflect about the consistency of the analyses performed.

As for the focus of research on mathematical modelling in the observed period, it is possible to notice that, although some aspects have appeared in all SIPEM's editions, new theoretical assumptions have supported the arguments of researchers in the area.

The discussion of epistemological aspects related to the characterization of what mathematical modelling is, and how its practice occurs in the classroom, has been recurrent in the presented studies. In 2006, in this theme, there was certain effort to elucidate characteristics of modelling activities, whereas in the edition of 2015, the studies addressed it with foundations of philosophical perspectives.

In this context, arguments by Wittgenstein would support authors' considerations regarding mathematics, hypotheses in modelling activities, and implications for classroom modelling practice. In addition, a phenomenological analysis would explain the role of simplifications when problems of reality are addressed in mathematics classes.

The training of teachers on mathematical modelling has maintained the interest of researchers during SIPEM's editions. Although some studies present teacher training as a research context and not as an object of analysis per se, discussions on the role of teachers, how they could achieve professional formation that offered consistent subsidies, and the implementation of modelling activities in their classroom are the themes evidenced as relevant to the working group throughout the analysed editions.

There is also differentiation from other investigations on teacher training, mainly because they support the need to conceive teacher training that surpasses the current models of training, including those already discussed by the community of mathematical modelling in mathematics education. The structuring of some support to teachers provided by educational curricular materials was a theme discussed in the past two editions. However, we believe that it is necessary to consider the use of these materials in order not to decharacterize the open and dynamic nature of modelling activities, which could result from the use of such materials.

The theoretical approach, which brought assumptions of social, philosophical, and semiotic nature to the field of mathematical modelling, is theme in consolidation in the scope of WG10. In the event of 2015, 8 of the 12 studies reported refer to some discussion in this context.

This concentration of research on these themes occurs due to the strengthening of groups in the state of Paraná, which have deepened into these themes through master and doctoral degree studies. This nucleation of research seems to have weakened in other regions such as the states of Bahia and São Paulo, at least in relation to the reports presented at SIPEM. This fact may be attributed to several factors, such as the migration of researchers to/from different locations, as in the case of Rio Grande do Sul state, which counts on new doctors in 2015, who used to belong to the state of São Paulo in the 2012 edition, to the institutions in which they undertook their PhDs.

Discussions on the studies within the field have also been a focus of research, allowing meta-understandings of mathematical modelling and its configuration in classrooms at different levels of schooling. In this context, "research" on mathematical modelling is characterized as investigation that seeks responses to questions that are posed on teaching or learning in situations where mathematical modelling is used, on curriculum and its relations with mathematical modelling activities, on teacher training, and on the "mathematics education" that is thought and intended for students at different levels of schooling. These answers must necessarily be placed within a coherent theoretical and epistemological framework and developed according to a well-defined research methodology.

In the 2006 edition of SIPEM, we were able to identify the concern of the working group to confirm the unfolding of these answers in the use of mathematical modelling in teachers' practice. In this direction, throughout the analysed editions, WG 10 indicates the need for research on modelling to reflect, directly or indirectly, the implementation of modelling in Brazilian schools and universities.

11.5 Final Considerations

According to Fiorentini and Lorenzato (2007), in a comprehensive way, one can say that the object of investigation and study in mathematics education, in general, and in mathematical modelling, in particular, consists of the multiple relations and determinations between teaching, learning, and knowledge. This does not mean that a given research cannot prioritize the study of one of these elements of the triad, or one of these relationships, but, at the same time, it cannot totally ignore the other elements.

In this sense, we agree with Klüber and Almeida (2015, p. 58)

> in the scope of WG10, the discussions have been conducted in order to understand internal aspects of the reported studies, as well as in the search for more comprehensive aspects regarding methodology and the constitution of themes of interest to our community. Vigilance on the divergences and convergences between research themes is thus estab-

lished. This aspect highlights the confrontation of perspectives, culminating in the need to maintain a critical posture so that the area does not close, becoming endogenous, and also does not become too broad as to lose its identity.

What we can conclude with respect to the research reported in SIPEM is that they cover different levels and different modalities.

As far as levels are concerned, most research is exploratory, associated with broader themes, and is generally associated with research in masters' or doctoral degree programmes. Some studies, however, are explanatory and retake previous themes, deepening them theoretically.

In terms of modalities, they are subdivided into individual research, guiding-oriented partnerships, inter-institutional research, funded research, research linked to the training of researchers.

In general, the surveys reveal a dialogue with the production of the area and, more specifically, with the history of WG10. They thematize aspects that emerge from the interior of modelling practices. These aspects were focused through several theories, some with a philosophical basis, some with theories of language and others with mathematics education. Examples are the interlocutions with the semiotic perspective of Charles Peirce, the arguments based on Ludvig Wittgenstein and the formative aspects pointed out by B. Bernstein.

What the reports presented in the SIPEM indicate is that the research in mathematical modelling in mathematics education has been developing aspects of the central themes. What still needs to be consolidated, however, is a more proficient and frequent dialogue between the different research groups established in the different Brazilian regions. In this sense, a criticism that could be pointed out is the little interlocution between researchers.

Regarding the methodological aspects, Brazilian research uses qualitative methodologies to analyse the collected data. A dialogue with international research could signal to the use of quantitative analyses as is the case, for example, of research developed by Kaiser and Brand (2015).

What Brazilian research could also contemplate refers to the characterization of mathematical modelling activities, especially regarding the possibility of approaching problems and the development of models, as is the case, for example, in the study developed by Reit and Ludwig (2015).

Regarding the agenda of what can be proposed for Brazilian research in mathematical modelling, we point out some aspects to be investigated:

- Strengthening of research in groups, establishing a dialogue between researchers from different Brazilian states
- Dialogue with international researchers in order to structure joint productions
- Investment in studies that, to some extent, dialogue with basic education
- Studies that contextualize the insertion of modelling in the classroom in face of the public educational policies related to teacher training
- Extension of the debate on the limits between mathematical modelling in mathematics education and other areas included in the context of the SIPEM, such as teacher training of mathematics teachers

- Implementation of regional, national, and international inter-institutional projects concerning teacher training on mathematical modelling and other themes

In general terms, this chapter presents the current configuration of research on mathematical modelling in Brazil, taking into account the research reported in the International Seminar for Research in Mathematics Education. On the one hand, advances can be observed regarding the characterization and use of mathematical modelling in the Brazilian context. On the other hand, there are indications that advances can still be consolidated aiming at the development of students in classes mediated by mathematical modelling activities.

References

Almeida, L. M. W., Araújo, J.L., & Bisognin, E. (2012). Reports of do 5th International Seminar for Research in Mathematics Education. *Brazilian Society of Mathematics Education*. pp. 13–15. Retrieved 15 June, 2016, from http://www.sbembrasil.org.br/sbembrasil/images/Boletim19.pdf

Almeida, L. M. W., Sousa, B. N. P. A., & Tortola, E. (2015). *Proceeding of the 6th International Seminar for Research in Mathematics Education*, Pirenópolis, Goiás, Brazil.

Almeida, L. M. W., Silva, K. A. P., & Veronez, M. R. D. (2015). Sobre a Geração e Interpretação de Signos em Atividades de Modelagem Matemática. *Proceeding of the 6th International Seminar for Research in Mathematics Education*, Pirenópolis, Goiás, Brazil.

Araújo, J. L. (2010). Brazilian research on modelling in mathematics education. *ZDM, 42*, 337–348.

Araújo, J. L., & Campos, I. S. (2015). Espaço de Negociação: Inserindo Elementos Sociais na Compreensão de Lucro. *Proceeding of the 6th International Seminar for Research in Mathematics Education*, Pirenópolis, Goiás, Brazil.

Barbosa, J. C. (2007). Sobre a pesquisa em modelagem matemática no Brasil. *Proceeding of 5th National Conference on Modelling in Mathematics Education*, Ouro Preto, Minas Gerais, Brazil.

Barbosa, J. C., Araújo, J. L., & Caldeira, A. (2006). Reports of do 3th International Seminar for Research in Mathematics Education. *Brazilian Society of Mathematics Education*. pp. 8–11. Retrieved 15 June, 2016, from http://www.sbembrasil.org.br/files/sipemIII.pdf

Barbosa, J. C., Caldeira, A. D., & Araújo, J. L. (2009). Reports of do 4th International Seminar for Research in Mathematics Education. *Brazilian Society of Mathematics Education*. pp. 2–10. Retrieved 15 June, 2016, from http://www.sbembrasil.org.br/sbembrasil/images/RelatóriogeralSIPEM.pdf

Barbosa, J. C., Caldeira, A. D., & Araújo, J. L. (Eds.). (2007). *Modelagem Matemática na Educação Matemática Brasileira: Pesquisas e práticas educacionais* (pp. 17–32). Recife: SBEM.

Bean, D. (2012). As premissas e os pressupostos na construção conceitual de modelos. In *Proceeding of the 5th International Seminar for Research in Mathematics Education*, Petrópolis, Rio de Janeiro.

Bernstein, B. (2000). *Pedagogy, symbolic control and identity: Theory, research, critique*. Lanham: Rowman & Littlefield Publishers, 230 p.

Bicudo, M. A. V. (2014). Meta-análise: Seu significado para a pesquisa qualitativa. *Revemat, 9*, 7–20.

Bicudo, M. A. V., & Klüber, T. E. (2011). Pesquisa em modelagem matemática no Brasil: A caminho de uma metacompreensão. *Cadernos de Pesquisa, 144*(41), 902–925.

Biembengut, M. S. (2009). 30 Anos de Modelagem Matemática na Educação Brasileira: Das propostas primeiras às propostas atuais. *Alexandria: Revista de Educação em Ciência e Tecnologia, 2*(2), 7–32.
Brito, D. S., Oliveira, C., & Milani, C. S. (2015). A 'Escolarização' do Espaço Vivido nas Atividades de Modelagem com Geometria: Uma Compreensão Sob a Perspectiva Fenomenológica. *Proceeding of the 6th International Seminar for Research in Mathematics Education*, Pirenópolis, Goiás, Brazil.
Candau, V. M. (2000). A didática hoje: Uma agenda de trabalho. In V. M. Candau et al. (Eds.), *Didática, currículo e saberes escolares*. Rio de Janeiro: DP&A.
Canedo Jr, N. R., & Istemann Jr, M. A. (2015). A Primeira Experiência em Modelagem Matemática de Alunos do Sexto Ano do Ensino Fundamental. *Proceeding of the 6th International Seminar for Research in Mathematics Education*, Pirenópolis, Goiás, Brazil.
de Almeida, L. M. W. (2006). Algumas reflexões sobre a pesquisa em Modelagem Matemática. *Proceeding of the 3th International Seminar for Research in Mathematics Education*, Águas de Lindóia, São Paulo, Brazil.
Deleuze, G. (2011). *Lógica do Sentido*. São Paulo: Perspectiva.
Fernandes, F. (1959). *Fundamentos empíricos da explicação sociológica*. São Paulo: Companhia Editora Nacional.
Fiorentini, D., & Lorenzato, S. (2007). *Investigação em educação matemática percursos teóricos e metodológicos* (2ª ed.). Campinas: Editores Associados.
Gouveia, J. A. A. (1971). A pesquisa educacional no Brasil. *Cadernos de Pesquisa, 2*, 1–10.
Kaiser, G., & Brand, S. (2015). Modelling competencies: Past development and further perspectives. In G. A. Stillman, W. Blum, & M. S. Biembengut (Eds.), *Mathematical modelling in education research and practice* (pp. 129–139). https://doi.org/10.1007/978-3-319-18272-8_1
Klüber, T. E., & Almeida, L. M. W. (2015). Reports of do 6th International Seminar for Research in Mathematics Education. *Brazilian Society of Mathematics Education*. pp. 51–56. Retrieved 15 June, 2016, from http://www.sbembrasil.org.br/sbembrasil/images/arquivos/relatorio_visipem.pdf
Klüber, T. E., & Burak, D. (2014). Sobre a pesquisa em Modelagem na Educação Matemática brasileira. *Revista Diálogo Educacional, 14*(1), 143–157.
Klüber, T. E., Tambarussi, C. M., Loureiro, D. Z., Wichnoski, P., & Oliveira, W. P. (2015). Rumos e avanços da Modelagem Matemática na Educação Matemática Brasileira: Um olhar sobre a pesquisa a partir dos resumos do V SIPEM. *Indagatio Didactica, 7*(1), 80–95.
Martins, H. H. T. S. (2004). Metodologia qualitativa de pesquisa. *Educação e Pesquisa, 30*(2), 289–300.
Milani, M. C., Kato, L. A., & Cardoso, V. (2015). Modelagem matemática e aprendizagem de geometria: Possíveis aproximações por meio de vídeos. *Proceeding of the 6th International Seminar for Research in Mathematics Education*, Pirenópolis, Goiás, Brazil.
Reit, X., & Ludwig, M. (2015). An approach to theory based modelling tasks. In G. A. Stillman, W. Blum, & M. S. Biembengut (Eds.), *Mathematical modelling in education research and practice* (pp. 81–92). https://doi.org/10.1007/978-3-319-18272-8_1
Sant'ana, M. F., & Sant'ana, A. A. (2015) Modelagem Matemática: Relação entre Formulação de Perguntas e Elaboração de Tarefas. *Proceeding of the 6th International Seminar for Research in Mathematics Education*, Pirenópolis, Goiás, Brazil.
Silva, M. S., Barbosa, J. C., & Oliveira, A. M. P. (2015) Princípios Operados por Professores na Recontextualização Pedagógica de Materiais Curriculares Educativos. *Proceeding of the 6th International Seminar for Research in Mathematics Education*, Pirenópolis, Goiás, Brazil.
Soares, D. S., & Borba, M. C. (2014). Os Caminhos da Modelagem no "Pós"-GPIMEM: Desenvolvendo uma árvore genealógica. *Revista Eletrônica de Educação Matemática, 9*, 74–99.
Souza, E. G., & Luna, A. V.A. (2015). Modos de Praticar Matemática em Modelagem Matemática. Way of Practicing Mathematics in Mathematics Modelling. *Proceeding of the 6th International Seminar for Research in Mathematics Education*, Pirenópolis, Goiás, Brazil.

Tambarussi, C. M., & Klüber, T. E. (2015). Formação de professores em Modelagem Matemática: Contribuições a partir do Programa de Desenvolvimento Educacional do Paraná, PDE. *Proceeding of the 6th International Seminar for Research in Mathematics Education*, Pirenópolis, Goiás, Brazil.
Weingarten, T., & Dalla Vecchia, R. (2015). Modelagem Matemática: Implicações do Desequilíbrio entre Sentido e Significado no Processo de Determinação do Problema. *Proceeding of the 6th International Seminar for Research in Mathematics Education*, Pirenópolis, Goiás, Brazil.

Chapter 12
Philosophy of Mathematics Education: A Panorama from Brazil

Maria Apaecida Viggiani Bicudo, Renata Cristina Geromel Meneghetti, Sônia Maria Clareto, and Tânia Baier

Abstract This chapter presents a panorama of the research on philosophy of mathematics education in Brazil, based on the studies presented and discussed in Working Group 11 (WG11), Philosophy of Mathematics Education of the Brazilian Society of Mathematics Education, during the VI International Seminar of Research on Mathematics Education (SIPEM VI) in 2015. It reports upon the hermeneutic analysis of each of the studies that was carried out, framed by the questions 'What are the article's purposes/objectives/questions/problems/interrogations?', 'What were the investigation procedures?', 'Who are the authors of the reference works cited?', and 'What does the article say, considering its proposal?' The analysis indicated a broad range of themes and problematizations were addressed in the papers, including reflections about mathematical objects, the conceptions of mathematics, the relationships between mathematics and the school environment, the scholarization of mathematics, the preparation of the mathematics teacher, the different cultural mediators in mathematics teaching, with special emphasis on computational environments and ICTs, and ethnomathematics. It also brought to light a uniqueness concerning the philosophy of mathematics education as it has been investigated across the world. This uniqueness manifests itself through the view about knowledge and reality presented from the Brazilian point of view, while informed by authors like Deleuze, Gadamer, Husserl, Merleau-Ponty, and Wittgenstein who shape the world view of this area.

M. A. V. Bicudo (✉)
Universidade Estadual Paulista, Campus de Rio Claro, São Paulo, Brazil

R. C. G. Meneghetti
Universidade de São Paulo, São Paulo, Brazil

S. M. Clareto
Universidade Federal de Juiz de Fora, Juiz de Fora, Brazil

T. Baier
Universidade de Blumenau—FURB, Blumenau, Brazil

A. J. Ribeiro et al. (eds.), *Mathematics Education in Brazil*,
https://doi.org/10.1007/978-3-319-93455-6_12

12.1 Introduction

This study presents a panorama of the research on philosophy of mathematics education in Brazil, based on the studies approved to be presented and discussed in Working Group 11 (WG11), 'Philosophy of Mathematics Education' during the VI International Seminar of Research on Mathematics Education (Seminário Internacional de Pesquisa em Educação Matemática, SIPEM VI) in 2015. With that in mind, a hermeneutic analysis of each of the studies was carried out. In light of the representativeness of WG11 in the Brazilian Society of Mathematics Education (Sociedade Brasileira de Educação Matemática, SBEM) and the importance of SIPEM as a vehicle to present Brazilian research to society, in what has become a fruitful environment for discussion and exchange of ideas, we understand that the studies presented to the group of philosophy of mathematics during the 2015 edition of SIPEM reflect important aspects of the current scenario of research on mathematics education. This is especially true for the studies addressing philosophical thought on mathematics education, or the philosophical approaches through which these studies are articulated.

The philosophy of mathematics education turns its eyes to mathematics education in the quest to understand its own way of doing and proceeding. As an interface between philosophy, philosophy of education, and philosophy of mathematics, it is from these fields of inquiry that it acquires its mode of proceeding, always inquisitive and attentive so as not to naturalize statements, concepts, and objectives. We understand that the object of philosophy of mathematics education is the critical and reflexive analysis of proposals and actions in mathematics education, in the various contexts where it takes place: in public institutions, families, the street, and the media. These analyses resort to the works of several authors renowned for their importance in these specific themes, understanding and interpreting their cases as a means to communicate with these proposals and actions and thus articulate a new discourse for mathematics education itself. The inquiries undertaken by philosophy of mathematics education also address the theory/practice articulation in the very same reality where it becomes effective or put to action—which is the reality of schools and of the mathematics classroom as much as of any other environment where mathematics education comes to be—where teacher, student, teaching contents, and mathematics themes are in action. These actions, which are outlined in their own context, stand on a perceptional, and therefore spatial, temporal, and historical ground, where educational postures, the teaching proposal, and the conception of mathematical object and knowledge unfold.

The Philosophy of Mathematics Group is a working group (WG11) that, from a philosophical perspective, investigates embryonic topics in the scope of mathematics education, and may establish a communication with other perspectives. The objective of WG11 is to collect investigations, studies, experiences, discussions, themes, and debates about mathematics as well as mathematics teaching and educational processes from different perspectives, such as the epistemological, ontological, and axiological viewpoints. Some of the topics investigated by the group are

'*What is there?*' (ontology), '*How do you know what is there?*' (epistemology), and '*How much is it worth and why?*' (axiology) as embedded in the theme mathematics education. The studies submitted, reviewed, and approved for presentation to this WG are the fruit of a research that derives support from the practices of debate established concerning these questions.

In addition to representatives of the numerous institutions located in several regions of Brazil (southeast, south, northeast, and midwest), the members of WG11 during SIPEM or the National Meetings on Mathematics Education (Encontro Nacional de Educação Matemática, ENEM) also included teachers, researchers, and educators of various specialization levels, including elementary and high schools, as well as universities and graduation courses. Importantly, it becomes clear that the number of participants of WG11 increased gradually, from 10 in the first edition of SIPEM in 2003 to 25 in 2015.

This diversity of representativeness also manifests as a wide variety of research lines of the studies presented in WG11, among which we highlight Phenomenology and Mathematics Education, Philosophical-scientific Thoughts, Teaching and Learning of Mathematics and the Interfaces with Other Disciplines, Studies on Mathematics History, Philosophy, and Education, Conceptions of Mathematical Knowledge and of Teaching and Learning of Mathematics, Conceptions and Perspectives in the Formation of Mathematics Teachers, Trends in Mathematics Education: a Critical Analysis of Conceptions Addressed, Philosophical Conceptions of Mathematics Environments for Online Learning, Philosophy of Language, Philosophy Fundaments in Ethnomathematics, Philosophy and Technology.

Such list of themes addressed in WG11 shows that it establishes a dialogue with topics discussed in other WGs of the Brazilian Society of Mathematics Education (SBEM), but always from the perspective of the philosophy of mathematics education, according to the way this group understands the discipline and describes it hereafter.

In addition to this variety of subjects, these studies also adopt different philosophical perspectives, which are frequently focused on more comprehensive research being conducted by the authors, covering the specific theme of a given article presented. Besides, the studies presented throughout the years translate a growing concern amidst members of the group with regard to resuming discussions held in previous meetings, both at SIPEM and ENEM, conducting the activities based on the presentation of studies that contribute more comprehensive and insightful notions in terms of what has been probed and requires further investigation by the WG. The effort to proceed with studies about issues understood as not fully discussed and that demand more engagement has become a constant purpose of this group.

Another characteristic observed in the activities of WG11 is the relationship that the studies under discussion establish with research previously presented and endorsed, like doctoral theses or master's degree dissertations. It is through this relationship that subthemes of philosophy of mathematics education are rearticulated, promoting the development of articles. Also, the group discusses articles whose authors declare that these studies were carried out as part of more comprehensive research projects.

12.2 Investigation Procedures

The present investigation was carried out using the hermeneutic-phenomenological approach, according to which the contents of a text being considered have to be understood without being naturalized. In this sense, the 13 articles published in the annals of the VI SIPEM by WG 11 were analysed under strict interpretation standards, based on the following questions: '*What are the article's purposes/objectives/questions/problems/interrogations?*', '*What were the investigation procedures?*', '*Who are the authors of the reference works cited?*', and '*What does the article say, considering its proposal?*'

In this chapter, each article evaluated was assigned a number. For instance, *The Ontological Conception of Mathematical Objects in the Theory of Objectification* was named 'article [1]'. Throughout this discussion, we will introduce the number assigned by us at the first mention of each article. The articles reviewed were distributed to the four authoresses of this chapter so that each of us analysed articles that were written by some other researchers, and reviewed the analyses made by other authoresses of the group. So, an article was reviewed by one authoress, while the review produced was analysed by another.

Phenomenological research includes successive reductions as a means to understand the comprehensive core ideas in the articulation of a text's discourse. In phenomenology, rather than summarizing or simplifying the description under analysis, the term *reduction* is the articulation of perceived meanings that interleave as a network of meanings converging to a core. This core is more comprehensive than the individual parts that form it, which are the perceived meanings, the different perspectives, and the phenomena analysed. Also, through successive articulations that represent the journey of thought with what reveals itself in the world-life where one is with the other, this core develops ideas that indicate perceived meanings, transcending these and being presented as an articulation of meanings. Therefore, in the first part of this analysis, we underlined the passages of a text we considered significant, based on the criterion that they should cover the questions germane to the article. These underlined passages were called *Meaning Units*, since they make sense to the investigator, who constantly dialogues with the question proposed, with what is said by the author of the article and its respective historical-cultural context and, then, proceeds with the interpretation. Therefore, the first stage of the present chapter was called *Ideographic Analyses*, when the *First Reduction* was carried out, that is, the articles were read as many times as necessary so as to detect their *Meaning Units*. Next, we attentively analysed these *Meaning Units* looking into what was contained in the text, transliterating those units the way we considered the most appropriate, and renamed them *Signification Units*. These units were given this name because they carry our interpretations of what is understood from each article based on the dialogue we established with it. After, we revealed our comprehensions in our own sentences and attentive reading and, then, through articulation, combined these sentences according to the meanings that we made of them. The union of meanings of the *Signification Units* generated a new sentence, called *Second Reduction Articulated Propositions*, since these express the articulation of the understanding and the interpretations we had and made, and were thus named to explicit this understanding. This nomenclature, which summarizes a new comprehension made clear in the text, was transferred to a table and given a code.

In the next stage, called *Third Reduction*, all *Signification Units* of all articles were collected in four tables named in accordance with the questions that directed the study of the articles, namely Question 1, Question 2, Question 3, and Question 4. These tables were printed and had a blank sheet of paper attached, where the *Articulated Propositions* were listed as we interpreted these tables, now working as a group. These articulated propositions were identified and written down on the said blank page.

Next, we constructed a *Table of Convergences* to support the analysis carried out, though it is not shown here due to space limitations. The first column listed the *Signification Units* that generated the *Third Reduction Articulated Propositions*, presented in the second column of the table. The third column contained the *Articulated Core Ideas*, which were expressed according to the articulated consideration of the data given in the second. After the attentive inspection of the *Articulated Core Ideas*, we proceeded to the articulation of the *Comprehensive Core Ideas*.

Each question was analysed following the steps described. For the first question, '*What is the purpose of the article, its objectives/questions/problems/interrogations?*' for example, three convergent core ideas were articulated, including the various objectives of the articles reviewed: (1) *The cyberspace environment in mathematics education*, (2) *The philosophical research on mathematics education carried out in elementary school and higher education institutions, in undergraduate courses*, and (3) *Constitutive themes of the philosophy of mathematics education*.

These *Convergent Core Ideas* were interpreted as a dialogue between the authors of the articles reviewed, the authors of the reference works cited in each article, and us. The interpretation effort, which is the core of this chapter, is described in the following section, called *Interpreting the analyses carried out*. This interpretation is given as an analytical, argumentative, and articulated text, emphasizing what the various studies reviewed propose to do, the investigation procedures that were proved to be important, the authors that were most often cited as reference, or, in other words, what the texts say in terms of their proposals.

12.3 Interpreting the Analyses

In this section, we present our interpretations of the *Articulated Core Ideas* concerning each of the four questions described above. Next, we name each of the topics and discuss our interpretations thereof.

12.3.1 Articulated Core Ideas When Investigating the Article's Purposes/Objectives/Questions/Problems/Interrogations

We examined each article looking into what it proposed, trying to understand its aim, or question, or problem, or even its verbalized interrogation. From this analysis, we articulated three *Convergent Core Ideas* that cover the various aims of these articles: *the cyberspace environment in mathematics education*; *the philosophical research on mathematics education carried out in elementary school and higher education institutions, in undergraduate courses*; *on the reality and nature of mathematical objects.*

The cyberspace environment in mathematics education, introduced for discussion in philosophy of mathematics education by WG11 in 2015, was based on studies published by researchers and professors dedicated to clarify *what to do* and *how to proceed* in mathematics teaching and learning scenarios that address the reality of cyberspace, using its resources and the respective informational screen that, in turn, provides a list of possible activities. These include the activities developed based on information and communication technologies (ICTs). Researchers investigate this reality, underscoring the ontological characteristics that those studies have made clear. They also examine the way communication is established between people, considering software logics, understanding that this communication mode also affords epistemological and anthropological comprehensions. The studies presented emphasized the dialogue enabled by situations in which the subject is with others near logic and near the possibilities created by the software. Such is the case of Article [5], *Communication in cyberspace: dialogues about mathematics* (Paulo & Ferreira, 2015), whose leading question was '*How does the dialogue about mathematics content become possible and take place in communities in social media like Facebook and Orkut?*'. Article [13], *A study on the mathematics demonstration by/with the computer* (Batistela, Barbariz, & Lazari, 2015) inquiries over the feasibility of an effective demonstration by a computer, from which the limitations or questions about computability vs. mathematics production are outlined. The article frames the following questions, '*How is the computer being used by mathematicians in the activity of proving?*' and '*Is mathematical proof carried out by/with a computer today?*' This question addresses the comprehension of the way through which mathematical deductive proofs are currently conducted in the life-world where technologies are present. The phrase 'life-world', as translated from the German word *Lebenswelt*, or *world of life* such as rendered by most authors whose first languages derive from Latin, is understood as the spatiality (modes of being in space) and temporality (modes of being in time) in which we live with other human beings, other living creatures and nature. Life-world is also seen to include all explanations of scientific, religious nature, or any other areas of human activity and recognition. The world is not a container, an object, but a space that extends itself as actions are carried out and whose horizon of comprehensibility widens as meaning becomes to each of us and to the community we live in (Bicudo, 2010, p. 23). Article [10], *The moving and the formal* (Figueiredo, 2015), declares that computer science is a *special field for the exploration of the connections* between the capacity to perceive changes, even movement, and the cognitive relationships with formal sciences.

The philosophical research on mathematics education carried out in elementary school and higher education institutions, in undergraduate courses, as we presented and investigated, underlines the study of the works of philosophers from a position of constant, critical, and inquisitorial dialogue. As they were being understood, we analysed these studies in light of the work scenarios in the school environment, seen as the epicentre of questions. Therefore, it is not about the application of philosophical theories to mathematics education defining *what has to be done and why it should be done this way* though these works reveal a philosophical stance by proposing, describing, analysing, and interpreting activities and taking the reality of schools as what is out there to be understood. Article [12], *Teacher, who invented mathematics? The course of a question that becomes a problem and a problem that defines curricula* (Clareto, 2015) frames one question already in its title. The question is alarming, and as it unfolds, it begins to be seen as a problem, that is, it is problematized in every school that mobilizes itself to face what is being asked. Article [11], *Concerns and trends in research on ethnomathematics* (Miarka, 2015), discusses the move towards the theorization of ethnomathematics and the theory that supports this area of research, including the ways through which it operates. Article [9], *Teachers-to-be and their conceptions about supervised trainee programs in mathematics* (Meneghetti & Oliveira, 2015), analyses the notions student teachers have of supervised trainee programmes considering the investigative character of the statements obtained and the issues surrounding mathematical knowledge and contents addressed in mathematics pre-service teacher education.

The core ideas collected *on the reality and nature of mathematical objects* address questions about the reality of mathematical objects, the way they are constituted and investigated by other authors, and background questions about mathematics education as observed in philosophical works. Several articles were presented in WG11 in this scope of discussions. Article [2], *Hermeneutics in mathematics education: comprehensions and possibilities* (Mondini, Mocrosky, & Bicudo, 2015), resumes the discussions held by WG11 in 2012, placing them on a widening horizon without judging lines of research, and much less deciding over any *hermeneutics* that may be seen as opposing the Heideggerian, Gadamerian, Habermesian, and other phenomenologies. The article attempts to reintroduce the *possibilities* of carrying out a hermeneutic study for debate, revealing comprehensions that aim to deepen the studies conducted by WG11, *un-veiling* the possible contributions for investigation and pedagogical practice in mathematics education. Article [3], *The language of Gadamer: its image in research on mathematics education* (Kluth, 2015), addresses language from Gadamer's perspective and, based on the studies cited as reference, frames the question, '*How does thought reveal itself in the movement of knowledge construction of algebra structures?*' aiming to clarify the hermeneutical investigation about this question. Article [10] declares that we are creatures of change, that is, beings whose world opens itself full of meanings of change, and frames the question, '*How can we understand the phenomenon of formal sciences?*'

The article highlights, as a doubt of philosophical nature, the human capacity of revealing meanings of change in the mathematical doing, for instance. In that article, ontological discussions stand out, emphasizing the formal and the moving. Yet, another article that poses questions about the ontology of mathematical objects is article [1], which investigates how the various theories conceive the ontology of mathematical objects, since not every modern theory of teaching in mathematics education discusses the ontological concept of its object of knowledge, more specifically, the object of the ontology of mathematical beings. Article [4], *That mathematics: is it a problem?* (Rotondo and Azevedo, 2015), also addresses mathematical contents of *distributing and measuring*, asking, '*How to operate and produce meanings with the concept of division?*' Article [6], *Curricular contents of mathematics in elementary schools in Brazil: a philosophical analysis* (Baier & Bicudo, 2015), is an attempt towards philosophical investigation, analysing, criticizing, and reflecting over the curriculum adopted in elementary education, stressing science and mathematics issues in light of their sociocultural historicity in the Western world. Again, we mention article [12] that, in its own core question, focused on the mathematical object. Article [13] investigates mathematical proof, considered one of the pillars of this science such as it exists in the western world. Article [7], *Directions towards a geometric philosophy of transformations* (Detoni and Pinheiro, 2015), looks into the dynamic geometry of movement, mathematics, and flow aiming to understand it from the historical and epistemological perspectives that cover a variety of conceptions of geometries.

12.3.2 Articulated Core Ideas in the Analyses of Studies About Understanding the Investigative Procedures Adopted by Authors

When trying to understand the investigative procedures adopted by the authors of the different studies reviewed, we articulated three *comprehensive core ideas* that were called *theoretical essay supported by bibliographic review*, *qualitative research that used experiences in everyday life at school as data*, and *qualitative research that used data obtained in the cyberspace.*

Theoretical essay supported by bibliographic review articulates methods described in studies by authors who are relevant to the theme under discussion and that focus on key concepts of different theories to reveal these concepts and develop argumentations highlighting the narrowing or increasing distances between authors. Such is the case of article [1], a *literature review* based on the theory of objectification, on a discussion about the ontological principle of this theory, and radical constructivism. The article makes, in the authors' words, *some considerations on the ontology of mathematical objects while fundamental principle to be taken into account (or not) by modern theories of learning in mathematics education.* These core ideas also include theoretical essays whose notions originate from literature reviews, as articles [2] and [3], as well as article [8], titled *Ethnomathematics and the ideas in language games* (Silva, 2015). Article [2] is a *text that presents an initial literature review about hermeneutics, supported on a phenomenological perspective, with the aim of revealing comprehension as human existence and contributing with research about mathematics education focused on comprehension of teachers, students, texts, concepts, meanings, and signification in the scope of this science*, while article [3] declares that *this theoretical essay intends to clarify language from a Gadamerian perspective*. Article [8] is also presented as a literature review, intending to *analyse the possibilities of family resemblance between mathematics, ethnomathematics, and the philosophical thought of Ludwig Wittgenstein in his book called 'Philosophical Investigations'*.

Qualitative research that uses experiences in everyday life in schools as data is a discussion about *Comprehensive Core Ideas* that examines articles developed at two moments, namely in schools (with teachers, students, and staff), and in cyberculture environments. Regarding the research *in* and *with* the teaching environment, article [12] is an investigation carried out in a school based on a question framed by a student during a mathematics lesson. The ways through which the school faces the question are reported and discussed, revealing the process in which the curriculum is being developed. Article [4] is an investigation conducted in a workshop of in-service teachers taking a continued education course, when a problematization of the formation of teachers evolves. The operator *division* was used as a guiding cue that, during the movement of the formation action, problematizes itself, that is, becomes a problem that is more complex that initially conceived aiming for the formation of these teachers. Using a semi-structured questionnaire answered before and after the training period, article [9] introduces an analysis of the conception students of teacher education programmes have about supervised trainee programmes. It was included in these comprehensive core ideas because it addressed undergraduate teacher education students directly who were taking the supervised trainee programme in elementary schools.

In *Qualitative research that uses data from the cyberspace environment*, article [5] follows members of Facebook and Orkut groups and analyses the ways they express themselves. The article also highlights the qualitative aspects of the mode of being in cyberspace, such as the dread felt when communicating with subjects in cyberspace. Article [7] evaluates the convenience of the traditional geometry practice as described in curricula in terms of graphics software environment. It reveals the articulation between theoretical studies about this theme and the dynamic meaning of graphics software.

12.3.3 Articulated Core Ideas in the Analysis Covering the Authors Studied

In our first investigative attempt on the subjects addressed by WG11 in SIPEM VI 2015, we also addressed the references cited in the articles, since these reference works reveal views of the world and knowledge present in the discussions held and in the production of knowledge of the researchers participating in this group.

As the common ground on which philosophical and mathematical thought flows across the participants of WG11, we introduce authors who, as we understand them, are relevant thinkers about these themes. Of the articles that address themes about mathematics and philosophy of mathematics, we highlight Bachelard (1968), Bourbaki (1950), Dedekind (1931), Euclid (2009), Galileu (1999), Hawking (1997), and Wussing (1969, 1989). Concerning philosophy authors, we mention Deleuze (1997, 2006), Deleuze and Guatari (2012), Gadamer (1999), Husserl (1997, 2008), Merleau-Ponty (1994, 2002), and Wittgenstein (1999).

Among the relevant authors cited in articles about mathematics education, we include D'Ambrosio (1979, 1984, 1998, 1999), Knijnik (2001), Knijnik and Wanderer (2008), and documents of the Brazilian legislation (Brasil, 1998, 2002). Also, one authoress is quoted as reference in several articles on various themes, especially concerning qualitative research procedures and questions about cyberspace: Bicudo (1993, 1998, 1999, 2000, 2010, 2011).

12.3.4 Articulated Core Ideas in the Analysis Showing What the Articles Say Considering Its Proposal

The successive reductions carried out led our articulations to four *Comprehensive Core Ideas* about what the texts under analysis say concerning the answers and the understandings made explicit by the authors when concluding their articles. These reductions were: *Constitutive themes of mathematics education, Constitution of mathematics objects, Philosophy of mathematics education focused on the elementary and higher education spaces, The cyberspace environment in mathematics education*, and *Themes of research presented in WG11 in the previous SIPEM.* As we understand it, the *Comprehensive Core Ideas* articulated in the reductions carried out while we tried to understand the object of our first question framed about the articles reviewed, that is, '*What are the article's purposes/objectives/questions/problems/interrogations?*', are coherent with the *Comprehensive Core Ideas* articulated in our analysis of the fourth question proposed to the articles, '*What does the article say, considering its proposal?*'

The *Comprehensive Core Ideas* discussed herein clarify what WG11 considers important in philosophy of mathematics education in SIPEM VI. An examination of international literature panorama on the theme reveals that some items coincide with contributions made by various authors, like Paul Ernest, Merylin Frankenstein, Jean Paul van Bendengen, and Ole Skovsmose, among others who write about philosophy of mathematics education, as it is the case here considered in terms of *Constitutive themes of mathematics education*. However, the references cited by the members of the WG Philosophy of Mathematics Education in SIPEM 2015 were not restricted to the works quoted in studies developed in countries other than Brazil, though it reveals, in several aspects, the existence of a different way to understand reality and the constitution of mathematical objects.

The theme philosophy of mathematics education focused on the *elementary and higher education spaces, in undergraduate courses,* seemed to innovate in this field of study. We understand that, considering the studies presented in this WG, we address the questions about philosophy of mathematics education and investigate them in the reality of school temporality and spatiality. Another relevant theme that surfaced in the studies reviewed and, as it seems to us, is important in the reality of the life-world as lived by us is the *cyberspace environment in mathematics education.* The authors writing on this theme investigate ontological, logical, and epistemological questions in this environment. Articles investigating themes not fully clarified in SIPEM V were also presented, requiring further examination. These articles were collected under the item *Themes proposed for research in WG11of the previous edition of SIPEM.*

Constitutive themes of the philosophy of mathematics education collects ideas about the conception of mathematical objects, the mathematical *doing*, ethnomathematics, the constitution of mathematical objects, the conception of knowledge, and knowledge as a technique. One of the concerns highlighted is the way through which these mathematical objects are constituted, which agrees with the idea that articulates the clarification of the conception of mathematical objects and that of doing mathematics.

One of the ways to understand this conception includes looking at mathematical objects as fixed standards generated during historical-cultural development through the action-reflection process promoted by social practice (Gomes & Morey, 2015). Another way to understand mathematical concepts is based on a given operation, as discussed in article [4], which considers division as distribution in equal parts, treating the operation as a mathematical concept that carries a need, that is, the division into identical parts though with no concern for the experience of the people who conduct this distribution in sociocultural concepts. Article [8] also reveals the existence of an extra-linguistic mathematical reality, which lends meaning to its propositions, going against the notion of mathematical objects that exist or pre-exist in the mental sphere, at empirical level, or even in social subjectivity. This is *mathematical doing*, that is, the procedures adopted in the construction of mathematical knowledge are addressed in article [7], which declares that transformations are not mere new clothing to Euclidean geometry. Nevertheless, the article claims that the dynamic transformations previously admitted and tacitly and axiomatically declared now become the epistemological medium of these transformations, which are now seen as a group so that the notion of group revolutionizes geometry doing, since the properties of geometric objects, rather than being embedded in them, are, or are not, a group with other objects. Article [13] discusses the question that doing mathematics is a method, not a procedure that inflexibly indicates the ways to go; on the contrary, it is a process of doing that involves judgements and choices of what is significant or useful.

In this collection of studies, discussions about ethnomathematics also are carried out to understand mathematics and address the question of mathematical objects. Article [8] discusses the comprehensions about what it calls *mathematics* associated with different life forms understood as sets of language games that share similarities and that may, according to the authoress, intersect the comprehensions of ethnomathematics. The author of article [11] discusses the ways through which ethnomathematics has been conceived: as a study of innate mathematical thought and of practices that emerge from various cultures, as the possibility of criticism and deconstruction of the universality of mathematics when decoding cultural practices and testing pedagogical possibilities that work with these decoded concepts. The article also emphasizes that these ways of conceiving ethnomathematics reveal a naturalized view (i.e. one that has not been problematized concerning the presupposed understandings) and the notion that ethnomathematics adapts itself to use.

WG11 has held constant discussions about the *constitution of mathematical objects*, since they have not been understood by the group's members as not existing outside the human world, which, put simply, is the world where we live with one another in cultural environments that are socially organized and in contact with the natural world as seen through the possible interpretations. For example, the authors of article [1] resort to the theory of objectification as an explanatory theory that affords to assume that mathematical objects derive from social practice. Together with Nietzsche and Deleuze, the authoress of article [12], in emphasizing the question: '*Who invented mathematics?*' points to the forces and desires that devise a mathematics. The authoress inquires over the invention *of* mathematics and the invention *with* mathematics. The authoress of article [4] discusses the invention that takes place in the composition with remainders, an invention of relationships, properties, and concepts that form the scope of what is called mathematics. This effort towards searching for the modes mathematics constitutes itself in the world is made by the participants of WG11, even though with distinct views and possibilities.

The discussion about the usefulness of mathematical objects is also addressed. In article [1], the usefulness of these objects lies in their role of models, standards, and essential matrices in the development of any human activity requiring mathematical knowledge. Article [11] discusses the decolonizing potential of ethnomathematics, revealing its political bias, and the decolonization of the educational system. Both political and educational views address the question: '*What is ethnomathematics useful for?*', similarly to the question about the usefulness of mathematics.

Underlying the ways WG11 articulates the ideas presented above is the understanding that knowledge is something that constitutes us but is not to be owned by us, as discussed in article [2]. The same article presents argumentations about positivist science, revealing that, when rationality prevails, knowledge begins to be constructed through techniques that validate truth, blurring the search for questions whose answers cannot be visualized, as a consequence of the premises of theory considered from an objectified standpoint.

The *Comprehensive Core Ideas* ***The cyberspace environment in mathematics education*** addresses the comprehensions about the conception of computer, covering questions about formal language and programming language, and mathematical proving carried out by mathematicians who use the computer. These background questions help understand the way of doing mathematics when one uses the computer, creating possibilities to understand the communication spaces constituted in the cyberspace where mathematics is discussed, in a dialogue. Therefore, a space was created for the ways to understand mathematics when talking about the discipline in social media groups.

One of the ways to see the computer is presented in article [13]. According to the authors, the computer goes beyond the descriptions of sequences of operations carried out in three main elements, namely the data input unit, the data processing unit, and the output unit. On the contrary, the computer represents new opportunities to the mathematician, improving efficiency of calculations, generation and editing of graphic elements, which turns it into a more responsive communication medium. In article [10], the author highlights an important principle of computer science: whatever is static in computation does not exist to the machine. The underlying comprehension of this principle is the perception of perception of change. It has been argued that this perception takes on a special mode, to which computers connect directly: the perception of the transformations taken place as a result of the stimuli we send, which, when shown on the screen, are considered as perceived reaction. Also, the notions of interaction and interactivity, when associated with computers, are already embedded as possibilities in these transformations.

Concerning the work of mathematicians next to the computers, the discussions make clear, at the current state of investigations and debates, that, certainly, the computer has proved to be important in the work with mathematics and mathematics education in several aspects, like the marking of tests, for example. Besides, another aspect that was emphatically highlighted was the creative experience of the mathematician, that action that synthetizes reasoning, intuitions, perceptions, and that comes out as the solution for the problem. In this experience, intuition plays a central role in the work of the mathematician and, as we understand it, it is this experience that differentiates the work of the computer from the work of the mathematician in mathematics production. Article [13] alludes to Turing (quoted from Copeland & Shagrir, 2013), who describes intuition as *a spark that shines for an instant and affords a glimpse of a possibility of adjustment. Creativity is responsible for the approach taken to this adjustment.*

The space opened to communication by the computational structure affords modes of expression between people who communicate verbally, in writing, imagery, and mathematics. Article [5] presupposes that thought and its expression constitute themselves simultaneously in the life experience of the lived body, according to the concept developed by Merleau-Ponty (1994, 2002). By focusing on the reality of cyberspace where one may be with the other establishing dialogues, the article understands that, in the dialogue about mathematics thought may advance, call the other in, and thus share what is perceived, understood, and interpreted. In terms of the situations in which comprehension of this science stands out, some studies cited in this chapter reveal that discourse, understood as an articulation of intelligibility, may express itself in social media groups.

The philosophical research on mathematics education carried out in the teaching environment of elementary and higher education is a set of *comprehensive core ideas* that, as we see them, contribute distinctiveness to the research on philosophy of mathematics education that we have been carrying out. This difference lies in the work conducted in the school environment based on philosophical studies that improve the clarification of ontological questions (modes of being of mathematical objects), epistemological questions (modes of knowing and producing knowledge of mathematical objects), and axiological questions (modes of valuing actions and attributing values, as it is the case, for instance, of evaluation and ethical posture), in addition to the planning and actualization of investigation activities. These include the didactic-pedagogical activities developed by the researcher with professionals in training, elementary school students, and undergraduates.

In a study carried out in the classroom environment, the authoresses of article [4] use an item of teaching content and problematize it. The study considers division as distribution: a mathematical concept that, as said above, carries the need for equal parts and clarifies the operation as a mathematical object, explaining the way it operates. The study also shows that the operation takes place far away from several lives, even though it produces lives by imposing ways to operate that are often memorized and begin to dominate the ways a person proceeds beyond the school walls. The inquiry expressed in the article and that maintains itself as an inquiring dialogue in its clarification analyses how to turn mathematics into a problem. Therefore, the intention expressed by the authoresses of the article is not to consider mathematics objectively as given, but to work it in teaching and learning activities, problematizing mathematics itself. Also, the authoresses understand that a trace of occupation in this mode of proceeding is mathematics as being instituted, which fills it with objects of a world that is distant from the students who are there in a scenario of producing (themselves) with mathematical knowledge. This generates tension, and when the authoresses carry out the activities always in an inquiring way, the subjects with whom they are in a situation of teaching and learning mathematics start the process of producing ways of operating invented amidst of disquietude. So, they do not resort to the reproduction of instituted modes. We understand that, in this process, the researchers who authored the article take the conceptions with which they work in the mathematics classroom as a means to understand the activity of conducting the generation of knowledge and to understand the other with whom they are, listening to them in their disquietude.

The understanding of this WG11 is that mathematics is actualized in the school, that is, it realizes itself through the assembly of several forces, not only by what is determined by established science or by a previously conceived and inflexible curriculum in light of what is happening. These forces generate from the enrolment of bodies with materiality and historically constructed modes of knowing that intertwine with modes of being *with* teachers *with* students *with* textbooks with … with … with (Clareto, 2015). The vigour and diversity of these forces are so intense that it becomes impossible to name where they come from. However, we understand that these forces are perceived and that it is necessary, in the school environment, to work with them. We understand that the curriculum of activities to be carried out in the school has to consider all the historicity of the construction of science in the western world, from the modern age and the current times. Nevertheless, as proposed in article [6], one cannot avoid accepting the ideas surrounding the elementary concepts of the theory of fractals, emphasizing its connections with the theory of chaos so as to include uncertainty and the instabilities that prevail in the world we live in, where chaos and order coexist in a close relationship.

Also, the theme *curriculum* was one of the subjects highlighted by WG11 and addressed in one of the studies presented, which focus on the work with geometry at school. Article [7] introduces a debate about *The New Scientific Spirit*, in which Bachelard (1968) positions the dialectics of the realism/rationalism in geometric philosophy, in what becomes a tool for the epistemological comprehension of new geometries, that is, the understanding of the notion of group. Since a group is a set of invariants that manifest due to a transformation, epistemologically speaking a significant turn is observed towards the definition of what geometry is (geometries are) when it is the physical invariants that would lend rational value to the principles of permanence. Bachelard (1968) also declares that the western geometry tradition is Euclidean, and that, as ancient and widely practised as it is, this geometry may circumscribe itself in completed notions. Each object found its own concept and may be placed as an object among objects. We understand that our work at school, based on what has been discussed so far, has to make this tradition relative. This may take place when, with the advent of non-Euclidean geometric thoughts, we accept the task of reconstructing concepts. It is important to underscore the epistemological question framed by Bachelard, since it implies discussing how one geometry—or a certain way to make geometry—realizes itself *de-realizing* another, exposing it to criticism. When taken to the pedagogical field, this question exposes the differences between more traditional teaching practices as we follow one geometry and one practice that will propose geometrizing itself in a given possible way.

Another topic addressed in article [9] discusses the conception and importance of the trainee period in teacher education programmes based on a questionnaire prepared by the researchers and that the students answered. For the authoresses, students understand the trainee period as an opportunity for them to gain experience about the school environment. In addition, the respondents understood that this is one of the few situations during the course when they can get near schools and know different realities. The article places emphasis on this experience in the formation of teachers, suggesting that the training period should take place already in the beginning of the teachers' education course, not at the end. It is also understood that a review of the activities proposed by supervisor professors and/or those allowed by the schools during training periods should be conducted, since these activities sometimes restrict the role of the student to a few moments during classes, when this student is mostly considered only an observer of the work conducted by the professor during the training period.

Finally, we discuss the *Themes of research presented in WG11 in the previous SIPEM*, which concerned ethnomathematics and hermeneutics. The questions that remained unanswered addressed the conceptions of ethnomathematics and the respective approaching or distancing from mathematics and the difference between the hermeneutics as presented by Gadamer and depth hermeneutics. In the article presented in that edition of SIPEM, the latter was mentioned as referring to the hermeneutics defined by Thompson.

Articles [8] and [11], which investigated ethnomathematics, clarified the subject and included themes covered in *Constitutive themes of mathematics education*, the reason why they will not be discussed again here.

Hermeneutics was discussed in two articles, [2] and [3]. The concerns expressed by the authoresses included clarifying the conception of hermeneutics and, for that, investigated the Heideggerian, Gadamerian, and hermeneutical views. The first article elucidated depth hermeneutics. Hermeneutics was understood from two standpoints: in its exegetic meaning, when it is methodical and normative, as it was seen between ancient times and the nineteenth century, and the philosophical hermeneutics, which was studied by Heidegger, Gadamer, and Ricoeur. In this case, hermeneutics is understood as a window to the world and, therefore, to the historic and the cultural, inasmuch as the act of understanding is seen as the constitutive of human beings who, by existing, understand themselves and their cultural works. This comprehension, which carries an interpretation and comprehension, supersedes the definition of hermeneutics as a method of interpretation, as some authors call it. The hermeneutic experience is that of linguisticity, in which language meets its mode of being in conversation and in mutual understanding. It is a movement that always brings the subject, the other, and the world into an interpretative hermeneutic circle, widening horizons and other possible comprehensions. It is in this frame that the appraisal made by Habermas is placed (as quoted by Negru, 2014 and Batista, 2012), an author who systematically investigates the distortions of communication, that is, of the linguistic world. For Habermas, communication is distorted and distortions should be elucidated by depth hermeneutics. This way of understanding hermeneutics is adopted by Thompson (2009), who, in the last chapter of his book, introduces a methodology to adopt with depth hermeneutics. For WG11, as observed during the discussions, any attempt to develop one interpretative method has to face the impossibility of comprehension since it limits its own method. The circle closes in the logicity of the method and does not open itself to the comprehension of the world and of ourselves.

In this sense, hermeneutics was understood as a possibility for research on mathematics education, as a means to understand the phenomenon of human language and its various cultural manifestations and expressions. Using the Gadamerian hermeneutics, article [3] investigates comprehension of algebraic structures.

12.4 A Comprehension of Philosophy of Mathematics Education

Philosophy of mathematics education, as the philosophy of any activity, turns itself to this very activity, aiming to understand it from the standpoint of its objectives and its logic. In this sense, it focuses on the investigation of the activities of this area, like teaching and learning mathematics, the importance of teaching mathematics to all elementary school students. Questions like '*What is mathematics?*' '*How does mathematics relate with society?*' '*What is the nature of learning (mathematics)?*' '*What is the status of mathematics education as a field of knowledge?*' were framed by Paul Ernst and by the group of organizers of the Topic Studies Group on Philosophy of Mathematics Education of the International Congress of Mathematics Education, 13th edition, in 2016.

The works presented and discussed in WG11 in SIPEM 2015 addressed several of these questions and strengthened the mode of proceeding mentioned during problematizations, analyses, and reflections about mathematical objects, the conceptions of mathematics, the relationships between mathematics and the school environment, the schoolarization of mathematics, the formation of the mathematics teacher, the different cultural mediators in mathematics teaching, with special emphasis on computational environments and ICTs, and ethnomathematics.

The analysis of what was carried out aiming at the essence of each article discussed by the group brings to light a uniqueness concerning the philosophy of mathematics education as it has been investigated across the world. This uniqueness manifests itself through the view about knowledge and the reality presented by several Brazilian authors that emerges from consistent studies and reflections about the works of authors like Deleuze, Gadamer, Husserl, Merleau-Ponty, and Wittgenstein. This view underlies the innovative investigations on research on philosophy of mathematics education as being realized in the very environment of elementary and higher education, besides investigations conducted hermeneutically. Concerning the latter, a quote extracted from the back face of *The Philosophy of Mathematics Education* (Ernest et al., 2016), with which we end this article:

> [...] A case study is provided of an emergent research tradition in one country. This is the Hermeneutic strand of research in the philosophy of mathematics education in Brazil. This illustrates one orientation towards research inquiry in the philosophy of mathematics education. It is part of a boarder practice of 'philosophical archeology': the uncovering of hidden assumptions and buried ideologies within the concepts and methods of research and practice in mathematics education [...]. (Ernest et al., 2016, back cover.)

Based on the analysis of the works presented in the group of Philosophy of Mathematics Education (WG11) in the latest edition of SIPEM, it was possible to draw an updated panorama of research on philosophy of mathematics and of the dialogue that these studies establish with works from both international and Brazilian authors as well as other lines of research on mathematics education, contributing a philosophical reflection to develop the ideas shaping this field of knowledge.

At last but not least, in being a group on philosophy of mathematics education, WG11 works philosophically, which means that the studies conducted by its members focus on several themes on the mathematics education investigated in the other working groups of SIPEM, since the attribute of philosophy is to reflect on what is taking place or what has been carried out. We can mention some of these themes as being: learning and teaching mathematics, mathematics education, informatics and distance education, the school reality where mathematics education takes place, and mathematics education carried out in the teaching environment of elementary and higher education.

References

Bachelard, G. (1968). *O novo espírito científico*. Rio de Janeiro, RJ: Tempo Brasileiro.

Baier, T., & Bicudo, M. A. V. (2015, November 15–19). Conteúdos curriculares de matemática na educação básica: Uma análise filosófica. *Proceedings of the VI SIPEM, VI Seminário Internacional de Pesquisa em Educação Matemática*, Brasília, DF.

Batista, M. (2012). Hermenêutica filosófica e o debate Gadamer-Habermas. *Crítica e Sociedade: Revista de Cultura Política, 2*(1), 101–118.

Batistela, R. F., Barbariz, T. A. M., & Lazari, H. (2015, November 15–19). Um estudo sobre demonstração matemática por/com computador. *Proceedings of the VI SIPEM, VI Seminário Internacional de Pesquisa em Educação Matemática*, Brasília, DF.

Brasil. (1998). Ministério da Educação. Secretaria de Educação Fundamental. *Parâmetros Curriculares Nacionais: Matemática*. MEC/SEF, Brasília, DF.

Brasil. (2002). Ministério da Educação. Secretaria de Educação Média e Tecnológica. *PCN + Ensino médio: orientações educacionais complementares aos Parâmetros Curriculares Nacionais: Ciências da Natureza, Matemática e suas Tecnologias*. MEC/SEMTEC, Brasília, DF.

Bicudo, M. A. V. (1993). A Hermenêutica e o Trabalho do professor de Matemática. *Cadernos da Sociedade de Estudos e Pesquisas Qualitativos, 3*(3), 63–96.

Bicudo, M. A. V. (1998). O papel do Educador. *Nuances—Revista do Curso de Pedagogia, 4*, 20–24.

Bicudo, M. A. V. (1999). *Pesquisa em Educação Matemática: Concepções e Perspectivas*. São Paulo, SP: Editora Unesp.

Bicudo, M. A. V. (2000). *Fenomenologia, confrontos e avanços*. São Paulo, SP: Cortez.

Bicudo, M. A. V. (Ed.). (2010). *Filosofia da Educação Matemática: Fenomenologia, concepções, possibilidades didático-pedagógicas*. São Paulo, SP: Editora Unesp.

Bicudo, M. A. V. (2011). Realidade Virtual: Uma abordagem filosófica. *Ciências Humanas e Sociais em Revista, 33*, 114–127.

Bourbaki, N. (1950). The architecture of mathematics. *The American Mathematical Monthly, 57*(4), 221–232.

Clareto, S. (2015, November 15–19). Professor quem inventou a matemática? Travessias de uma pergunta que se torna problema e um problema que inventa currículo. *Proceedings of the VI SIPEM, VI Seminário Internacional de Pesquisa em Educação Matemática*, Brasília, DF.

Copeland, B. J., & Shagrir, O. (2013). Turing versus Gödel on computability and the mind. In B. J. Copeland, C. J. Posy, & O. Shagrir (Eds.), *Computability: Turing, Gödel, church and beyond*. London, UK: The MIT Press.

D'Ambrosio, U. (1979). Adequate mathematics for third world countries: Consideranda and strategies. In M. El Tom (Ed.), *Developing mathematics in third world countries* (pp. 33–46). Amsterdam: North Holland Publishing Company.

D'Ambrosio, U. (1984, August 24–30). Socio-Cultural Bases for Mathematical Education. *Proceedings of the 5th International Congress in Mathematics Education*, Adelaide.

D'Ambrosio, U. (1998). *Etnomatemática: Arte ou técnica de explicar e conhecer* (5a. ed.). São Paulo, SP: Ática.

D'Ambrosio, U. (1999). A história da matemática: Questões historiográficas e políticas e reflexos na Educação Matemática. In M. A. V. Bicudo (Ed.), *Pesquisa em Educação Matemática: Concepções e perspectivas* (pp. 97–115). São Paulo, SP: UNESP.

Dedekind, R. (1931). *Gesammelte mathematische Werke. Zweiter Band*. Braunschweig, Germany: Verlag Vieweg & Sohn. (Original work published in 1923).

Deleuze, G. (2006). *Diferença e repetição*. (Luiz B. L. Orlandi, Roberto Machado, Trans.) (2nd ed.). Rio de Janeiro, RJ: Graal.

Deleuze, G. (1997). Mistério de Ariadne segundo Nietzsche. In G. Deleuze (Ed.), *Crítica e Clínica* (p. 34), (Peter Pál Pelbar, Trans.). São Paulo, SP: Editora.

Deleuze, G., & Guattari, F. (2012). *Mil Platôs: Capitalismo e Esquizofrenia* (Vol. 5, p. 34) (Peter Pál Pelbar, & Janice Caiafa, Trans.). São Paulo, SP: Editora.

Detroni, A. R., & Pinheiro, J. M. L. (2015, November 15–19). Direções para uma filosofia geométrica das transformações. *Proceedings of the VI SIPEM, VI Seminário Internacional de Pesquisa em Educação Matemática*, Brasília, DF, Brazil.

Ernest, P., Skovsmose, O., Van Bendegem, J. P., Bicudo, M., Miarka, R., Kvasz, L., et al. (2016). *The philosophy of mathematics education. ICME-13 topical surveys*. Hamburg, Germany: Springer. https://doi.org/10.1007/978-3-319-40569-8

Euclides. (2009). Os Elementos (Traduzido por Irineu Bicudo). São Paulo: EDUNUESP.

Figueiredo, O. A. (2015, November 15–19). O movente e o formal. *Proceedings of the VI SIPEM, VI Seminário Internacional de Pesquisa em Educação Matemática*, Brasília, DF.

Gadamer, H. G. (1999). *Verdade e método—Traços fundamentais de uma hermenêutica filosófica*. (Flávio Paulo Meurer, Trans.). Petrópolis, RJ: Vozes.

Galileu, G. (1999). *O Ensaiador. Coleção Os Pensadores*. (Helda Barraco, Trans.). São Paulo, SP: Nova Cultura. (Original work published in 1623).

Gomes, S. C., & Morey, B. B. (2015, November 15–19). A concepção ontológica dos objetos matemáticos na teoria da objetivação. *Proceedings of the VI SIPEM, VI Seminário Internacional de Pesquisa em Educação Matemática*, Brasília, DF.

Hawking, S. (1997). *Breve História do Tempo Ilustrada*. Curitiba, PR: Albert Einstein.

Husserl, E. (1997). *Die Urstiftung und das Problem der Dauer. Der Ursprung der Geometrie*. Diederichs, München: Steiner, UCH.

Husserl, E. (2008). *A crise da humanidade europeia e a filosofia*. Porto Alegre, RS: Edipucrs.

Knijnik, G. (2001). Educação matemática, exclusão social e política do conhecimento. *Bolema: Boletim de Educação Matemática, 14*(16), 12–28.

Knijnik, G., & Wanderer, F. (2008). Discursos produzidos por colonos do sul do país sobre a matemática e a escola de seu tempo. *Revista Brasileira de Educação, 13*(39), 555–599.

Kluth, V. (2015, November 15–19). A linguagem em Gadamer: Sua imagem em uma pesquisa da educação matemática. *Proceedings of the VI SIPEM, VI Seminário Internacional de Pesquisa em Educação Matemática*, Brasília, DF.

Meneghetti, R. C. G., & Oliveira, M. S. F. D. (2015, November 15–19). Futuros professores e suas concepções sobre os estágios supervisionados de matemática. *Proceedings of the VI SIPEM, VI Seminário Internacional de Pesquisa em Educação Matemática*, Brasília, DF.

Merleau-Ponty, M. (1994). *Fenomenologia da percepção*. São Paulo, SP: Martins Fonte.

Merleau-Ponty, M. (2002). *A prosa do mundo*. São Paulo, SP: Cosac & Naify.

Miarka, R. (2015, November 15–19). Preocupações e tendências da pesquisa em etnomatemática. *Proceedings of the VI SIPEM, VI Seminário Internacional de Pesquisa em Educação Matemática*, Brasília, DF.

Mondini, F., Mocrosky, L. F., & Bicudo, M. A. V. (2015, November 15–19). A hermenêutica em educação matemática: Compreensões e possibilidades. *Proceedings of the VI SIPEM, VI Seminário Internacional de Pesquisa em Educação Matemática*, Brasília, DF.

Negru, T. (2014). O debate entre Gadamer e Habermas e a universalidade da hermenêutica. *Revista Redescrições: Revista do GT Pragmatismo e Filosofia Americana, 2*(3), 53–59.

Paulo, R. M., & Ferreira, M. J. A. (2015, November 15–19). Comunicação no ciberespaço: Diálogos acerca de matemática. *Proceedings of the VI SIPEM, VI Seminário Internacional de Pesquisa em Educação Matemática*, Brasília, DF.

Rotondo, M. A. S., & Azevedo, F. O. (2015, November 15–19). A tal da matemática: Um problema. *Proceedings of the VI SIPEM, VI Seminário Internacional de Pesquisa em Educação Matemática*, Brasília, DF.

Silva, J. M. P. (2015, November 15–19). Etnomatemática e as ideias de jogos de linguagem. *Proceedings of the VI SIPEM, VI Seminário Internacional de Pesquisa em Educação Matemática*, Brasília, DF.

Thompson, J. B. (2009). *Ideologia e Cultura Moderna: Teoria social crítica na era dos meios de comunicação de massa*. Petrópolis, RJ: Vozes.

Wittgenstein, L. (1999). *Investigações Filosóficas* (José Carlos Bruni, Trans.). São Paulo, SP: Nova Cultural.

Wussing, H. (1969). *Die Genesis des abstrakten Gruppenbegriffes: Ein Beitrag zur Entstehungsgeschichte der abstrakten Gruppentheorie*. Berlin, Germany: Veb Deutscher Verlag der Wissenschaften.

Wussing, H. (1989). *Lecciones de historia de las matemáticas*. Madrid, Spain: Siglo XXI de España Editores.

Chapter 13
Every Citizen Needs to Know Statistics! What Are We Doing? Brazilian Research in Statistics Education

Mauren Porciúncula, Suzi Samá, Cristiane de Arimatéa Rocha, and José Ivanildo Felisberto de Carvalho

Abstract This chapter aims to present an overview of the research discussed in Working Group 12 (WG12) on teaching probability and statistics, of the Brazilian Society of Mathematics Education, as it has developed during the International Seminars of Research on Mathematics Education (SIPEM). It begins by considering the emergence of statistics education as a research field in its own right in the international context and by reflecting on how Brazilian researchers became part of this movement. In sequence, the other sections present the research trajectory, highlighting the importance of studies in the field of probability and statistics education for Brazilian mathematics education. This part of the paper is organised into two main sections. The first seeks to provide the reader a general idea of the research in statistics education in Brazil, from the papers presented in the SIPEM from 2003 to 2015, which the research organised into four categories: teaching strategies; teacher training; analysis of learning and analysis of curricula and textbooks. It then goes on to consider in more detail the projects discussed at the most recent editions of SIPEM, which collectively contemplated an analysis of the knowledge of mathematics teachers to teach combinatorics; a diagnosis of the initial ideas of the teachers about the knowledge of probability; a discussion about the knowledge necessary to the future teachers of mathematics for the understanding of the concept of probability; and an evaluation of a pedagogical proposal for learning statistical inference.

M. Porciúncula (✉) · S. Samá
Universidade Federal do Rio Grande, Rio Grande, Brazil

C. de Arimatéa Rocha · J. I. F. de Carvalho
Universidade Federal de Pernambuco, Recife, Brazil

A. J. Ribeiro et al. (eds.), *Mathematics Education in Brazil*,
https://doi.org/10.1007/978-3-319-93455-6_13

13.1 Trajectory of the Research in Statistics Education and the Creation of WG12

Since the 1970s, the need to break with the deterministic culture in mathematics classes has led many countries to defend the teaching of statistics from the early years of basic education (Cazorla, Kataoka, & Silva, 2010). Consequently, research work began to be carried out on teaching and learning on this field of knowledge, which began to be socialised in national and international events.

Created in 1991, as a result of international mobilisation, the International Association for Statistical Education (IASE) was meant to be a new education section of the International Statistical Institute (ISI) to promote the advancement of statistics education in the educational, industrial and governmental spheres (Cordani, 2015). IASE promotes satellite conferences in the ISI every 2 years, and the last edition took place in 2015 in Brazil, bringing together several researchers from WG12. Another important action in favour of statistical education occurred in 1982, when the International Statistical Institute (ISI) Education Committee held the first International Conference on Teaching Statistics (ICOTS). Since then, an edition of ICOTS has been held every 4 years. This conference aims to bring together professionals from different areas of knowledge to share and discuss the results of their research related to statistics education. In 2006, this Conference was held in Brazil. According to Cazorla et al. (2010), 15 out of the 94 Brazilian researchers that participated in the conference were from WG12.

Following this international scenario, in Brazil, at the end of the 1990s, statistics and probability began to be officially incorporated into the curriculum structure of basic education with the publication of the National Curricular Parameters (PCN). These parameters (Brasil, 1997, 1998, 2000) aimed to guide the organisation of school curricula throughout Brazil.

This historical moment was the beginning of the first research works in Brazil in the area of statistics education (Santos, 2015). WG12 was created in that context in 2000, during SIPEM I. This working group emerged with the objective of studying and understanding how people teach and learn statistics, which also involves the cognitive and affective aspects of teaching and learning, as well as the development of teaching methods and materials for statistical literacy. To contemplate such approaches, statistics education uses theoretical and methodological resources from other areas, such as psychology, pedagogy, philosophy, neuroscience, computing, among others, besides mathematics itself. To study the teaching of statistics focusing on these sciences, WG12 currently counts on researchers from different teaching and research institutions in Brazil and several knowledge areas, predominantly mathematics, pedagogy and psychology.

The constitution of this WG by SBEM has great social relevance for two main reasons: (i) with the legal inclusion of the concepts of statistics and probability in basic education, their teaching became mathematics teachers' responsibility. Therefore, it is necessary to develop, integrate and disseminate research that subsidises the training of teachers to teach the concepts of this area, especially in basic education; (ii) Besides legislation, we also need to consider that the media, such as television and the internet, use statistical results to give credibility to the information transmitted to the population. These statistics are presented in the form of graphs, such as the evolution of the dollar quotation; position measures, such as average sales in trade; tables, such as those of educational statistics; as well as results of opinion polls, experiments and scientific studies. In this way, statistics is increasingly present in daily life. Therefore, it is very important that every citizen is literate in statistics to make their own decisions consciously.

This section aimed to situate the reader in relation to the context in which WG12 is inserted in Brazil, and its relevance. We present below the research works submitted to the first editions of SIPEM, demonstrating what is being done in the field of probability and statistics education in Brazil, in line with the demands of the current society.

13.2 Statistics Education Research Presented at WG12 from SIPEM II to SIPEM V

Over the years, WG12 has been configured as a catalyst for research in statistics education in Brazil, with the first socialisation of the studies at SIPEM II in 2003.

We found a diversity of statistical concepts and an interest on the different levels of education. The concepts investigated include construction and interpretation of graphs, statistical measures (such as mean and variance), combinatorics, basic concepts of probability and hypothesis tests. In order to promote the learning of these different concepts, at different levels of education, various methodologies and theoretical references were used.

In the studies with a relatively small number of subjects, reports of teaching experiences, case studies, action research, participatory research, exploratory research and different qualitative analyses, such as content analysis from data collected by observations and/or interviews, were used. On the other hand, in studies with greater number of subjects, statistical tests, regression and correlation analyses, hypothesis tests and multivariate statistics were the methodological strategy of data analysis adopted. Also, some studies made use of both qualitative and quantitative approaches, and other studies have used bibliographic research.

Different theoretical references provide the background for WG12 research. A few works originated from the education area, such as David Ausubel's theory of meaningful learning. According to him, it is fundamental to establish a relationship between the content to be taught and the reality in which the student is inserted. It is necessary to create a motivating environment to be suitable for learning.

Other studies have used theoretical frameworks such as the knowledge onto-semiotic approach (Godino, 2009) to study the types of mathematical objects, such as language, situations, concepts, procedures and arguments, and possible semiotic conflicts that may compromise the understanding and meaning of basic concepts of statistics and probability. Some other studies are based on theoretical frameworks that discuss the knowledge categories of mathematics teachers, such as the theory of didactic-mathematical knowledge underlying OSA, and the theory of mathematical knowledge for teaching (MKT) proposed by Ball, Thames, and Phelps (2008).

In relation to the aims of the research presented, the largest number of papers published were on Teaching Strategies (19), followed by Teacher Education (10), Learning Analysis (3) and Analysis of Curricula and Textbooks (5). These themes provide the reader with a general idea of the research works on statistics education since the SIPEM II, when WG12 became part of this event. We present each of these approaches, organised by theme, in subsections, and some of the research works that constituted them.

13.2.1 What Teaching Strategies Is WG12 Proposing? There Are Many …

Proposing teaching strategies has been a main goal of the research work presented from SIPEM II to V by WG12 members. There are many specificities about this topic, and we have chosen to organise subsections according to their focus: Teaching through projects; Digital technologies: Problem solving through games; and Understanding statistical measures.

13.2.1.1 Teaching Through Projects

Six papers address this teaching strategy. Some are about teaching through projects, others are statistical projects or mathematical modelling. Although the nomenclature differs from one researcher to the other, all defend a similar pedagogical proposal.

In this perspective, Mendes (2003) emphasises the importance of working with statistical concepts with real problems in order to promote student learning. This strategy allows the student to take an active role in the construction of knowledge, as well as to live an investigative activity on topics of his/her interest (Mendonça & Lopes, 2009). Jacobini and Wodewotzki (2006) point out that the academic and political education of the student, resulting from the learning by modelling, makes it possible to promote the statistical literacy of students linked to the social conscience. During the opening of SIPEM II, Mendes (2003) suggested a transposition of the proposals of the American Statistical Association (ASA) into our national reality, seeking to introduce the Quantitative Literacy Project, in order to adapt the statistical projects to the PCN as a way of working on cross-cutting themes.

In the same direction, Silva and Groenwald (2003) emphasise the importance of inserting the environmental dimension in mathematics curriculum of basic education, through teaching through projects, so that the student experiences a contextualised statistics. The interdisciplinary work, enabled by teaching through projects, is a way of promoting the "development of a full citizen able to contribute to the formation of a democratic, solidary and just society" (Groenwald & Seibert, 2003: 158).

The proposal of teaching through projects was also used in the training of future teachers of the initial years of elementary school. Biajone and Carvalho (2006), dissatisfied with the negative attitudes of preservice teachers regarding the discipline of statistics, used teaching through projects as a pedagogical proposal. Researchers have found that the effectiveness of this approach is "directly related to the nature of the content, its objectives and the cultural complexity of the target audience for its destination" (p. 177). For the authors, it is necessary to make adjustments in the dynamics of the classroom and in the process of evaluation, in order to achieve meaningful learning, consistent with the formative needs of the future professional.

13.2.1.2 Digital Technologies in Statistics Education

Viali and Sebastiani (2009) sought to motivate and involve students in the learning of statistical concepts, through a contextualised approach, allied to the use of technological resources. In this paper, high school students conducted a data survey and used a spreadsheet in order to organise and present their results through tables and graphs. From the results and the evaluation of the approach, the authors concluded that the activities developed promoted a greater participation of the students, improving the learning of the statistical concepts.

In another project, Viali, in partnership with Cunha, investigated the contributions of a learning unit for the application of statistical concepts with the use of the spreadsheet, with elementary school students. Based on the observations collected during the research, and on the analysis of the materials produced by the students, the authors verified "an evolution in the statistical literacy level of the participating students" and "positive changes in both individual behavioural terms and coexistence as a group" (Cunha & Viali, 2012, p. 1).

Another important contribution to the promotion of student statistical literacy through the use of technologies was implemented by (Cazorla and Gusmão 2009; Lopes, 2006a). The authors present a teaching sequence that integrates a set of activities that aims to provide theoretical and methodological support to a Virtual Environment to Support Statistical Literacy—AVALE. This interactive and free virtual environment was developed in order to assist the teaching and the learning statistics and probability in basic education.

Seeking to boost the probabilistic literacy of the high school students, Ferreira, Kataoka, and Karrer (2012) have used simulation through software R integrated with other activities in a paper and pencil environment. The purpose of the simulation was to enable students themselves to analyse whether the concepts of probability that they had at the beginning of the activity were confirmed after the simulation. Although the use of the R software favoured the comparison of the results obtained initially, the work carried out in sequence, in the paper and pencil environment with the tree of possibilities, played a prominent role in the understanding of the concepts of theoretical probability.

13.2.1.3 Problem Solving Through Games and Understanding of Statistical Measures

Brazilian official documents, such as the National Curriculum Parameters (PCN) , advocate problem solving as a centrepiece for teaching mathematics. In this direction, Lopes (2006b, 2009b, 2012) proposes problem solving through games as a possibility for teaching and learning basic concepts of probability and statistical measures, such as mean, median and mode. The author emphasises that in the methodology of problem solving the students assume an active attitude in facing challenges, assuming a posture of researcher and constructor of their own knowledge. In this process, the "mistakes are used for the reflection of the students about their action and it is also important for the teachers, because it helps them to understand and interpret the development of the students' ideas" (Lopes, 2006b, p. 12). From the manifestations of the high school students that participated in the games proposed by the author, it was possible to conclude that the students became more active in the construction of their own knowledge. In addition, the researcher found that the proposal of problem solving, through games, promoted greater interest of the students on the concepts worked in class.

Problem solving was also the teaching strategy discussed by Carvalho, Monteiro, and Campos (2009) when working on the interpretation of graphs in many contexts. In graph interpretation, people "evoke previous knowledge related to facts or experiences of their lives that influence the interpretations of graphically presented data" (Carvalho et al., 2009, p. 2). The authors emphasise the importance of working in the educational environment with the reading and interpretation of graphs since they find that they develop students' reasoning in situations involving relations between variables.

Studies on the understanding of different statistical measures as a way of promoting learning, summarising and representing a set of data were also found in SIPEM. For example, Novaes and Coutinho (2009) present some ways of determining quartiles in order to promote students' learning on variability. The choice to work with the concept of quartiles is based on the hypothesis, raised by the authors, that the low mathematical complexity involved in determining this statistical measure "can constitute a facilitating agent for the construction of meanings and, consequently, for the development of the statistical literacy" (p. 1).

Concerning probability, Ferreira et al. (2012) investigated the learning of probabilistic concepts through the application of the teaching experiment "Carlinha's Random Rides". The results showed advances in the understanding of probabilistic concepts and the work with the tree of possibilities. Also in relation to probability, Bittencourt and Viali (2006) presented a proposal for teaching the normal distribution and the central limit theorem through practical examples such as: generation of random numbers, relation of the mean with the standard deviation and selective process results to be added into undergraduate courses. They emphasise that the understanding of these concepts, beyond simple memorisation, is fundamental to the success of teaching and learning.

13.2.2 Not Only Teaching Strategies Were Proposed, Learning Analysis Were Also Carried Out!

The analysis of the learning of statistics and probability by students has also been the objective of the research presented by WG12 in the SIPEM. From 2003 to 2015, several papers contemplated such approach. Sometimes the focus was strictly on learning; other times, this analysis of learning occurred after a teaching intervention.

Research works that relate to the verification of attitudes and skills regarding statistics were present in many SIPEM editions, revealing different findings. Many of them counted on the contribution of cognitive theories, as well as descriptive statistics, correlation and multiple regression and multivariate statistics, for learning analysis.

In the context of a study conducted with 319 university students, Vendramini and Brito (2003), by applying some instruments and scales, found positive and statistically significant associations between: statistical performance and attitude towards statistics; and performance in statistics tests of and mathematics tests. These results confirm that the teachers should be concerned, when planning their way of teaching, with the correlation between what an attitude can cause in a student and his/her consequent performance. This aspect was evidenced through a multiple regression that showed that the more positive the attitudes towards statistics, the better the performance.

Furthermore, studies involving the elaboration of scales to measure interaction and attention strategies were also presented. After validation of a scale, through a factorial analysis with 236 students, Oliveira, Kataoka, and Silva (2009) identified a high level of interaction between students and a low level of interaction between student and teacher. In relation to attention, the scale allowed to verify an association with lower learning outcomes. Knowing this profile of students, considering interaction and attention was considered of utmost importance to the teacher when planning their classes and teaching strategies.

Another important set of studies about the learning of statistics includes the verification of students' knowledge. These studies have made use of instruments containing items that involve specific content. Some of these are created by the researchers themselves, others adapt instruments already validated by international researchers. Most of them make use of quantitative analyses, as well as statistical tests to obtain and compare the results. A study with this characteristic, presented in the SIPEM III, had as its objective to verify if the contents of statistics indicated in the PCN were being delivered in high schools, as well as the level of knowledge of students of the last year. Through an instrument containing fifteen items on interpretation, construction of graphs, mean, median and mode, standard deviation, sample and probability, the study obtained as main results how little some statistical measures, such as mode and median were known, and how high was the error rate for simple questions of proportion (Echeveste, Bittencourt, Bayer, & Rocha, 2006). In general, 11 of the 15 questions had a success rate of less than 50%. The study also showed that the grades reached by students in private schools were significantly higher when compared to those in public schools.

13.2.3 And About the Teachers! How Is the Research About the Training of These Professionals That Teach Statistics?

The papers presented in the SIPEM by WG12 also focused on teacher training. In this theme, in the period from 2003 to 2012, the contributions presented were focused on: initial training of teachers teaching statistics; continuing education of teachers teaching statistics; knowledge, conceptions and identity of the teacher.

The manuscripts on the teachers initial training deal with different aspects, such as: investigations of the feasibility of formative processes, for example, the validation of a didactic sequence to teach basic concepts of probability (Cazorla & Gusmão, 2009; Lopes, 2006a); the analysis of situations that allow the discussion of the processes involved in the interpretation of graphs (Monteiro, 2006) and the teaching of statistics through work with projects (Biajone & Carvalho, 2006). The first of these studies was carried out with preservice mathematics teachers for basic education, and the last two with preservice teachers for the initial years of elementary school.

Such studies prioritise not only aspects related to statistical and probabilistic concepts, but also didactic-pedagogical aspects, including: activities that enable the mobilisation of mathematics, statistics and other knowledge areas; activities of graphs interpretation; practices that encourage the widening of the meaning of statistics in the teaching training course for the initial years of elementary education; activities that promote the broad development of probabilistic reasoning through analysis and sequencing applications, as well as discussions on conflict points such as conditional probability, independent events, or the generalisation of the binomial distribution. The results of these research works reaffirm the need for training practices that address the teaching of statistics throughout initial teacher training.

Other research focused on the continued training of teachers. They contemplate the approach of processes identified in the praxis of statistics and probability teaching, by teachers who teach mathematics at the end of elementary and middle (Muniz & Gonçalves, 2006), and the processes of teachers training in mathematics focusing on the contents of descriptive statistics, through the simultaneous use of various representations of the same data set (Coutinho, Silva, & Almouloud, 2009).

These investigations took place, respectively from the development of an extension course for teachers; and in continuing education processes for teachers of the public network, offered by a higher education institution for more than 10 years. Both reaffirm the need for didactic aspects for the teaching of statistics and probability, or for the construction of statistical literacy of teachers.

In particular, Muniz and Gonçalves (2006), during the development of a course for teachers, found that there are weaknesses in the concepts of chance and randomness. Coutinho et al. (2009), in the development of a project to investigate the perception of variability by in-service mathematics teachers at the elementary school, note the importance of continuing education. Such research demonstrates the need for training processes for in-service teachers, revolving around specific knowledge, as well as pedagogical aspects.

The papers published from SIPEM II to V also discussed and investigated teachers' knowledge of statistics, probability and combinatorics, and the factors that influence the construction of the identity of these teachers.

The research works point out some challenges for teachers, namely: gaps regarding variation reasoning (Silva & Almouloud, 2003); challenge of considering the curricular demands of statistics in teaching practices in elementary and middle school (Bayer, Echeveste, & Félix, 2003); the insertion of formative practices that address a deepening of the knowledge of combinatorial structures (Rocha, 2012); and the understanding of the power relations existing in the construction of the identity of the teachers who teach statistics, delimited by the duality between researchers in the areas of statistical education and statistics (Pamplona & Carvalho, 2009).

13.2.4 A Look at the Curriculum and the Textbook: What Do Researchers Point To?

Throughout the editions of the SIPEM, it is possible to find research that analyses curricular documents and textbooks. These works reveal that the insertion of statistics into the basic education curriculum in Brazil is very recent, having only occurred since the end of the 1990s. We believe that this movement may have influenced the researchers' perspective, not only for curricular documents, but also to understand the approach of textbooks for teaching statistics and probability.

Despite a small number of research involving the curriculum and textbook, such studies present relevant questions. Some of these studies (Girard, 2003; Lopes, 2009a, 2009b) discussed the curriculum, and others (Coutinho & Gonçalves, 2003; Gitirana & Anjos, 2009; Viali & Oliveira, 2009) addressed the mathematics textbook for basic education in Brazil.

Regarding curriculum research, Coutinho and Gonçalves (2003) identified divergences between the conceptions and practices of foreign teachers, when related to Brazilian curricular proposals. Girard (2003) emphasises the insertion, in French curricula, of statistics and probability in an articulated way. The author draws attention to the fact that the prescription in the curriculum does not solve the difficulties with teaching and learning these concepts. Like Girard (2003), even today, we believe that such question remains as something to be constructed, as a challenge for us, researchers and teachers. Lopes (2009a, 2009b) strengthens this issue when analysing the curricular documents and textbooks of Brazil and the United States of America.

The challenge of working effectively with statistics and probability can become even more complex if textbooks do not keep up with curriculum innovations. In the study of probability, Viali and Oliveira (2009) did not find the presentation of relations between probability and statistics in the books destined to high school.

The analyses of Brazilian textbook present in SIPEM led both Gitirana and Anjos (2009) and Viali and Oliveira (2009) to conclude that textbooks present a deficient approach, both from the point of view of the concepts addressed and the way they are presented, making it difficult for students to construct knowledge of statistical and probabilistic concepts.

13.3 Contributions of SIPEM VI: From Teaching Strategies to Teacher Training

In 2015, during SIPEM VI, the papers presented in WG12 contemplated the diversity that has characterised the WG over the years, focusing upon the teaching and learning of statistics, probability and combinatorics, both in basic and in higher Education.

A study conducted by Porciúncula and Samá (2015), from the Research Group of Statistical Education of the Universidade Federal do Rio Grande—FURG evaluated a teaching strategy based on Ausubel's theory of meaningful learning (Ausubel, 1968; Moreira, 2014). This research was developed in the course of statistical inference, in the psychology course. There were 32 students enrolled.

Based on the meaningful learning theory, student problem-solving was adopted as the pedagogical strategy and all classes were planned around problem-solving situations. At the end of the course, students were asked to write a text that contemplated, among other aspects, an evaluation of the pedagogical proposal of the course.

To analyse the written records, the authors used a qualitative approach, through the analysis of the discourse of the collective subject (DSC), proposed by Lefèvre and Lefèvre (2005a, 2005b). It consists in the construction of discourses through the junction of fragments of the individual manifestations of the students, subjects of this research. Porciúncula and Samá (2015) found that the authorship of the students, when elaborating problem situations with themes they chose, provided the awakening of their interests; teamwork made it possible for students, through interaction, to incorporate new statistical concepts into their cognitive structures; learning from error, insofar as they came to perceive it as something positive, was used as a learning factor; and that continued, rather than summative assessment, minimised the fear of having to demonstrate their knowledge in a single moment.

In this edition of SIPEM, Felisberto de Carvalho and Macedo (2015) presented the design and evaluation of a teaching strategy involving a didactic sequence for the development of probability knowledge with 42 students, enrolled in the course of statistics and probability for mathematics preservice teachers. In the course, notions concerning probabilistic meanings—classical and frequentist, and the concept of conditional probability were discussed. In this paper, the authors also used the theoretical framework discussed by Ball et al. (2008), which contributed both to the elaboration of the didactic sequence and to the analysis of the teachers' knowledge in initial formation in the development of the didactic sequence.

The results of this research indicated the difficulties and limits in understanding the distinction between chance and probability. The researchers realised that at the moment of the calculation of the probability a correct mapping of the sample space was not carried out by a majority of the participants. Another finding was that the classical meaning of probability has strong validation by students, to the detriment of the frequentist meaning. The authors also point out several obstacles involving the conditional reasoning, such as the understanding of reduced sample space and the insufficiency to present a mathematical demonstration.

The teaching strategy addressed in the study by Felisberto de Carvalho and Macedo (2015) can provide an experience of situations that mobilise the knowledge necessary to teaching probability, and that contribute to teachers being effective in their teaching practice in this field of knowledge.

Concerning the discussions on teacher education, from four studies presented in SIPEM VI, two studies discuss the knowledge of mathematics teachers, namely: Rocha, Lima, and Borba (2015) presented the results of two projects carried out within the scope of the Group of Studies in Combinatory Reasoning of the Education Centre—Generation of the Universidade Federal de Pernambuco—UFPE; and Felisberto de Carvalho, Pietropaolo, and Campos (2015) analysed a questionnaire in order to diagnose the didactic-mathematical knowledge of the teachers of the final years of elementary school on probability.

Rocha et al. (2015) aimed to analyse the knowledge of mathematics teachers to teach combinatorics, based on the results of two master's dissertations (Lima, 2015; Rocha, 2011), in which teachers were interviewed teaching mathematics, specifically the combinatorial and the fundamental counting principles (FCP). Both Rocha (2011) and Lima (2015) used as theoretical basis the domains of mathematical knowledge for teaching, developed by Ball et al. (2008).

According to Rocha et al. (2015), teachers participating in both studies, when analysing different combinatorial situations, mobilise and express different domains of knowledge to teach combinatorics or the FCP, also known as multiplicative principle. Respondents pointed out different resolution strategies, such as listing, the tree of possibilities, the double entry table, multiplication and the FPC, to solve the different types of problems worked on in the combinatorics. The authors also point out that teachers reflect on the statements of combinatorial problems and suggested strategies and were able to point out possible difficulties and/or readiness for learning that students may have.

Rocha et al. (2015) advocate for investment in new research addressing the different domains of knowledge in the practices of initial and/or continued teacher education, especially for those who teach combinatorics at different levels of education.

Felisberto de Carvalho et al. (2015) also investigated the knowledge of teachers, but in this case, the focus of the research was on elementary notions of probability. These researchers present and discuss a questionnaire oriented to analyse the initial knowledge on elementary probabilistic notions of mathematics teachers of the final years of elementary school. This questionnaire is the first element of the interventional design of a doctoral research, applied to teachers participating in a group of the Teacher Education Observatory programme in Brazil.

In this study, the authors used as theoretical basis the didactic-mathematical knowledge theory of Godino, Ortiz, Roa, and Wilhelmi (2011), which underlies the ontosemiotic approach to mathematical knowledge and instruction (Godino, 2009).

Felisberto de Carvalho et al. (2015) point out teachers' difficulties with probabilistic knowledge, either at the time of explaining what is random or in defining probability—for example, or in calculations involving the quantification of probabilities. In general, this diagnostic tool points out that this group of teachers, many of whom have been practising for more than a decade, present gaps in the different knowledge distinguished by the teacher's didactic—mathematical knowledge theory.

13.4 The Consolidation of Statistics Education in Brazil! Contributions to Continue Promoting the Learning of Statistics by Brazilian Citizens…

The focus on statistics education, in an international and national scope, derives, initially, from the demands of training undergraduate students in this area; followed by the need to break with the deterministic culture in mathematics classes, adopted when many countries started to adopt teaching of statistics from basic education.

As Brazilian legislation came to include statistics in basic education, which was predominantly the responsibility of mathematics and pedagogy teachers, the need to know more about this field of knowledge—statistics education—intensified, in so far as their concepts were being incorporated into the curriculum.

Specifically, in this text, we have sought to provide the reader a general idea of the research in statistics education in Brazil, from the papers presented in the SIPEM from 2003 to 2015. Through the themes that emerged from the analysis of the texts published in the annals of the events, it was possible to perceive the comprehensiveness of the research that has been developed in Brazil: teaching strategies; teacher training; analysis of learning and analysis of curricula and textbooks.

In addition to these topics, described in general terms, this text also sought to socialise with readers the recent researches published in the last SIPEM. They contemplate an analysis of the knowledge of mathematics teachers to teach combinatorics; a diagnosis of the initial ideas of the teachers about the knowledge of probability; a discussion about the knowledge necessary to the future teachers of mathematics for the understanding of the concept of probability; and an evaluation of a pedagogical proposal for learning statistical inference.

Undoubtedly, the findings from all the research works presented in this chapter have contributed to the consolidation of research groups and graduate programmes in this area. By integrating and surpassing the academic environment, these results in the field of research are reflected in the pedagogical practice of teachers in basic and higher education.

References

Ausubel, D. P. (1968). *Educational psychology: A cognitive view*. New York: Holt, Rinehart & Winston.

Ball, D., Thames, M. H., & Phelps, G. (2008). Content knowledge for teaching: What makes it special? *Journal of Teacher Educacion, 5*, 389–407.

Bayer, A., Echeveste, S., & Félix, V. (2003). Estatística no Ensino Fundamental e Médio: como professores de Matemática estão se preparando para este desafio. *Anais II SIPEM*, Santos, Sao Paulo, Brasil 2.

Biajone, J., & Carvalho, D. L. (2006). Estatística em projetos na pedagogia: promovendo aprendizagens e (re)significando atitudes. *Anais III SIPEM*, Águas de Lindóia, Sao Paulo, Brasil, 3.

Bittencourt, H. R., &Viali, L. (2006). Contribuições para o ensino da distribuição normal ou curva de Gauss em cursos de graduação.*Anais III SIPEM*, Águas de Lindóia, Sao Paulo, Brasil, 3.

Brasil. (1997). Ministério da Educação. Secretaria de Ensino Fundamental. *Parâmetros Curriculares Nacionais: anos iniciais do Ensino Fundamental (1° e 2° ciclos Matemática). Volume III*. Brasília: MEC/SEF. Retrieved from http://portal.mec.gov.br/seb/arquivos/pdf/livro03.pdf.

Brasil. (1998). Ministério da Educação. Secretaria de Ensino Fundamental. *Parâmetros Curriculares Nacionais: anos finais do Ensino Fundamental (3° e 4° série Matemática)*. Brasília: MEC/SEF. Retrieved from http://portal.mec.gov.br/seb/arquivos/pdf/introducao.pdf.

Brasil. (2000). Ministério da Educação. *Parâmetros Curriculares Nacionais. Ensino Médio. Parte III Ciências da Natureza, Matemática e suas Tecnologias*. Brasília: MEC. Retrieved from http://portal.mec.gov.br/seb/arquivos/pdf/ciencian.pdf.

Carvalho, L. M. T. L., Monteiro, C. E. F., & Campos, T. M. M. (2009). Interpretação de gráficos como uma atividade que envolve fatores visuais e conceituais. *Anais IV SIPEM*, Taguatinga, Federal District, Brasil, 4.

Cazorla, I. M., & Gusmão, T. C. (2009). Uma análise semiótica dos passeios aleatórios da Mônica: atividade para ensinar conceitos básicos de probabilidade. *Anais IV SIPEM*, Taguatinga, Federal District, Brasil, 4.

Cazorla, I. M., Kataoka, V. Y., & Silva, C. B. (2010). Trajetória e Perspectivas da Educação Estatística no Brasil: um olhar a partir do GT12. In C. E. Lopes, C. Q. S. Coutinho, & S. A. Almouloud (Eds.), *Estudos e Reflexões em Educação Estatística*. Campinas, Sao Paulo: Mercado das Letras.

Cordani, L. K. (2015). Caminhos da Educação Estatística ao Longo do Tempo: uma leitura pessoal. *International Journal for Studies in Mathematics Education, 8*(3), 157–182.

Coutinho, C. Q. S., & Gonçalves, M. C. (2003). O livro didático e a formação do professores de Matemática para o ensino de probabilidades. *Anais II SIPEM*, Santos, Sao Paulo, Brasil, 2.

Coutinho, C. Q. S., Silva, M. J. F., & Almouloud, S. A. (2009). Professores de matemática e a análise da variabilidade de dados representados simultaneamente por histograma e box-plot. *Anais IV SIPEM*, Taguatinga, Federal District, Brasil, 4.

Cunha, M. L., & Viali, L. (2012). Contribuições de uma unidade de aprendizagem para a compreensão de conceitos estatísticos com o recurso da planilha. *Anais V SIPEM*, Petropólis, Rio de Janeiro, Brasil, 5.

Echeveste, S., Bittencourt, H. R., Bayer, A., & Rocha, J. (2006). Um estudo sobre o nível de conhecimento dos alunos do 3° ano do Ensino Médio sobre Estatística. *Anais III SIPEM*, Águas de Lindóia, Sao Paulo, Brasil, 3.

Felisberto de Carvalho, J. I., & Macedo, R. C. (2015). Conhecimentos Necessários para o Ensino de Probabilidade: discussão de uma sequência didática desenvolvida com estudantes de matemática-licenciatura. *Anais VI SIPEM*, Pirenópolis, Goiás, Brasil, 6.

Felisberto de Carvalho, J. I., Pietropaolo, R. C., & Campos, T. M. M. (2015). Discussão de um Diagnóstico Inicial sobre Noções Probabilísticas na Perspectiva do Conhecimento Didático – Matemático. *Anais VI SIPEM*, Pirenópolis, Goiás, Brasil, 6.

Ferreira, R. S., Kataoka, V. Y., & Karrer, M (2012). Sequência de ensino "passeios aleatórios da Carlinha": contribuições da árvore de possibilidades. *Anais V SIPEM*, Petropólis RJ, Brasil, 5.

Girard, Jean Claude. (2003) A ligação estatística-probabilidade nos novos programas franceses de ensino médio: a posição da Modelização e da simulação. Anais II SIPEM, Santos, SP, Brasil, 2.

Gitirana, V., & Anjos, D. (2009). Invariantes do Conceito de Média: uma análise dos livros didáticos dos anos finais do ensino fundamental. *Anais IV SIPEM,* Taguatinga, Federal District, Brasil, 4.

Godino, J. D. (2009). Categorías de Análisis de los conocimientos del profesor de matemáticas. *Unión Revista Iberoamericana de Educación Matemática, 20,* 13–31.

Godino, J. D., Ortiz, J. J., Roa, R., & Wilhelmi, M. R. (2011). Models for statistical pedagogical knowledge. In C. Batanero, G. Burrill, & C. Reading (Eds.), *Teaching statistics in school mathematics-challenges for teaching and teacher education: A joint ICMI/IASE study* (pp. 271–282). Berlin, Germany: Springer.

Groenwald, C. L. O., & Seibert, T. E. (2003). Projetos de trabalho e o ensino de estatística no ensino fundamental. *Anais II SIPEM,* Santos, SP, Brasil, 2.

Jacobini, O. R., & Wodewotzki, M. L. L. (2006). Temas contemporâneos nas aulas de matemática: um caminho para combinar aprendizagem e reflexões políticas. *Anais II SIPEM,* Águas de Lindóia, Sao Paulo, Brasil, 3.

Lefèvre, F., & Lefèvre, A. M. C. (2005a). *Discurso do Sujeito Coletivo: um novo enfoque em pesquisa qualitativa* (2nd ed.). Educs: Caxias do Sul.

Lefèvre, F., & Lefèvre, A. M. C. (2005b). *Discurso do Sujeito Coletivo.* Retrieved from http://www.fsp.usp.br/~flefevre/Discurso_sujeito_coletivo.htm.

Lima, A. P. (2015). *Princípio Fundamental da Contagem: conhecimentos de professores de Matemática sobre seu uso na resolução de situações Combinatórias.* (Dissertação - Mestrado) - Programa de Pós-graduação em Educação Matemática e Tecnológica - EDUMATEC, UFPE. Recife, Pernambuco, Brasil.

Lopes, C. A. E. (2006a). A estatística e a probabilidade na educação básica e a formação dos educadores matemáticos. *Anais III SIPEM,* Águas de Lindóia, Sao Paulo, Brasil, 3.

Lopes, C. A. E. (2009a). A implementação curricular da Estatística e da Probabilidade na Educação Básica.*Anais IV SIPEM,* Taguatinga, Federal District, Brasil, 4.

Lopes, J. M. (2006b). O ensino de probabilidade através de jogos e da metodologia de resolução de problemas. *Anais III SIPEM,* Águas de Lindóia, Sao Paulo, Brasil, 3.

Lopes, J. M. (2009b). Uma proposta para o estudo de conceitos básicos de probabilidade. *Anais IV SIPEM,* Taguatinga, Federal District, Brasil, 4.

Lopes, J. M. (2012). O estudo da média, da mediana e da moda por meio de um jogo e da resolução de problemas. *Anais V SIPEM,* Petropólis, Rio de Janeiro, Brasil, 5.

Mendes, C. R. (2003). Alfabetização estatística: possibilidade de implementação através de projetos interdisciplinares. *Anais II SIPEM,* Santos, Sao Paulo, Brasil, 2.

Mendonça, L. O., & Lopes, C. E. (2009). A Educação Estatística no Ensino Médio por meio da Modelagem Matemática. *Anais IV SIPEM,* Taguatinga, Federal District, Brasil, 4.

Monteiro, C. E. (2006). Estudantes de pedagogia refletindo sobre gráficos da mídia impressa. *Anais III SIPEM,* Águas de Lindóia, Sao Paulo, Brasil, 3.

Moreira, M. A. (2014). *Abandono da narrativa, ensino centrado no aluno e aprender a aprender criticamente.* Retrieved from http://www.if.ufrgs.br/~moreira/Abandonoport.pdf.

Muniz, C. A., & Gonçalves, H. J. L. (2006). A Educação Estatística no ensino fundamental: discussões sobre a práxis de professoras que ensinam matemática no interior de Goiás.*Anais III SIPEM,* Águas de Lindóia, Sao Paulo, Brasil, 3.

Novaes, D. V., & Coutinho, C. Q. S. (2009). Quartis: uma análise didática de alguns dos diferentes métodos para sua determinação. *Anais IV SIPEM*, Taguatinga, Federal District, Brasil, 4.

Oliveira, M. H. P., Kataoka, V. Y., & Silva, C. B. (2009). Estratégias de atenção e de interação no processo de autorregulação da aprendizagem de estatística: validação de uma escala.*Anais IV SIPEM,* Taguatinga, Federal District, Brasil, 4.

Pamplona, A. S. & Carvalho, D. L. (2009). Relações de Poder em Comunidades de Prática: os professores de estatística na licenciatura em matemática. *Anais IV SIPEM,* Taguatinga, Federal District, Brasil, 4.

Porciúncula, M. M. S. & Samá, S. (2015). A Teoria da Aprendizagem significativa e o Ensino de Estatística na Graduação. *Anais VI SIPEM,* Pirenópolis, Goiás, Brasil, 6.

Rocha, C. A. (2011). *Formação docente e o Ensino de Combinatória: diversos, olhares, diferentes conhecimentos.* Dissertação. Pós-graduação em Educação Matemática e Tecnológica. Universidade Federal de Pernambuco. Recife, Pernambuco, Brasil. Retrieved from http://repositorio.ufpe.br/xmlui/bitstream/handle/123456789/3774/arquivo2888_1.pdf?sequence=1&isAllowed=y.

Rocha, C. A. (2012). Ensino de Combinatória: expectativas de professores que atuam no Ensino Fundamental. *Anais V SIPEM,* Petropólis, Rio de Janeiro, Brasil, 5.

Rocha, C. A., Lima, A. P. B., & Borba, R. E. S. (2015). Conhecimentos de Professores para Ensinar Combinatória: contribuições de pesquisas. *Anais do VI SIPEM,* Pirenópolis, Goiás, Brasil, 6.

Santos, R. M. (2015). *Estado da Arte e história da pesquisa em educação estatística em programas brasileiros de pós-graduação.* Tese. Universidade Estadual de Campinas. Campinas, São Paulo, Brasil.

Silva, C. B., & Almouloud, S. A. (2003). A compreensão de média e variabilidade como conceitos essenciais para a noção intuitiva de intervalo de confiança. *Anais II SIPEM,* Santos, Sao Paulo, Brasil, 2.

Silva, C. K., & Groenwald, C. L. O. (2003). Estatística e Meio Ambiente: uma abordagem para o ensino fundamental e médio. *Anais II SIPEM,* Santos, Sao Paulo, Brasil, 2.

Vendramini, C. M. M., & Brito, M. R. F. (2003). Implicações das habilidades matemáticas e das atitudes na aprendizagem dos conceitos de estatística. *Anais II SIPEM,* Santos, Sao Paulo, Brasil, 2.

Viali, L., & Oliveira, P.I.F. (2009). Uma análise de conteúdos de probabilidade em livros didáticos do Ensino Médio. *Anais IV SIPEM,* Taguatinga, DF, Brasil, 4.

Viali, L., & Sebastiani, R. G. (2009). A Estatística no Ensino Médio: uma abordagem alternativa. *Anais IV SIPEM,* Taguatinga, Federal District, Brasil, 4.

Chapter 14
Difference, Inclusion and Mathematics Education in Brazil

Miriam Godoy Penteado, Fabiane Guimarães Vieira Marcondes, Clélia Maria Ignatius Nogueira, and Leo Akio Yokoyama

Abstract This chapter presents an overview of participation of the working group on Difference, Inclusion and Mathematics Education of the Brazilian Society of Mathematics Education (SBEM) at the sixth International Seminar of Research on Mathematics Education (SIPEM VI). This is the most recently created working group, launched nationally in 2013, with its first meeting taking place at SIPEM VI, where 14 papers were presented. In this chapter, we report upon the four themes around which these papers were organised: (1) communication and language; (2) content and curriculum; (3) teachers, teacher knowledge and teaching practices and (4) questioning the concept of normality. The ideas addressed in each of these themes are presented in this chapter. Discussion of the concerns raised in the research studies suggested that the fourth theme could be seen as issues that cut across all the others, leading to the perspective that everyone is different and, further, that a problematisation of the concept of normality should guide the definition of content, methodology and assessment at all levels of mathematics education.

14.1 Introduction

Inclusion is a subject that has permeated discussions in many social spheres, mobilising public policies that aim to allow people with disabilities to attend and interact in various social situations. These discussions are the result of long and difficult battles of people with disabilities throughout history. In Brazil, different stages of

M. G. Penteado (✉)
Universidade Estadual Paulista, Rio Claro, SP, Brazil

F. G. V. Marcondes
Instituto Federal de São Paulo, São José dos Campos, SP, Brazil

C. M. I. Nogueira
Universidade Estadual do Oeste do Paraná e Universidade Estadual de Maringá, Cascavel, Maringá, PR, Brazil

L. A. Yokoyama
Colégio de Aplicação da Universidade Federal do Rio de Janeiro, Rio de Janeiro, Brazil

A. J. Ribeiro et al. (eds.), *Mathematics Education in Brazil*,
https://doi.org/10.1007/978-3-319-93455-6_14

these battles can be inferred from the national legislation that determines the rights and duties of citizens and can be considered as guiding public policies. We begin by highlighting some of the legal aspects that have characterised the trajectory of educational inclusion in our country.

In 1978, an amendment to the Federal Constitution of Brazil gave citizens with disabilities the right to free special education. In 1988, the new Constitution stated that the education of citizens with disabilities should preferably be in the regular education system. The 1996 Guidelines and Bases of National Education replaced the term "disabled" with "learners with special needs", seeking to focus on needs rather than disability. In the 1990s, at the UNESCO World Conference on Education for All, it was proposed that all students should be integrated into the school environment, with diversity being respected by offering answers to potentialities and needs. In 1994, 92 countries (Brazil included) reaffirmed the right to education of every individual (UN. Organisation of the United Nations, 1948), by agreeing to the Salamanca statement about the education of all disabled children, which sets out the following principles: recognise difference, meet specific needs, promote learning, recognise the "school for all" and educate teachers for the new educational demands.

The Guidelines and Bases of National Education (Brasil, 1996) and Resolution 02 of the National Education Council (Brasil, 2001a) addressed the enrolment of students with disabilities in regular schools and maintain Specialised Educational Assistance (AEE) to complement the education of students with disabilities.

The document Curricular Adaptations (Brasil, 1998) recognised diversity in the school population, as well as the need to respond to this diversity by offering adequate responses to each group through planned and organised processes, always aiming at education for all. In 2001, the Inter-American Convention (Declaration of Guatemala), which reaffirmed the same rights and freedoms of persons without disabilities for people with disabilities (Brasil, 2001b), was launched and signed in Brazil.

In 2008, the National Policy on Special Education in the Inclusive Perspective PNEE (Brasil, 2008) was established as guiding document for states and municipalities, transforming their educational systems into inclusive systems. PNEE shifts the focus from the individual's difficulties to the conditions of the environment in which they live. Decree 6949/2009 ratified this understanding: "people with disabilities are those who have long-term disabilities of a physical, mental, intellectual or sensorial nature, which, in interaction with various barriers, may obstruct their full and effective participation in society on an equal basis with other persons" (Brasil, 2009, p. 1).

According to PNEE, educational inclusion is a political, cultural, social and pedagogical action, defending the right of all students to learn and participate together, without any kind of discrimination.

In 2015, Law no. 13146 of July 6, the Brazilian Law of the Person with Disability (Statute of the Person with Disability), was signed.

Article 27 states that

> Education constitutes the right of the person with disability, ensuring an inclusive educational system at all levels and lifelong learning in order to achieve the maximum possible development of their physical, sensory, intellectual and social skills and abilities, according to their characteristics, learning interests and needs (Brasil, 2015a).

This paragraph reaffirms as the duty of all to ensure quality education for the person with disability and to protect them from all forms of violence, neglect and discrimination.

Currently, Brazilian education is explicitly considered within an inclusive perspective and, as a consequence, between 2003 and 2010, the number of students with disabilities in regular schools increased by 234%, with the 2010 school census indicating that 95% of children with disabilities from ages 6 to 14 are in school.

Although this is progress, it does not mean that the service is in line with what the legislation indicates or that all members of the school community feel prepared to meet the educational needs of all students, ensuring them a good quality education according to the guiding principle of inclusive school.

This situation has been the subject of discussion and research by educators from different areas, particularly in mathematics education. Over the past decade, the interest for researching the education of disabled people and their relationships with mathematics education has increased. At several Brazilian universities, research groups are structured around the theme of mathematics education and inclusion, and there is already a significant bibliographic production.

In 2013, Working Group 13 (WG 13) entitled Difference, Inclusion and Mathematics Education was created within the Sociedade Brasileira de Educação Matemática (Brazilian Society of Mathematics Education—SBEM). Working Group 13 aims at holding meetings among researchers to promote joint projects and share research results—in particular, projects that take into account the present demand on mainstream schools to assume their responsibility for addressing disabled learners from an inclusive perspective.

> This group aims to approach studies that contribute to a mathematics education that favours a deep understanding of the teaching and learning processes, focusing on theoretical, methodological, pedagogical and epistemological issues. The concerns of this WG include the discussion of the adequacy of school practices, educational policies, teacher education, academic performance, and experience with mathematics outside the school context (SBEM, 2016).

According to the current government policy, special education addresses people with disabilities, those with global developmental disorders, as well as high abilities/giftedness. Although these populations have been the focus of interest in the initial establishment of WG 13, the group intends to move beyond the issue of special education by addressing other groups of learners who are marginalised by the current system and educational policies.

Once becoming official in the SBEM, WG 13 prepared its first meeting at the VI Seminário Internacional de Pesquisa em Educação Matemática (International Seminar of Research on Mathematics Education—SIPEM). According to the regulations of the event, 14 papers were chosen out of the 19 submitted. WG 13 coordinators read the articles, identified the focus of each one of them with the purpose of establishing a grouping. The themes that emerged from this process were: (1) Communication and Language; (2) Content and Curriculum; (3) Teachers, Teacher Knowledge and Teaching Practices and (4) Questioning the Concept of Normality.

In the following sections, we present the ideas addressed in the discussions of each of these themes and we conclude by summarising the main concerns of the groups.

14.2 Communication and Language

In this group, the following works were presented: (1) *The Inclusion of Deaf Learners in Mathematics Lessons: Stories Narrated by Interpreters of Libras*[1]; (2) *Between two Languages: Teaching and Learning of Mathematics to Included Deaf Learners*; (3) *Deaf Children in a Mathematical Landscape of Investigation* and (4) *Deaf-Hearing Dialogues: Paths to Inclusion.* All investigations focused on communication, in particular, with reference to Libras. The first two papers focus on studies that were carried out in the classroom, emphasising the role of the interpreter of Libras and used observations and interviews as principal methodologies. The third study focuses on deaf children and their engagement in landscapes of investigation. The meetings with children were video recorded and an interview was conducted with their teacher. The locus was a specialised institution that provides care for children with disabilities. The fourth focuses on the creation of signs for distance learning mathematics classes and counted upon the collaboration of deaf adults.

In general, the papers presented indicate that the education of deaf learners, even in a school that proposes inclusive education, is faced with a barrier, as oral communication is the main method adopted by the teacher. According to Borges and Nogueira (2015), the communication with deaf learners becomes reduced in an environment that uses a language that is not accessible to them. As a consequence, their learning becomes partially obstructed.

Silva and Fernandes (2015) refer to Alrø and Skovsmose (2010) in emphasising the problematic consequences for the learning of the deaf learners due to issues of communication. Alrø and Skovsmose point to a range of particular qualities of learning based on the dialogic qualities of a communication. Communication in mathematics is further complicated by the fact that mathematics has its own language, with terms that do not correspond to the conventional signs in Libras.

Indeed, the construction of mathematical signs in Libras is an important issue to be investigated. In order to do so, it is relevant to involve both the deaf and the hearing. Furthermore, both the mathematical and the linguistics aspects of the process of communication must be covered, and sharing of the newly created signs with other communities of deaf people is also needed.

Silva and Fernandes (2015) argue that interaction between the deaf and the hearing is important in order to encourage the construction of signs supporting the learning of mathematics by deaf learners.

[1] Língua Brasileira de Sinais (Brazilian Sign Language).

> "[...] We listen to their needs and generate alternatives that help to understand the mathematical content that appears" (p. 2).

In an inclusive school, communication between the deaf and the hearing in the classroom is mediated by an interpreter.[2] As this is still a relatively new phenomenon, more research is needed with respect to his or her role in the classroom.

In discussing the dialogical processes between deaf learners and a hearing teacher, Lacerda (1996) highlights some of the complex factors in this relationship. One factor is a superficial knowledge of signs among deaf learners and among the interpreters, another factor is that often the interpreter does not participate in any previous curricular planning with the teachers, and, as a consequence, many improvisations in the classroom may occur. Considering the difficulties that deaf learners have in reading and writing Portuguese, this situation becomes even more complicated.

Considering the working conditions of the interpreter, Jesus and Thiengo (2015, p. 3) point out that

> "[...] the presence of the interpreter does not ensure that methodological issues are considered, as he or she frequently does not have academic training to develop methodological strategies for teaching different contents to deaf learners".

In several studies, the search for different methodological strategies that consider the particularities of the deaf learner has been addressed. The adoption of certain trends in mathematics education, such as the implementation of mathematical investigations, sheds light on many difficulties in communication. Thus, Moura and Penteado (2015, p. 4) emphasise that much literature highlights that "the key element in the investigative process remains communication" and that it is "through communication that it is possible to organize a scenario in which you can put into practice various dialogic acts".

The establishment of a reliable and effective communication between deaf and hearing people and between deaf and deaf people is crucial. It is important to provide deaf children with an immersion in a language environment that enables them to acquire Libras as soon as possible. It is recommended that this takes place before the deaf children get to the school.

In many cases, the interpreter has no education in mathematics and he or she is not familiar with mathematical vocabulary. According to Borges and Nogueira (2015, p. 2), this lack of mathematics education greatly hampers the performance of the interpreter and, consequently, the teaching and learning of the discipline. Mathematics involves abstract knowledge, which brings the challenge of providing adequate exemplifications. Furthermore, mathematics uses words from natural language in order to name objects, "giving them meanings that can even not be imagined by a layman". This can generate "misunderstanding as they are held in colloquial meaning of the used word".

[2] Translator and Interpreter of Brazilian Sign Language. In Portuguese, this professional is named Tradutor e Intérprete de Língua Brasileira de Sinais (TILS).

Hence, research that seeks to create signs and proposes alternatives for the work of the interpreter becomes important. Jesus and Thiengo (2015, p. 10) highlight the "need for joint planning of lessons between the interpreter and mathematics teacher". Furthermore, they emphasise that "the teacher contributes to conceptual and methodological issues and the interpreter with knowledge related to deafness".

Such joint planning can minimise the interpreter's lack of knowledge of mathematics education, and help the teacher to prepare a lesson that meets the particularities of the deaf. With explicit reference to linguistic issues, it would be of value to explore the structure of Libras in terms of pedagogical possibilities.

14.3 Content and Curriculum

In this group, the following works were presented: (1) Adjusted Tasks to Introduce Combinatory Analysis to Visually Impaired and Deaf Learners from the Elementary School; (2) First Number Concepts of a Teenager with Down Syndrome using Multisensory Materials; (3) An Investigation with Deaf Learners from Elementary School: Mental Calculations in Question and (4) Constructing Takãra—Investigation and Mathematics Teaching Practice in the TAPI'ITÃWAA tribe. The first two projects were developed in the state of Rio de Janeiro, the third in the state of Paraná and the fourth in an indigenous tribe in the state of Mato Grosso. The first three involved the participation of students with disabilities (deaf, visually impaired and Down syndrome) in the development of activities with curricular contents. The fourth involved prospective teachers in an indigenous tribe.

In the case of learners with intellectual disabilities as Down syndrome, many find it difficult to understand basic mathematical concepts such as quantification of a set of up to ten elements. Reports from parents of these children state that they have no difficulties in relating with friends, practising team sports, or attending classes in science and arts. But, for most of them, mathematics has been an obstacle. One possible explanation is the deficit in the short-term verbal memory of a child with Down syndrome (Jarrold and Baddeley, 2001).

For some with intellectual disabilities, it is not trivial to acquire the concept of number related to quantification. Yokoyama (2015) comments on a teenager with Down syndrome who has no well-defined concept of number. After specific activities she begins to relate the procedure of counting to 10 with amounts of up to ten elements. Despite having daily experiences of counting in the mathematics classroom, this teenager did not relate counting directly to quantification.

Experiences as described in Yokoyama (2012) should be shared, so that more teachers can be motivated to develop new ways of teaching. This does not mean that a child with Down syndrome cannot learn with his or her classmates. But a more flexible curriculum is needed in order to acknowledge the differences. For example, it is not always possible for a person with intellectual disabilities to accompany the same speed of learning as those with typical development. However, this does not mean that they are unable to learn, just that they may learn at a different pace.

Considering deaf learners, schools should provide the teaching of Libras and written Portuguese as second language. However, since the educational inclusion of deaf people in Brazil is still incipient, many children are presented for both languages (written Portuguese and Libras) almost simultaneously, making it difficult, for example, to learn numerals and number words in Libras and in written Portuguese.

Hearing listeners, even before enrolling in school, tend to have memorised the sequence of number at least up to 10. This happens on account of social interaction, children's songs, rhymes, etc. They know that numbers are used for counting (Nogueira, 2016). This is not always the case for deaf children.

The lack of prior number knowledge of deafs in relation to hearing children was studied by Nunes, Evans, Barros, and Burman (2011), who concluded that this happens because deaf children do not have experiences involving counting in their daily lives and the school has to take this into account when organising the teaching of mathematics.

When dealing with the teaching of the decimal number system to deaf learners, Nogueira, Borges, and Frizzarini (2013) emphasise "the need for different methodological strategies, particularly to fill in the gaps in prior knowledge of deaf children caused by impaired interaction with the environment" (p. 173).

In order to promote these actions, public schools can use a specialised educational service. This is provided by a specialist teacher who, together with classroom teacher and maybe other educators, makes an individual development plan to meet the special needs of the learners.

With respect to initiatives for promoting access to official assessments, the following is observed:

(1) The National Secondary Education Examination,[3] which allows access to universities, is written in Braille for blind students by a team specialised in transcribing texts. (2) Deaf students have a right to be tested through Libras in order to have access to public universities; however, there is no mathematical vocabulary in Libras provided at the national level. In basic education, the deaf student learns the signs for mathematical notions through negotiation with the teacher and interpreter. There are different signs, depending on the region in which the student is going to school. This makes it impossible to draw up a video-test in Libras that could function at the national level. (3) There are no tests taking into account students with intellectual disabilities or global developmental disorders. If they want to apply for the university, the maximum they have as support is a person who reads the test for them.

Concerns with indigenous mathematics education are also addressed by WG 13. The study presented in SIPEM VI shows the importance of considering and valuing indigenous culture, their habits, customs and their own mathematical knowledge. Such approach provides favourable conditions for the dissemination of indigenous culture and skills to younger generations. This is also a way of providing space for the participation of elderly in assisting and assessing the school work.

---

[3] Exame Nacional do Ensino Médio (ENEM).

Although there are some ongoing actions taking place, there is still a huge demand. In order to promote education from the perspective of inclusion, it is necessary to consider: curriculum flexibility, different methodologies and sharing of good experiences.

14.4 Teachers, Teacher Knowledge and Teaching Practices

The articles that characterise this group are: (1) *From Edges to the Centre: Reflections of Mathematics Teachers about School Inclusion*; (2) *Mathematics Teacher Education under the Inclusive Perspective: State of Art from 2006 to 2015* and (3) *The Teaching of Mathematics for Deaf Learners from High School: An Analysis of the Teachers' Practices from the Federal District*. They were developed in the state of São Paulo, Rio de Janeiro, Federal District and Goiás, with the theme of inclusion and teacher education, as well as the presence of the interpreter teacher in mathematics classes.

The researches aimed at discussing the inclusion of students with disabilities, developmental disorder, high skills and giftedness, through the analysis of the narratives of mathematics teachers and specialised teacher by the oral history method. One aim was through a bibliographical revision on education of teachers who teach mathematics and the inclusive perspective to come to know the scientific productions in different contexts and documents that guide this discussion in Brazil. Another aim was to verify the relevance and the difficulties of the educational interpreter of sign language that acts in the mathematics classrooms in high school. In order to do so, observation and questionnaires with open and closed questions were applied to the educational interpreters and the mathematics teachers. All studies reveal the importance of teacher education in order to work in inclusive mathematics landscapes of investigation.

Inclusive education raises concerns with respect to the identity of the teacher and his or her perspectives regarding the teaching of mathematics taking into account the differences among learners. Such differences could refer to disabilities, cultural diversities or social-political issues. It is essential that teachers have an understanding of "the historical, socio-cultural and ideological contexts that have deployed discrimination and oppression practices in education" (Ballard, 2003, p. 59).

As part of teacher education, it should be possible in a broader way to discuss the concept of difference and inclusion. According to Rosa (2015), the idea of integration seems to dominate discourses with respect to schools. Thus, the idea seems to be to integrate learners into an already existing educational system without changing it. Another feature of this discourse is that the disabled persons are considered unable to learn. Their disability is seen as a personal failure, as a lack of this or that functionality. It is necessary to involve teachers in practices that focus on what learners can do instead of lamenting what they cannot (Healy and Powell, 2013).

Castro, Pinto, and Ramos (2015) state that their analysis of research journals shows a recent increase in research on inclusion and teacher education, but there is still a shortage. Regarding mathematics, most articles focus on teachers' education with respect to primary education. According to the authors, the shortage of articles that have mathematics as area of interest confirms that inclusion in mathematics lessons has been discussed mostly from the perspective of special education.

Thus, it is necessary to raise the following questions: How are curricula in mathematics teacher education addressing inclusive education? How is this topic approached in the education of teachers for secondary school? What are the spaces for this discussion in in-service teacher education?

With respect to compulsory education, research with focus on the education of deaf learners points out difficulties in the relationship between teachers and interpreter. Teachers recognise the importance of the interpreter of Libras, however fail in not planning the lessons together with him or her. Often the teachers justify this attitude by referring to the fact that there are too many learners in the class, and furthermore that they need to focus on the requirements for tests. They do not seek help from specialised educational services, which might be available in the school. As already mentioned, the interpreter rarely has an education background in mathematics, and, as a consequence, he or she has difficulties with specific mathematical signs that might not even exist in Libras. It is necessary that schools encourage the establishment of partnerships among educators who work in the classroom in order to plan strategies that help the interpreter doing the work.

New teacher knowledge is needed. It becomes necessary that teachers' mathematical content knowledge comes to include an understanding of relationships between different concepts. This gives the teacher the security of reorganising the curriculum. Such broader content knowledge also assists teacher in preparing assessments. Thus, it becomes important to move beyond standardised tests, which do not favour learners who, for example, do not master spoken or writing Portuguese.

It is necessary to discuss the accountability for the learners' education at school. Responsibility and planning should be shared among the interpreter, the teacher, the special education professional and other educators in the school. The way it is organised now, this broad assumption of responsibility is not taken.

14.5 Questioning the Concept of Normality

The articles that characterise this group are: (1) *The Zero of Deaf Learners: Zero is Absence, Zero is a Place, Zero is Failure, Zero is a Companionship, and Zero is Round*; (2) *Reflections of Prospective Mathematics Teachers about the Challenges of Teaching Mathematics in Inclusive Lessons* and (3) The *Rural School: Searching for Identity of Mathematics Learners from a Teacher Education Course of Rural Education*. They were developed in the state of São Paulo and Mato Grosso do Sul, focused on the theme of deafness, inclusion in teacher education and education in the field. The methods used were interviews with deaf students of a mainstream public school, analysis of the manifestations of teachers and prospective mathematics teachers from Brazil and England on inclusive scenarios and meetings with prospective teachers for rural schools.

These studies try to approach the communication of mathematical ideas in unconventional ways. They address how teachers and prospective teachers reflect on the challenge of inclusive classrooms and on the articulation of demands from those who live in the field (rural area). The studies aim at making teachers capable of acting together in their communities and realities.

Universal access to school highlights the paradox of inclusion/exclusion. Groups considered outside the homogenising school standards, although accepted, are excluded through selections, policies, naturalisation of school failure, among other reasons. This is because, traditionally, the school is seen to privilege the group defined as normal.

Historically, the organisation of special education is based on the concept of normality/abnormality, defined by Secretary for Continuing Education, Literacy, Diversity and Inclusion (SECADI), which is part of the Ministry of Education and Culture (MEC) (MEC/SECADI). The idea that disability is a pitiable condition and a personal disadvantage that need to be overcome results in an institutional and personal prejudice against learners with disabilities. This can have a marked effect on their academic trajectories.

In their research, Healy and Fernandes (2011) indicate that when blind learners use their hands as tools to indicate what they want to say, a variety of interpretations does arise. However, those who can see are accustomed to more static approaches and do not always recognise other ways of expressing mathematics. Sometimes, a sighted person finds it difficult to recognise as valid a non-conventional way of expressing mathematical ideas.

Blind learners see with their hands; deaf learners talk with their hands; and learners with Down syndrome use tactile and visual representations as memory resources. All such cases may express valid and innovative abstractions that exceed the expectations of teachers (Fernandes and Healy, 2013; Healy and Fernandes, 2011; Yokoyama, 2012). Deaf learners can make narratives with their hands in order to express their understanding of a mathematical concept. This is an unconventional way for exploring mathematical properties. Mathematical ideas can be expressed in narratives that would not make much sense in a traditional mathematics classroom. This indicates new directions for mathematics education that can contribute to inclusive practices (Marcondes and Healy, 2015).

This reflection can be extended to other contexts such as rural education, which refers to schools that meet the education of, for example, farmers, natives, settlers and camped of agrarian reform, rural workers, forest peoples and others who live from work in rural areas (Brasil, 2010).

In this case, normality tends to be defined through urban education. Teacher education for rural schools shows, however, an attempt to react to the normalisation set by urban education. Thus, in rural education there are efforts to establish a school characterised through the settlement area of agrarian reform (Neto and Silva, 2015). There is a search for a rural identity. It is important to keep the history and tradition of the peasant population. However, there is no broad support for this approach, and, in many cases, an urban school setting becomes installed in the rural school. According to Neto and Silva (2015), popular knowledge seems to be subjugated or supplanted, as it does not represent standardised knowledge. The authors observe a dichotomy in rural education. It does not wish to deny the rural characteristic, struggling to establish themselves as such. However, simultaneously, it takes as a reference what is considered "normal" and, in this way, its practices become tied to the traditional urban school model. This dichotomy emerges as, on the one hand, the need to break rules in order to escape of the normality set up by urban life, and, on the other hand, a concern to ensure quality education according to already established urban standards.

It is necessary to deconstruct what is considered normal. Thus, normality is an arbitrarily selected identity in terms of which other identities are evaluated and prioritised (Silva, 2012). According to Neto and Silva (2015), the normal is what one wants to achieve; it represents the desired, the sought for, towards which every effort and energy become directed. But learners are different "because of intellectual, physical, cultural, social and linguistic characteristics, among others, structuring the traditional model of education" (Brasil, 2015b, p. 26).

Considering what has just been said, one could ask if it is possible to imagine a new school that questions such homogenising standards—a school where different learners share time and learning. Such a new school assumes "breaking of rules" as they have been imposed, and, in this way, makes room for new practices.

In order to legitimise practices that are previously devalued/silenced, one needs to consider the exchange of experiences. Such exchange could help pointing out possibilities arising from "rule breaking". However, these practices should not be institutionalised, as this could turn into a new set of "homogenising standards" and a new normality.

14.6 Final Considerations

WG 13 is ready to meet the challenges inherent in the perspective of inclusion as equal access and as participation of any student in the school system. Thus, the working group has expanded its initial goal from researching mathematics education for the target audience of special education to including other educational segments and during SIPEM VI, research results related to rural schools and indigenous communities were also presented and discussed.

The perspective presented by the group is that everyone is different and that education should focus on the possibilities of the learners and not on impossibilities. The presence of learners with disabilities makes the weakness of the school explicit. Mathematics education has shortcomings in quality for everybody. A problematisation of the concept of normality should guide the definition of content, methodologies and assessments. Thus, it is important to avoid the adaptation of tasks initially organised to target a standard learner, and instead, to organise tasks that take into account the needs of a diversity of learners.

Most of the papers presented at the SIPEM VI (8 among 14) discussed the education of deaf learners. With respect to this education, it is necessary to emphasise the need to make the teaching of Libras compulsory at all levels of education. Furthermore, it becomes important that the interpreters' role be discussed both in terms of their initial and continuing education, as well as in term of their professional performance. The interpreters should be seen as educational agents. Their integration in school should be together with teachers and not restricted to the environment of technicians.

In relation to teaching performance, the group recognises that it is necessary that the school provides structures that foster collaborative work involving all those responsible for the education. This includes teachers of regular classrooms, interpreters, specialised teachers, as well as the school board. The professional development of the educational agents must be focused on an inclusive perspective. From this perspective, collaboration may inspire different methodological strategies that support the learning of a larger number of learners.

However, taking into account the contributions as a whole, the fact that research studies referring to indigenous and rural education were considered alongside those involving the target population of Special Education, evinces a consolidating of the idea that inclusion addresses all students who, at some point of their schooling, have difficulties of accessing everything that the school offers. The studies show that the concern of researchers of WG 13 is not only with students with disability.

Although the challenges of inclusive mathematics education are enormous, a movement trying to overcome such obstacles in Brazil has started.

It is important to provide a reframing of learning, assessment, content, and the very structures and functions of the educational system. Inclusive education is essential in order to create a society that respects difference and offers opportunities for development for all.

References

Alrø, H., & Skovsmose, O. (2010). *Diálogo e Aprendizagem em Educação Matemática*. Belo Horizonte, MG: Autêntica.

Ballard, K. (2003). The analysis of context: Some thoughts on teacher education, culture colonization and inequality. In T. Booth, K. Ness, & M. Stromstad (Eds.), *Developing inclusive teacher education* (pp. 59–77). London: Routledge/Farmer.

Borges, F. A., & Nogueira, C. M. I. (2015). Entre duas línguas: o ensino e a aprendizagem de Matemática de alunos surdos inclusos. In VI SIPEM - Seminário Internacional de Pesquisa em Educação Matemática. *Anais do VI SIPEM*. Pirenópolis, GO.

Brasil. (1996). Ministério da Educação e do Desporto. Lei 9.394, de 20 de dezembro de 1996. *Lei de Diretrizes e Bases da Educação Nacional*. Brasília, DF: MEC/SEF.

Brasil. (1998). Secretaria de Educação Fundamental. Secretaria de Educação Especial. *Parâmetros curriculares nacionais: Adaptações Curriculares*. Brasília, DF: MEC/SEF/SEESP.

Brasil. (2001a). Conselho Nacional de Educação. *Resolução n° 2/2001, de 11 de setembro de 2001*. Brasília, DF: CNE/CEB.

Brasil. (2001b). Presidência da República. *Decreto n. 3956, de 8 de outubro de 2001*. Brasília, DF.

Brasil. (2008). Secretaria de Educação Especial. *Política Nacional de Educação Especial na Perspectiva da Educação Inclusiva*. Brasília, DF: MEC.

Brasil. (2009). Presidência da República. *Decreto 6949, de 25 de agosto de 2009. Promulga a Convenção Internacional sobre o Direito das Pessoas com Deficiência e seu Protocolo Facultativo.* Brasília, DF.

Brasil. (2010). Presidência da República. *Decreto n. 7352 de 4/11/2010. Dispõe sobre a política de educação do campo e o Programa Nacional da Reforma Agrária (PRONERA).* Brasilia, DF.

Brasil. (2015a). Presidência da República. Lei no. 13.146, de 6 de julho de 2015. *Lei Brasileira de Inclusão/Estatuto da Pessoa com Deficiência.* Brasília, DF.

Brasil. (2015b). Ministério da Educação. *Orientações para implementação da política de educação especial na perspectiva da educação inclusiva.* Brasilia, DF: MEC/SECADI.

Castro, E. S., Pinto G. M. F., & Ramos L. C. S. (2015). Formação de Professores que Ensinam Matemática sob a Ótica Inclusiva: Estado da Arte de 2006 a 2015. In VI SIPEM - Seminário Internacional de Pesquisa em Educação Matemática. *Anais do VI SIPEM.* Pirenópolis, GO.

Fernandes, S. H. A. A., & Healy, L. (2013). Expressando generalizações em Libras: álgebra nas mãos de aprendizes surdos. *Cadernos CEDES, 33*, 349–368, Campinas, SP.

Healy, L., & Fernandes, S. H. A. A. (2011). The role of gestures in the mathematical practices of those who do not see with their eyes. *Educational Studies in Mathematics, 77*, 157–174.

Healy, L., & Powell, A. B. (2013). Understanding and overcoming 'disadvantage' in learning mathematics. In M. A. Clements, A. Bishop, C. Keitel, J. Kilpatrick, & F. Leung (Eds.), *Third international handbook of mathematics education.* Dordrecht, the Netherlands: Springer.

Jarrold, C., & Baddeley, A. D. (2001). Short-term memory in Down syndrome: Applying the working memory model. *Down Syndrome Research and Practice, 7*(1), 17–23.

Jesus, T. B., & Thiengo, E. R. (2015). A inclusão do aluno surdo nas aulas de matemática: histórias narradas por intérpretes de Libras. In VI SIPEM - Seminário Internacional de Pesquisa em Educação Matemática. *Anais do VI SIPEM.* Pirenópolis, GO.

Lacerda, C. B. F. (1996). *O processo dialógico entre aluno surdo e educador ouvinte: examinando a construção de conhecimentos.* Tese (Doutorado em Educação) Universidade Estadual de Campinas. Campinas, SP.

Marcondes, F. G. V., & Healy, L. (2015). O zero dos alunos surdos: O zero é ausência, o zero é um lugar, o zero é fracasso, o zero é companhia e o zero é redondo. In VI SIPEM - Seminário Internacional de Pesquisa em Educação Matemática. *Anais do VI SIPEM.* Pirenópolis, GO.

Moura, Q., & Penteado, M. G. (2015). Crianças surdas em um cenário para investigação matemática. In VI SIPEM - Seminário Internacional de Pesquisa em Educação Matemática. *Anais do VI SIPEM.* Pirenópolis, GO.

Neto, V. F., & Silva C. M. (2015). A escola do Campo: A busca por uma Identidade, a partir dos Enunciados de Alunas da Área de Matemática, Licenciatura em Educação do Campo. In VI SIPEM - Seminário Internacional de Pesquisa em Educação Matemática. *Anais do VI SIPEM.* Pirenópolis, GO.

Nogueira, C. M. I. (2016). Educação especial, inclusão e educação matemática nos anos iniciais de escolarização. In R. E. S. R. Borba & M. C. S. Cruz (Eds.), *Ciclo de Palestras* (Vol. 2). Recife, PE: EdUFPE.

Nogueira, C. M. I., Borges, F. A., & Frizzarini, S. T. (2013). Os surdos e a inclusão: uma análise pela via do ensino de Matemática nos Anos Iniciais do Ensino Fundamental. In C. M. I. Nogueira (Ed.), *Surdez, inclusão e matemática.* Curitiba, PR: CRV.

Nunes, T., Evans, D., Barros, R., & Burman, D. (2011). Promovendo o Sucesso das Crianças Surdas em Matemática: Uma Intervenção Precoce. *ANAIS do XIII CIAEM-IACME.* Recife, PE.

Rosa, E. A. C. (2015). Das Bordas ao Centro: Reflexões de Professores que Ensinam Matemática Sobre a Inclusão Escolar. In VI SIPEM - Seminário Internacional de Pesquisa em Educação Matemática. *Anais do VI SIPEM.* Pirenópolis, GO.

SBEM. (2016). *Diferença, Inclusão e Educação Matemática.* Retrieved November 21, 2016, from http://www.sbembrasil.org.br/sbembrasil/index.php/grupo-de-trabalho/gt/gt-13

Silva, T. T. (2012). A produção social da identidade e da diferença. In T. T. SILVA (Ed.), *Identidade e diferença: a perspectiva dos estudos culturais* (13th ed., pp. 73–102). Petrópolis, RJ: Vozes.

Silva, E. L., & Fernandes, S. H. A. A. (2015). Diálogo surdo-ouvinte: caminhos para a inclusão. In VI SIPEM - Seminário Internacional de Pesquisa em Educação Matemática. *Anais do VI SIPEM*. Pirenópolis, GO.

UN. Organisation of the United Nations (1948). Declaration of Human Rights. United Nations General Assembly in Paris, December 1948, fromhttp://www.un.org/en/ga/search/view_doc.asp?symbol=a/res/217(III)

Yokoyama, L. A. (2012). *Uma abordagem multissensorial para o desenvolvimento do conceito de número natural em indivíduos com síndrome de Down*. Tese de Doutorado, Universidade Bandeirante de São Paulo, São Paulo, SP.

Yokoyama, L. A. (2015). Primeiras Noções Numéricas de uma Adolescente com Síndrome de Down Através de Materiais Multissensoriais. In VI SIPEM - Seminário Internacional de Pesquisa em Educação Matemática. *Anais do VI SIPEM*. Pirenópolis, GO.

Index

A. J. Ribeiro et al. (eds.), *Mathematics Education in Brazil*,
https://doi.org/10.1007/978-3-319-93455-6

MIX
Papier aus verantwortungsvollen Quellen
Paper from responsible sources
FSC® C105338

If you have any concerns about our products,
you can contact us on
ProductSafety@springernature.com

In case Publisher is established outside the EU,
the EU authorized representative is:
Springer Nature Customer Service Center GmbH
Europaplatz 3, 69115 Heidelberg, Germany

Printed by Libri Plureos GmbH
in Hamburg, Germany